Recent Progress in Human Genome

Recent Progress in Human Genome

Edited by Caleb Elliot

SYRAWOOD
PUBLISHING HOUSE
New York

Published by Syrawood Publishing House,
750 Third Avenue, 9th Floor,
New York, NY 10017, USA
www.syrawoodpublishinghouse.com

Recent Progress in Human Genome
Edited by Caleb Elliot

International Standard Book Number: 978-1-64740-393-5 (Hardback)

Cataloging-in-publication Data

Recent progress in human genome / edited by Caleb Elliot.
p. cm.
Includes bibliographical references and index.
ISBN 978-1-64740-393-5
1. Human genome. 2. Genomics. 3. Human genetics. I. Elliot, Caleb.
QH431 .H86 2023
611.018 166--dc23

TABLE OF CONTENTS

PREFACE

Human genome refers to the complete set of genetic data in an organism. It is the full set of nucleic acid sequences in human beings that are encoded as deoxyribonucleic acid (DNA), inside the twenty three pairs of chromosomes in cell nuclei along with the DNA molecule present inside the mitochondria. The mitochondrial genome and the nuclear genome are often considered separate. Noncoding sequences of DNA are also found in the human genome, along with coding DNA sequences. Understanding the development of the human genome can help in providing data on human evolution. This formed the basis of the Human Genome Project. It was focused on decoding chemical sequence that underlies the human genome and identifying all the genes that are there in the genome. This book unravels the recent studies in the field of human genome sequencing and research. Researchers and students engaged in the study of the human genome will assisted by it.

The information contained in this book is the result of intensive hard work done by researchers in this field. All due efforts have been made to make this book serve as a complete guiding source for students and researchers. The topics in this book have been comprehensively explained to help readers understand the growing trends in the field.

I would like to thank the entire group of writers who made sincere efforts in this book and my family who supported me in my efforts of working on this book. I take this opportunity to thank all those who have been a guiding force throughout my life.

Editor

1

Trapping DNA Replication Origins from the Human Genome

Toshihiko Eki [1,2,*], Yasufumi Murakami [2,3] and Fumio Hanaoka [2,4]

1 Molecular Genetics Laboratory, Division of Bioscience and Biotechnology, Department of Environmental and Life Sciences, Toyohashi University of Technology, 1-1 Hibarigaoka, Tempaku-cho, Toyohashi, Aichi 441-8580, Japan

2 Cellular Physiology Laboratory, RIKEN, 2-1 Hirosawa, Wako, Saitama 351-0198, Japan

3 Department of Biological Science and Technology, Faculty of Industrial Science and Technology, Tokyo University of Science, 2641 Yamazaki, Noda, Chiba 278-8510, Japan; E-Mail: yasufumi@rs.noda.tus.ac.jp

4 Faculty of Science, Gakushuin University, 1-5-1 Mejiro, Toshima-ku, Tokyo 171-8588, Japan; E-Mail: fumio.hanaoka@gakushuin.ac.jp

* Author to whom correspondence should be addressed; E-Mail: eki@ens.tut.ac.jp

---

Abstract: Synthesis of chromosomal DNA is initiated from multiple origins of replication in higher eukaryotes; however, little is known about these origins' structures. We isolated the origin-derived nascent DNAs from a human repair-deficient cell line by blocking the replication forks near the origins using two different origin-trapping methods (*i.e.*, UV- or chemical crosslinker-treatment and cell synchronization in early S phase using DNA replication inhibitors). Single-stranded DNAs (of 0.5–3 kb) that accumulated after such treatments were labeled with bromodeoxyuridine (BrdU). BrdU-labeled DNA was immunopurified after fractionation by alkaline sucrose density gradient centrifugation and cloned by complementary-strand synthesis and PCR amplification. Competitive PCR revealed an increased abundance of DNA derived from known replication origins (*c-myc* and *lamin* B2 genes) in the nascent DNA fractions from the UV-treated or crosslinked cells. Nucleotide sequences of 85 and 208 kb were obtained from the two libraries (I and II) prepared from the UV-treated log-phase cells and early S phase arrested cells, respectively. The libraries differed from each other in their G+C composition and replication-related motif contents, suggesting that differences existed between the origin fragments isolated by the two different origin-trapping methods. The replication activities for seven out of 12 putative origin loci from the early-S phase cells were shown by competitive PCR. We mapped 117 (library I) and 172 (library II) putative origin loci to the human genome; approximately 60% and 50% of these loci were assigned to the G-band and intragenic regions, respectively. Analyses of the flanking sequences of the mapped loci suggested that the putative origin loci tended to associate with genes (including conserved sites) and DNase I hypersensitive sites; however, poor correlations were found between such loci and the CpG islands, transcription start sites, and K27-acetylated histone H3 peaks.

Keywords: DNA replication; replication origins; sequence analysis; human genome; competitive PCR

1. Introduction

DNA replication is a fundamental process for maintaining and transmitting genetic information to proliferating cells. In eukaryotes, genomic DNA replication starts bi-directionally from sites called "replication origins". Because the initiation step of DNA replication is crucial for regulation of cell proliferation, the structures of the origins and the proteins involved in their functions have been extensively studied. In *Saccharomyces cerevisiae*, where DNA replication has been studied in detail, the nucleotide sequence motifs that are conserved in the replication origins and are called autonomously replicating sequences (ARSs) have been elucidated. In this species, the origin recognition complex (ORC) and the proteins associated with it have been shown to play essential roles in origin replication initiation. The ORC specifically recognizes a 17-bp AT-rich consensus sequence (ARS consensus sequence, ACS), where the initiation proteins assemble in a stepwise manner to initiate DNA replication [1]. In fission yeast, the distinct core sequences of the origins have not been identified although it is known that the origins in *S. pombe* contain AT-rich regions [2–5]. In metazoans, the individual origins of replication within, for example, the Chinese hamster *DHFR* locus [6,7] human *c-myc* [8,9], and *lamin* B2 genes [10] have been extensively studied. Although some of these studies [8,9] suggested that DNA replication initiates from a broad region (*i.e.*, an initiation zone) in the genome, no particular sequences were identified. Microarray- and high throughput sequencing-mediated methods have been recently applied to map the replication origins in the sequenced mammalian genomes (including the human genome) [11–19] (also reviewed in [20,21]). These genome-wide mapping studies on replication initiation sites have identified a number of replication origins in mammals. Some of these studies have suggested that the human origins preferentially associate with relatively G+C-rich regions [11,19] (unlike that observed with yeast origins), and CpG islands and transcription levels near the origins can influence the replication initiation events [12,15,16]; this suggests potential regulation of replication initiation coupled with transcriptional regulation during development and differentiation [13,18,22–24]. Although one study has claimed that a 36-bp human consensus sequence supports autonomous DNA replication [25], we still do not know if any conserved sequences or motifs encode replication origins. In addition, studies conducted some time ago have suggested that replication initiation sites are closely associated with the nuclear structure [26–29] (also reviewed in [30]) of the cell; however, little is known about the distribution of nuclear attachment sequences around the origins, such as the nuclear scaffold associated region (SAR) and the nuclear matrix associated region (MAR).

Over the past decades, several methods for origin mapping have been developed for identifying replication origins in eukaryotic chromosomes [20,31]. These methods fall into three categories: (1) analysis of nascent DNA (e.g., isolation of newly replicated DNA); (2) analysis of DNA structure (e.g., 2D-gel electrophoresis analysis or bubble trap [32]); and (3) ARS assays on cellular DNA. With one exception [25], the difficulty involved in conducting ARS assays in mammalian cells has resulted in previous studies on origins being performed mainly by nascent DNA analyses or structural analysis of DNA. Since the former methods are suitable for detecting unidentified origins, unlike a gel electrophoresis-based analysis, a nascent DNA-based approach has been used for recent comprehensive studies on replication origins [11–16,18,19]. In analyses of nascent DNA molecules, the bromodeoxyuridine (BrdU)-labeled DNAs and the λ exonuclease-resistant short DNAs (*i.e.*, nascent DNA with a 5' RNA primer) from the replication origins have frequently been used to assay for origin activities and/or clone such DNAs from the origins. Both methods have their drawbacks, which include contamination of nicked BrdU-labeled DNA in the former method, and variable efficiency of λ exonuclease activity and RNase contamination in the latter. Nevertheless, Karnani *et al.* compared both methods and showed that their usefulness for mapping replication origins in human cells was comparable [14]. In the former method, when exponentially proliferating (log-phase) cells were labeled with BrdU, vast amounts of BrdU-labeled DNAs were generated; these were distinct from the origins themselves, and were likely to be broken during purification leading to contamination of the BrdU-labeled short DNAs produced from the origins. Therefore, when analyzing nascent DNA for mapping and cloning replication origins using the BrdU-labeling method, it is very important to effectively arrest the replication fork movements near the origins upon labeling with BrdU. In other words, the ability to efficiently arrest replication forks near their origins is crucial for origin mapping using BrdU-labeled nascent DNAs.

In this study, our intention was to systematically clone the BrdU-labeled nascent DNAs from arrested replication forks that had arrested near the origins and map the replication origins in the human genome at the nucleotide sequence level in order to elucidate the features of the origin sequences. To achieve this, first, we developed a novel origin-trapping method by forming UV-induced DNA lesions in a nucleotide excision repair-deficient human (GM8207) cell line. Replication fork progression was effectively blocked in the exponentially proliferating GM8207 cells by UV irradiation and Trioxsalen-mediated DNA crosslinking. We performed competitive PCR assays using the nascent DNAs from the treated cells and showed that DNAs from well-characterized *c-myc* and *lamin* B2 origins were enriched by the UV- and crosslink-mediated origin-trapping methods. Second, for isolating the early-firing origins from highly synchronized cells, we labeled the nascent DNAs with BrdU in early S phase in the presence of replication inhibitors.

BrdU-labeled DNAs prepared by the two different methods were cloned for construction of genomic DNA libraries (*i.e.*, library I from the UV-irradiated log-phase cells and library II from the cells synchronized in early S phase) and for their sequence analyses. In particular, the nascent DNA abundance of 12 potential origin loci identified from the synchronized cells was examined by competitive PCR to access the origin activities of these loci from library II.

The nucleotide sequences from libraries I (85 kb) and II (208 kb) were compared with each other or with control sequences from the human genome to elucidate the features of the isolated putative origins. The results of this analysis indicated that the two libraries differed in their G+C composition and their replication-related motif contents. One hundred and seventeen and 172 putative origin loci from libraries I and II, respectively, were mapped to the human genome. In addition, the flanking sequences of the mapped origin loci were analyzed by comparison to 236 genomic control loci to examine the association of the origin loci with genes, DNase I hypersensitive sites, transcription start sites, CpG islands and an epigenetic marker of K27-acetylated histone H3 peaks.

Although our comprehensive study has been conducted on a relatively small scale, the origin-trapping methods reported here could be useful for studying DNA replication origins in a number of eukaryotes whose genomes have not been sequenced. That the methods require little or no specialist genome analysis tools (such as microarrays or next generation sequencers) should be a distinct advantage for many researchers.

2. Results and Discussion

2.1. Scheme for Isolation and Cloning of Nascent DNAs from Origins of Replication

In this study, we designed and used two different methods for isolating the nascent DNA molecules arising from DNA replication origins (Figure 1). The approach shown in Figure 1A depends upon arresting the replication fork movements near the origins by forming DNA adducts with UV irradiation in DNA synthesizing cells from a proliferating cell population. To evaluate this approach, we also employed the origin-trapping method by using a chemical crosslinker reported in the literature previously [33,34]. In parallel experiments, we aimed to facilitate accumulation of the nascent DNAs around the origins firing in early S phase by arresting the replication forks with replication inhibitors (Figure 1B). Synchronized M phase cells were allowed to progress to early S phase and were arrested using DNA synthesis inhibitors (*i.e.*, aphidicolin and hydroxyurea) and then labeled with BrdU to allow the nascent DNAs to be BrdU-labeled in the replication forks that were arrested near the early-firing origins. Nascent DNAs from origins initiating in early and late S phase could be trapped by the former method, whereas DNAs from the early-replicating origins were mainly isolated by the latter method. Trapped BrdU-incorporated nascent DNAs near the origins can be completely purified by alkaline sucrose density gradient centrifugation followed by immunoprecipitation with an anti-BrdU antibody and protein A/G-coated latex beads. To clone a very small amount of BrdU-labeled single-stranded DNA, the purified single-stranded DNA was converted into double-stranded DNA via tailing with poly dC at the 3'-ends by terminal deoxynucleotidyl transferase (TdT) and oligo dG-primed DNA synthesis of the complementary strands by DNA polymerase. After adaptor ligation, the DNA was PCR amplified and size fractionated by agarose gel electrophoresis for cloning into a plasmid. In this study, libraries I and II were constructed from the nascent DNAs isolated from UV-irradiated log-phase and synchronized cells, respectively (Figure 1).

Figure 1. Scheme for trapping and isolating nascent DNAs from origins of replication in human cells. (**A**) The origin-trapping method where UV- and chemical crosslinking-mediates replication fork arrest in exponentially proliferating cells; (**B**) The method for trapping the early-initiating origins is conducted by arresting the cell populations in early S phase using replication inhibitors. BrdU-labeled nascent DNAs from the origins are shown in red.

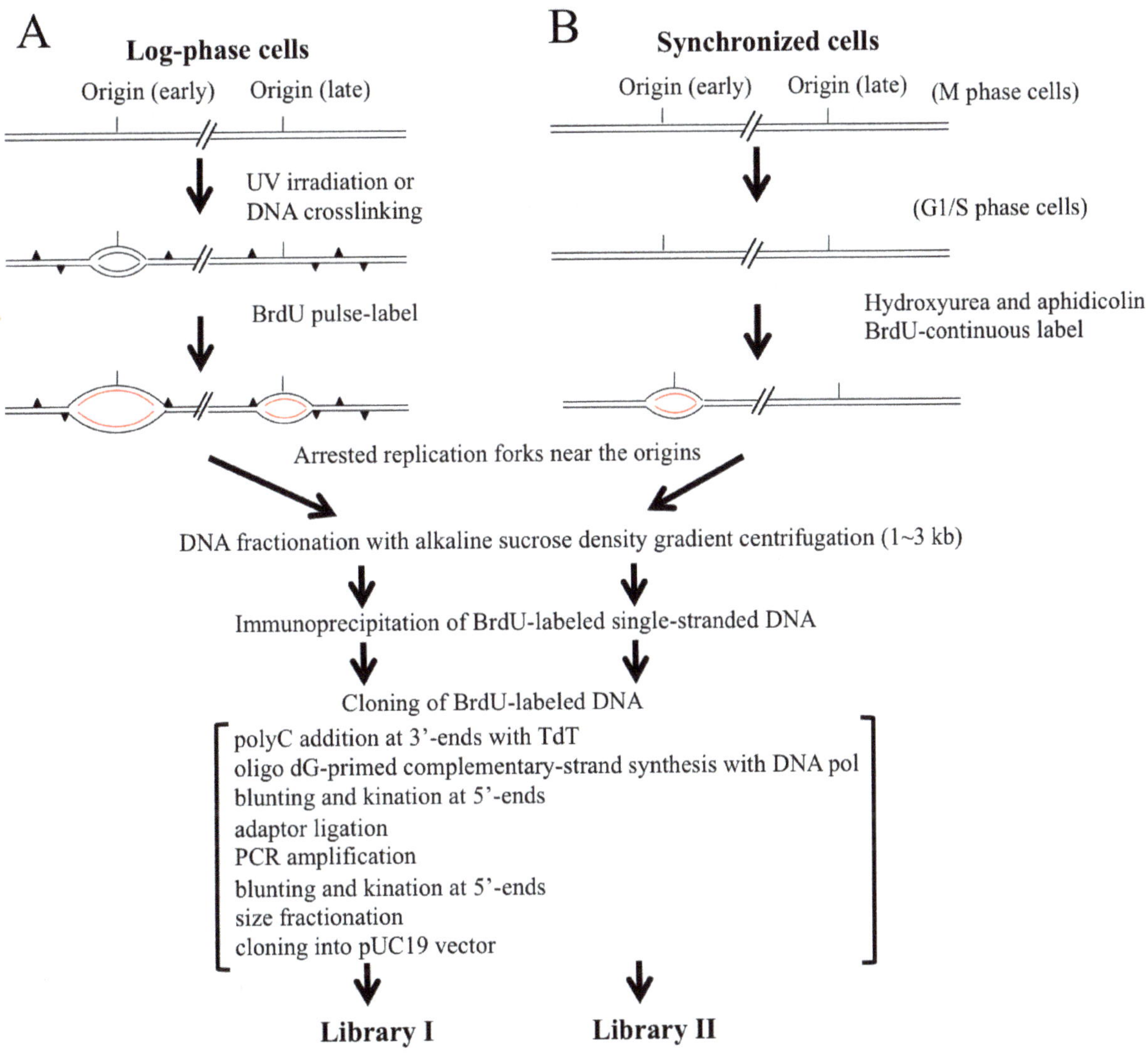

2.2. Trapping Nascent DNAs in Human Cells by UV Irradiation and Chemical Crosslinking

We first applied the UV-mediated origin-trapping method to the human SV40-transformed fibroblast cell line GM8207, which is deficient in nucleotide excision repair [35]. In GM8207 cells after UV irradiation, recovery from the arrested replication fork movement by DNA repair mechanisms is poor unlike in other human cell lines such as HeLa cells (data not shown). In addition, GM8207 cells are suitable for isolation of nascent DNA molecules, because BrdU-incorporation into UV-damaged genomic DNAs by DNA repair mechanisms can be minimized. We examined the effects of UV irradiation during cellular DNA synthesis by pulse-chase experiments followed by analytical alkaline sucrose density gradient centrifugation. DNA (approximately 0.5–2 kb) was labeled with [^{3}H]thymidine (TdR) during a 15-min pulse in GM8207 cells synchronized in S phase, after which the labeled DNA was allowed to maturate to bulk DNA during the 90-min chase in the non-irradiated cells (Figure 2B).

However, DNA elongation was totally inhibited in the cells treated with 300 J/m^2 UV prior to pulse-labeling, as was expected (Figure 2A). To examine if genomic DNAs are broken by UV irradiation, the cells whose parental DNA strands were uniformly labeled with [^{14}C]TdR were synchronized in S phase and irradiated with or without 300 J/m^2 UV. After 45 min in culture, labeled DNAs were observed in the bulk DNA fractions indicating that the DNA strand breakage after UV irradiation was negligible (Figure 2C). BrdU-labeled DNA of approximately 0.5–2 kb was pooled for further DNA purification (yellow zone in Figure 2A).

Figure 2. Inhibition of DNA synthesis in GM8207 cells after UV irradiation. (**A**) Distribution of pulse-labeled nascent DNAs in the UV (300 J/m^2)-irradiated cells after a 0-min (red), 30-min (orange), or 90-min (pink) chase along the alkaline sucrose density gradient fractions; (**B**) Distribution of pulse-labeled nascent DNA in the non-irradiated cells after a 0-min (purple), or a 90-min (light blue) chase; (**C**) Distribution of pre-labeled parental DNA with [^{14}C] TdR in the gradient fractions at 45 min after UV-irradiation (0: black; 300 J/m^2: red). Arrows: 3 kb size marker.

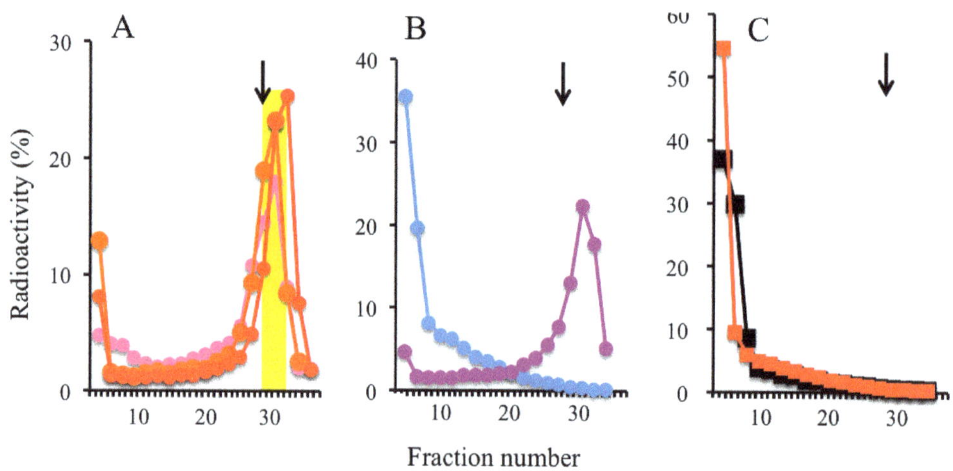

We also compared the previously reported DNA crosslink-mediated origin-trapping method that uses Trioxsalen [33,34] with our UV-mediated method. Inter-crosslinking of double-stranded DNA also effectively arrests the progression of replication forks as well as UV-induced DNA adducts on the template DNA for replication. Dimitrova *et al.* [34] isolated and cloned nascent DNAs from mouse cells by Trioxsalen-mediated crosslinking and showed that some of the isolated fragments were A+T-rich and were bound to protein factors. Although these researchers showed enrichment of the ARS sequences from yeast cells using this method, it is not known if the mouse origins were concentrated by the method, or if the isolated DNAs acted as origins *in vivo*.

We initially examined the inhibitory effects of Trioxsalen treatment (*i.e.*, variations in the times used for irradiation for photo-crosslinking) on DNA synthesis in GM8207 cells. The cells in mid-S phase were photo-irradiated using the conditions indicated in the Experimental Section, with or without Trioxsalen (1.25 μg/mL) (Figure 2). The sizes of the pulse-labeled DNAs were >10 kb in the irradiated and non-irradiated cells without Trioxsalen (Figure 3B), and in the non-irradiated cells with Trioxsalen (blue line in Figure 3A); however, a labeled nascent DNA peak of approximately 3 kb was observed from the cells irradiated with Trioxsalen (3 min, twice) (red line in Figure 3A). As shown in Figure 3C, nascent DNA formation was arrested at around 3 kb, even after a 90-min chase in the Trioxsalen-treated cells. In contrast, labeled DNA was found to maturate to bulk DNA in the non-treated cells after the chase (Figure 3D). The peak fractions of the BrdU-labeled DNAs (approximately 1–4 kb) were pooled for additional DNA purification and use in nascent DNA abundance assays (yellow zone in Figure 3C).

Figure 3. Inhibition of DNA synthesis in GM8207 cells by Trioxsalen treatment. (**A**) Distribution of pulse-labeled nascent DNA in the cells treated with Trioxsalen (photo-irradiation conditions: 3 min, twice in red; 2 min, once in orange; no irradiation in blue) in the alkaline sucrose density gradient fractions; (**B**) Distribution of pulse-labeled nascent DNA in the cells without Trioxsalen; (**C**) Distribution of pulse-labeled nascent DNA in the cells crosslinked with Trioxsalen (3 min, twice) after 0-min (red), 30-min (orange), and a 90-min (pink) chase; (**D**) Distribution of pulse-labeled nascent DNA in the non-crosslinked cells after 0-min (purple) and 90-min (light blue) chase. Arrows: 3 kb size marker.

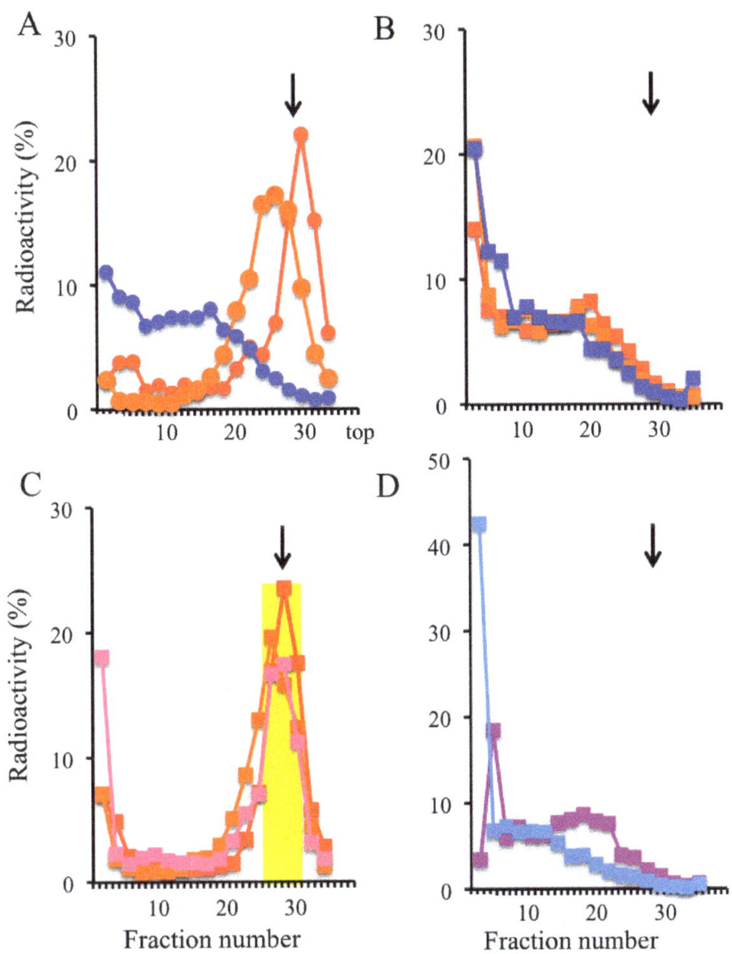

2.3. Trapping Newly Synthesized DNAs in Cells Arrested in Early S Phase with Aphidicolin and Hydroxyurea

To identify the early-initiating origins in the genome, we isolated nascent DNAs from GM8207 cells arrested in early S phase after double cell synchronization (Figure 1B). After release from the aphidicolin block, cells that had been synchronized in M phase with colcemide were released and allowed to enter G1, after which they were continuously labeled with BrdU in the presence of aphidicolin and hydroxyurea during early-S phase. Under these conditions, the replication forks in most of the cells would be arrested near their replication origins. To monitor DNA synthesis in the cell population after release from the mitotic block, [^{3}H]TdR incorporated into the cells and measured in the absence of the replication inhibitors at 1-h intervals after release (Figure 4A), showed that the synchronized population of cells started to replicate their DNA from about 5 h post-release. To isolate the nascent DNA, the synchronized cells were continuously labeled with BrdU in the presence of high concentrations of aphidicolin (5 μg/mL) and hydroxyurea (1.5 mM) from 5 h to 9 h after release, after which the labeled DNA was purified. The results of the analytical alkaline sucrose density gradient centrifugation showed that nascent DNA of approximately 3–5 kb had accumulated in the cells that were arrested in early S phase (Figure 4B). The labeled DNA peak in the gradient fractions was found at a position that was slightly larger than the 0.5–3 kb observed in the UV-treated or crosslinked cells. The size difference of the nascent DNA observed in the UV-radiated and synchronized cells may be related to the different modes of action of the treatments used for replication fork arrest because DNA crosslinking arrests replication forks more tightly than treatment with DNA synthesis inhibitors. Nascent DNA from the peak fractions was pooled for cloning (yellow zone in Figure 4B). The mean sizes of the DNA inserts in the libraries prepared from the UV-irradiated and synchronized cells differed from each other (Section 2.5); this probably reflects the different sizes of the pooled DNA obtained from the different treatments of the cells.

Figure 4. Analyses of DNA synthesis in GM8207 cells synchronized in early S phase. (**A**) DNA synthesis in the synchronized cells in the absence of aphidicolin and hydroxyurea after release from the mitotic block. [^{3}H]TdR incorporated into the cells was measured at 1-h intervals after release. Arrowed bar: labeling period with BrdU. Labeling was carried out in the presence of aphidicolin and hydroxyurea; (**B**) Distribution of labeled DNA molecules in the fractions obtained from alkaline sucrose density gradient centrifugation. Arrow: 3 kb marker.

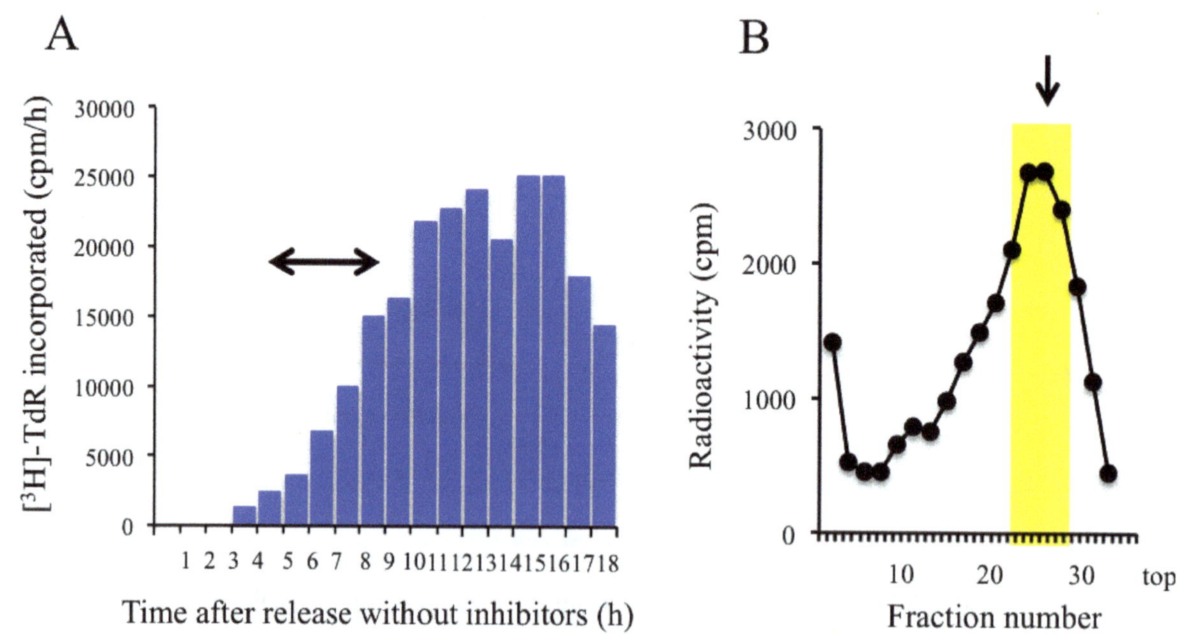

2.4. Isolation of Nascent DNAs and Evaluation of Enrichment of the Origin-Derived DNAs Using Competitive PCR

To isolate the BrdU-labeled DNA from the UV irradiated cells, we used alkaline sucrose density gradient centrifugation and immunoprecipitation with an anti-BrdU antibody, according to a previously reported procedure [34] with the following modification (*i.e.*, use of protein A/G-coated latex beads for minimizing non-specific binding of single-stranded DNA to the beads). We evaluated the purification coefficient of the nascent DNA from the GM8207 cells treated with Trioxsalen (3 min, one treatment). Parental strands from the bulk DNA were uniformly labeled with [^{14}C]TdR, while the nascent DNAs were labeled with BrdU and [^{3}H]cytidine. After fractionation by alkaline sucrose density gradient centrifugation, the nascent DNAs arrested at around 3 kb were pooled and purified by immunoprecipitation. The total radioactivity counts from the [^{3}H] and [^{14}C] recovered in each step are summarized in Table 1. The results indicate that the nascent DNA fragments were concentrated more than 3,000-fold in the immunopurified fraction.

Table 1. Purification of BrdU-labeled nascent DNAs.

Fraction	^{3}H total count (cpm)	^{14}C total count (cpm)	^{3}H/^{14}C ratio	^{3}H-labeled DNA abundance (fold)
Total DNA	12,997	127,230	0.102	1
Pooled DNA	4,811	293	16.4	161
Immunopurified DNA	3,320	<10	>332	>3,250

We also examined enrichment of the nascent DNAs from the known origins of replication using competitive PCR methodology [36]. Replication origins in the human *c-myc* [8] and *lamin* B2 [10] loci have been extensively studied and can, therefore, be used as positive controls. Two non-origin loci were also selected for the assay. One locus (β-globin_40k) is a site in the human β-globin gene locus that is approximately 20 kilobases from the replication initiation zone [37,38]. The other is the sWXD1449 locus (accession no. L77324) that is located on human chromosome Xq27; this region has been shown to be replicated during the very late stages of S phase by Hansen *et al.* [39]. Primer sets for the four loci (*i.e.*, *c-myc*, *lamin* B2, β-globin_40k, and sWXD1449) were prepared for competitive PCR (supplementary Table S1) and the abundance of the DNAs from each locus was determined for both the nascent and genomic DNA fractions. For nascent DNA abundance assays, the adjacent DNA sequences to the origins were frequently used as negative controls. Several primer sets for the non-origin loci including the flanking regions of *c-myc* and *lamin* B2 origins have been tested, however, we could only successfully prepare two primer sets for two non-origin loci (β-globin_40k and sWXD1449) and competitors that had sufficient stability for use in the competitive PCR assay. In this assay, the relative abundance of DNA from an origin locus in the nascent DNA, relative to a genomic DNA locus, was higher than that for a non-origin locus. PCR amplicons from the competitive PCR assays were separated by gel electrophoresis (Experimental Section); the results of which are shown in Figure 5 and summarized in Figure 6 and supplementary Table S2. The relative abundances of the origin-derived DNAs in the purified nascent DNA fractions from the UV-irradiated (Figure 6A,B) and crosslinked (Figure 6C,D) cells were approximately 8–20-fold higher than those from the non-origin loci-derived DNAs; this indicated that the nascent DNAs from the origins were enriched in the DNA fractions purified from the UV- and crosslinker-treated cells. Thus, the origin-trapping methods by UV irradiation and chemical crosslinking are effective at isolating replication origins from human cells.

Figure 5. Competitive PCR assay for assessing the relative abundance of DNA from *c-myc*, *lamin* B2, β-globin_40k, and sWXD1449 loci. Competitive assays for the four loci (indicated on the left of the figure) were carried out using the purified BrdU-labeled DNA fractions from the crosslinked cells (left columns) and genomic DNA (right columns) as a template in the presence of the indicated number of the corresponding competitor DNA molecules (from left to right at the top of the lanes: 40, 130, 400, 1,200, 3,600, 10,800, and 32,400 molecules). PCR products derived from target DNA (T) and competitor DNA (C) are indicated. The product from the competitor DNA is 20 bp longer than the target DNA-derived product. The corresponding results from quantitation of the PCR products are shown in Figure 6C and supplementary Table S2.

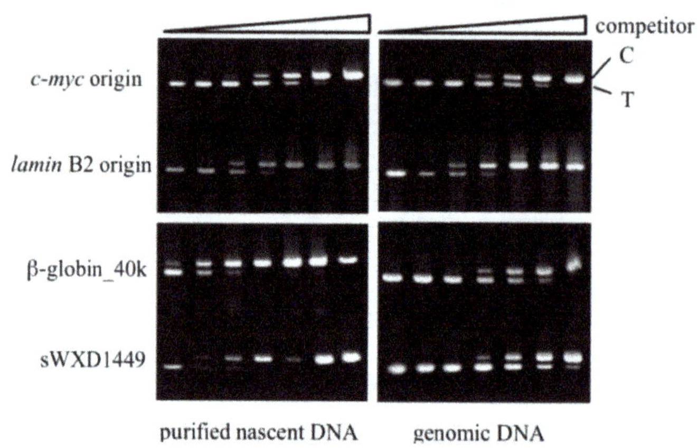

Figure 6. Abundance of the origin-derived DNAs in the DNA fractions purified from UV irradiated and crosslinked cells. The relative DNA abundances from the origin and non-origin loci compared with the β-globin_40k (**A**, **C**, **D**) or the sWXD1449 (**B**) locus in the purified BrdU-labeled DNA fractions from UV-irradiated (**A** and **B**) or crosslinked (**C** and **D**) cells are shown. Further details appear in supplementary Table S2.

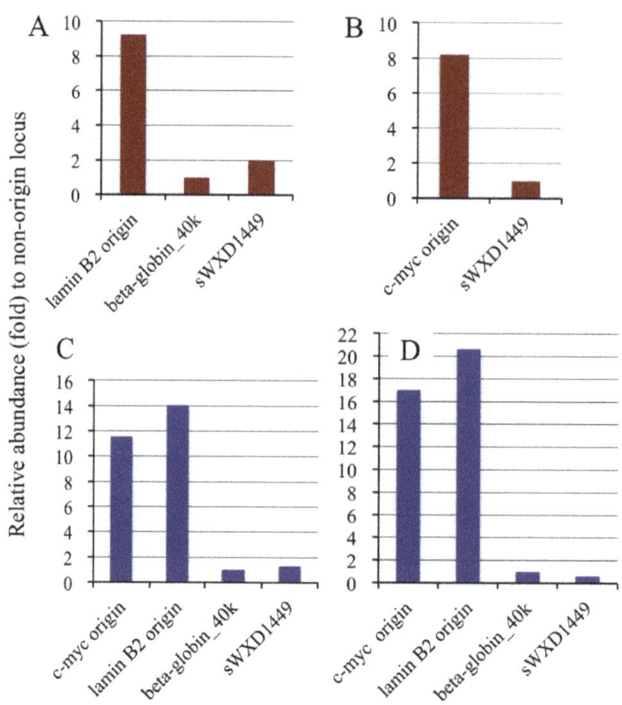

2.5. Cloning of BrdU-Labeled Nascent DNA and Sequence Analysis of the DNA Clones

Based on the procedures described above, we have independently isolated and cloned BrdU-labeled nascent DNAs from UV (300 J/m^2)-irradiated proliferating GM8207 cells and cells synchronized in early S phase. Because the quantity of the labeled nascent DNA was very small, and BrdU-labeled DNA is both heat- and photo-labile (e.g., heat denaturation at 95 °C for 5 min completely disrupted the BrdU-labeled DNAs), we avoided exposing the DNA samples to high temperatures and light during the cloning process. Nascent DNAs were cloned into plasmids via addition of poly dC tails at their 3' ends, oligo dG-primed complimentary strand synthesis, adaptor ligation, and PCR amplification (as described in the Experimental Section). Complementary strand synthesis by DNA polymerase was totally dependent upon both the poly dC tailing reaction and the oligo dG primer (data not shown). For DNA cloning from the UV-irradiated samples, we used T7 DNA polymerase for complementary strand synthesis at 37 °C; however, the resultant clones frequently contained non-specifically primed DNA inserts that were presumably caused by the repetitive sequences in the DNA samples. Therefore, despite the need to avoid using a high temperature reaction (65 °C for 4 min), we used Taq DNA polymerase for DNA synthesis for cloning the DNA samples from the synchronized cells. In addition, we implemented a size fractionation process for the amplified DNAs to remove the contamination derived from clones with short inserts. In this study, the two genomic libraries (I and II) were prepared from the UV-irradiated proliferating cells and the cells arrested in early S phase, respectively.

By subsequent sequencing of the DNA inserts in both directions, a large number of nucleotide sequences were obtained from each of the libraries, library I (133 sequences) and II (221 sequences) (Table 2; supplementary Table S3). Clones from library II contained longer DNA inserts than those from library I (Table 2, 640 bp *versus* 941 bp), which was made from the UV-irradiated cells; this result is consistent with the sizes of the nascent DNAs (0.5–2 kb in Figure 2A *versus* 3–5 kb in Figure 4B) that were used for library preparation, although we cannot exclude the possibility that a size bias occurred during the cloning process. Library I contained more chimeric clones than library II (23 *versus* 7 out of the total number of clones analyzed in each library).

Table 2. Characteristics of the sequenced clones from libraries I and II.

	Library I	**Library II**
DNA source	UV-treated proliferating cells	Synchronized cells in early S phase
No. of determined sequences	133	221
No. of independently sequenced inserts	132	187
Total length of the sequences (bp)	85,168	207,895
Mean length of the sequences (bp)	640 ± 328	941 ± 279

Note: Number of independently sequenced inserts: this denotes where two independently sequenced DNA inserts in a chimeric clone were counted as two, and two sequences from a single DNA insert (e.g., two sequences reads in both direction for a long-insert DNA) were counted as one. The mean length for each of the library sequences is shown with the associated standard deviation.

2.6. Competitive PCR Assays for Library II Loci

We next examined the abundance of each locus identified in the nascent DNAs using competitive PCR. This enabled us to examine if individual loci in the cloned DNAs were derived from an origin region. For this purpose, 107 sets of PCR primers were designed based on unique sequences in library II. After screening to obtain unique signals in both the PCR and Southern blot analyses using human genomic DNA, 27 sets of primers were finally obtained. Thereafter, out of 27 loci, 12 sets of competitive PCR primers were successfully generated (supplementary Table S1). The primer sets for three target loci (*i.e.*, AC4886-158, Z83847-223, and Y978-250) from the 12 sets were generated from unique regions in their flanking sequences. The abundance of each target locus in the nascent DNA fractions relative to genomic DNA was examined by competitive PCR. The data obtained indicated that DNA from the XPD4-C6 locus was as abundant as the *c-myc* origin-derived DNA (Figure 7). Despite somewhat low abundances overall, the DNA abundances of seven loci (*i.e.*, AC4886-158, Z83847-223, XPD1-C5, XPD2-G2, XPD4-C6, XPD4-D3, and XPD5-C6) were more than 4-fold higher than XPD4-F10 (Figure 7). This finding suggests that the early-replicating origins are most likely located close to these loci, and also that the origins can be concentrated effectively by the origin-trapping method using cell synchronization. Enrichment of the origin loci tested from library II showed no statistically significant trends; however, two possible reasons might account for this. First, the poor enrichment observed in library II may have been caused by incidental contamination with broken BrdU-incorporated non-origin fragments during nascent DNA purification. Alternatively, the five loci that exhibited poor abundances may be derived from inefficient origins activated by the cell synchronization treatment. Previous studies have shown that most origins are used in less than 10% of the cell cycles [32,40–43], suggesting that mammalian origin usage is highly flexible. Altered patterns of replication initiation have also been found in mammalian cells treated with DNA synthesis inhibitors [44] as described later. Therefore, cell synchronization treatment might activate a subset of inefficient origins. In addition, significant differences in the G+C composition and the sequence motif contents were not found between the seven putative origin loci and the other five loci investigated here.

Figure 7. Measurement of the relative abundances of the nascent DNAs from 12 library II loci using competitive PCR assays. The DNA abundance for each locus relative to that of XPD4-F10 is indicated. The locus name and the derived sequence ID (in parentheses) are shown. Details of the results appear in supplementary Table S4.

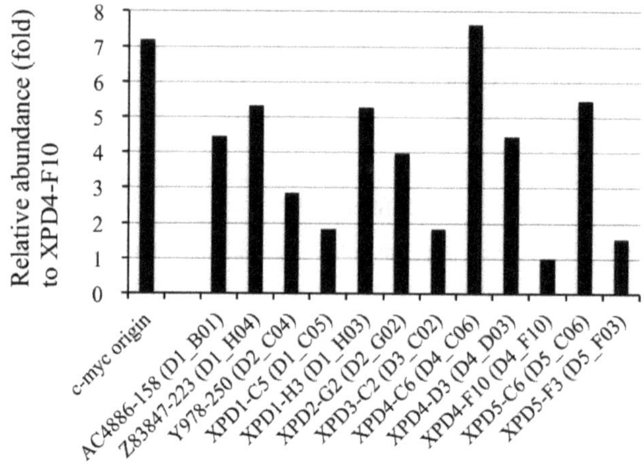

2.7. Sequence Analyses of the DNA Clones from Libraries I and II

The nucleotide sequences of the DNA clones from the two libraries were characterized using bioinformatics to examine the features of their primary structures. We examined the G+C composition of a contiguously assembled sequence from each library, as well as the composition of the repeat sequences, and five replication-related sequences (ACS [45], nuclear scaffold sequences [SAR and MAR] [46,47], topoisomerase II recognition sequences [48], and AT-rich elements). Because several studies [26–29,49] suggest that replication origins are associated with the nuclear scaffold, the composition of the SAR and MAR sequences was examined. Topoisomerase II sites have been shown to associate with MARs [46]. Furthermore, AT-rich regions have frequently been observed in the replication origins of the fission yeast genome [2–5] and in about 1% of the human genomic region [14]. In contrast, some studies suggest that human replication origins are associated with G+C-rich regions of the genome [11,19]; hence, the G+C composition and the content of the AT-rich elements were examined. Genome sequences (90–200 kb/locus) from 14 loci including each of the six loci from cytogenetic G- and R-band regions were selected from the human genome and used as controls (supplementary Table S5A). As shown in Figure 8A and supplementary Table S5B, the G+C contents of libraries I and II were 41.2 and 36.5%, respectively. The G+C content of library I is close to the average value of about 41% for the human genome [50]; however, library II has a lower value. The compositions of the repeat sequences and motifs in the libraries had similar values to the mean values of the 14 control loci. However, it should be noted that the compositions of the AT-rich elements (0.17 sites/kb) in library II (Figure 8C), and long interspersed elements (LINEs) in both libraries (0.55 and 0.57 sites/kb in the library I and II, respectively) (Figure 8D), were higher than the corresponding mean values (0.10 for AT-rich elements and 0.35 for LINEs) of the control loci (Figure 8C,D and supplementary Table S5B). In addition, the composition of the MAR sequences (3.27 sites/kb) in library II and the DNA repeat family (0.23 sites/kb) in library I were higher than the corresponding mean values of the control loci (2.48 and 0.14 sites/kb, respectively) (supplementary Table S5B). Although previous studies have suggested that a close association exists between the replication initiation sites and the nuclear scaffold [30], substantial differences in the compositions of the MAR and SAR sequences and topoisomerase II recognition sequences were not found between the sequences in the libraries and the control loci (Figure 8B,C). As for the replication-related motifs, Price *et al.* have reported that a 36-bp consensus sequence supports autonomous replication in human cells [25]; however, we did not detect this sequence in our libraries (data not shown). In addition, a recent genome-wide mapping study of λ exonuclease-resistant short DNAs revealed that human replication origins associate with G-quadruplex motifs [19]. Analysis of our putative origins with the QGRS Mapper program showed that 55.6 and 39.8% of the sequences in library I, and 62.9 and 43.4% of the sequences in library II, contained G4L1-15 and G4L1-7 motifs, respectively (supplementary Table S3). However, it is difficult to make direct comparisons between our data and the data from the genome-wide mapping study because these studies differed in their experimental conditions and the way in which the motif analyses were performed.

Figure 8. The G+C content and sequence compositions of repeat elements and replication-related motifs in libraries I and II. Supplementary Table S5 contains additional information. It should be noted that the human genome sequences used as controls contained no satellite elements. Bars represent the standard deviation of the mean.

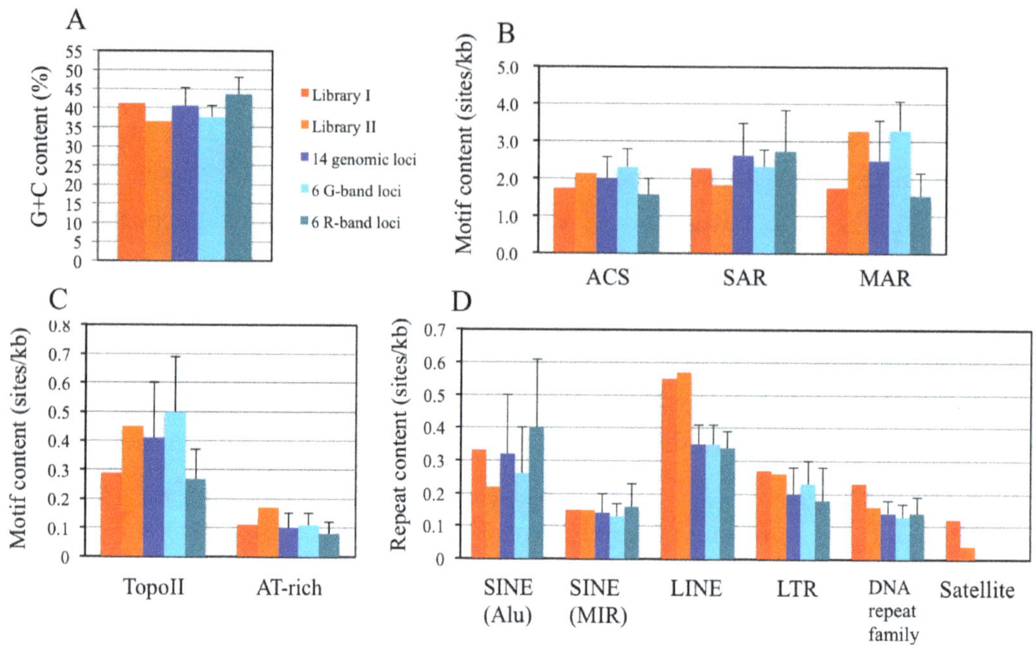

The sequence compositions of the two libraries could be clearly distinguished by their G+C content as well as by the composition of some of the motifs and repeat sequences (e.g., MAR, AT-rich elements and Alu repeats). Few Alu sequences and a low G+C content in library II are features generally observed in the G-band region of chromosomes, and are consistent with the corresponding mean values of the six G-band loci (Figure 8A and supplementary Table S5). Nevertheless, library I had a slightly higher G+C content (41%), and its motif compositions were similar to the mean values of the six R-band loci. As was mentioned in the Section 2.6, these differences are likely to reflect the different origin-trapping methods used for preparation of the libraries, especially library II, which is derived from cells treated with hydroxyurea and aphidicolin. Anglana *et al.* reported that depletion of nucleotide pools by hydroxyurea treatment modulated the initiation of replication in Chinese hamster cells [44]. Mesner and colleagues prepared origin-enriched libraries from both log-phase and synchronized human cells and found that the sequence compositions of the resultant libraries differed from each other; hence, they concluded that cell synchronization affects the replication initiation sites [17]. Thus, differences in the features of the sequences from libraries I and II might be explained by the presence of different initiation origins in the synchronized cells that had their nucleotide pools depleted with hydroxyurea.

2.8. In Silico Mapping of the Putative Replication Origin Sequences to the Human Genome and Analysis of their Flanking Sequences

The putative origin sequences from the libraries were further mapped to the human genome using BLAST/BLAT searches. One hundred and seventeen and 172 unique loci from 133 (library I) and 221 (library II) sequences were successfully assigned to their respective genomic locations on human chromosomes (supplementary Table S6A,B). Figure 9A is a histogram of the number of loci that mapped to each chromosome, while the ratios for locus abundance to chromosome size are shown in Figure 9B. The abundances of the loci fluctuate around a ratio of 1.0 (green dotted line, Figure 9B) for more than half of the chromosomes, indicating a correlation between the origin contents and chromosome size. Interestingly, the putative origin loci for chromosomes 6 and 21 from library I and for chromosomes 2, 4 and 21 from library II had ratios over 1.5 (supplementary Table S7). In contrast, loci on chromosomes 17–19 from library I and on chromosomes 10, 14, 19 and 20 for library II were underrepresented in comparison with the values predicted for their associated chromosome sizes (*i.e.*, they had ratios <0.5). In a related genome-wide study using human cells, Besnard *et al.* observed that the distribution of origin peaks had no correlation with chromosome length [19]. We also found no consistent correlations between the G+C compositions and/or the contents of the CpG islands and the abundance of loci mapping to these chromosomes. For example, although chromosome 19 has the highest G+C content (49%) and CpG island composition (19.7 sites per Mb) among the human chromosomes [50], fewer loci from both of the libraries mapped to this chromosome than to any of the other chromosomes (Figure 9A).

Figure 9. Assignment of the putative replication origin loci from libraries I and II to human chromosomes. (**A**) Histogram showing the number of loci from each library that mapped to individual chromosomes. The number of mapped loci from library I (blue) and II (red) are indicated; (**B**) Ratio of the abundance of mapped loci relative to the chromosome size. A green dotted line depicting a ratio of 1.0 indicates that the abundance of the mapped loci can, in theory, be predicted from the size of the chromosome (supplementary Table S7).

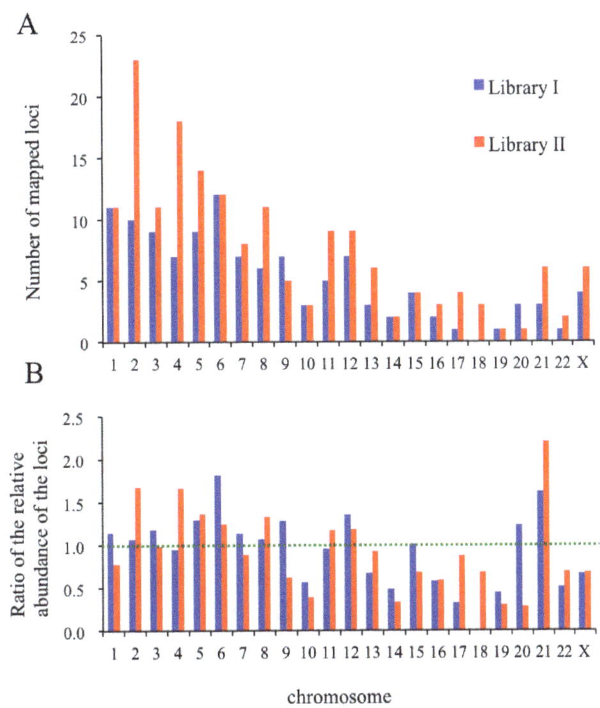

Next, we assigned each locus to a chromosome band. As shown in Figure 10A and in supplementary Table S8, more than 60% of the loci from both of the libraries mapped to G-band regions. The results from library II were unexpected because G- and R-band regions are believed to replicate in late and early S phase, respectively [51]. If this hypothesis about chromosome bands and replication timing is correct, the R-band loci should be enriched in the library prepared from the early S phase cells (library II). The discrepancy between previous cytogenetic observations and our results may be partially explainable based on the results from recent studies on chromosome bands. Studies on high-resolution banding and replication timing now suggest that a chromosome band is made up of several domains (or isochores) that differ in their replication timing [52,53]. Thus, a single G-band is likely to contain multiple domains that replicate in early S phase. Consequently, it is possible that some loci from library II are derived from chromosomal domains in the G-band regions.

Figure 10. Analyses of the flanking regions of the putative replication origin loci mapped to the human genome. The frequencies with which library I and II loci and 236 loci selected from the human genome (control human loci) mapped to G- or R-band regions (**A**) or to intra- or inter-genic regions (**B**). It should be noted that the ratio of the G- and R-bands in the "Control human loci" does not reflect the actual ratio of the bands in the human genome (see Section 3.9 for further explanation); (**C**) Frequency of loci with functional sites in their flanking regions. Information about human genomic loci was mainly obtained from the UCSC Human Genome Browser [54]. The number of loci with functional sites in their flanking regions was counted within the window sizes indicated. Digital DNaseI hypersensitive sites (DNaseI HS) and mammalian conserved sites were classified as being high (H) or medium (M) based on the strength of the signals. TSS: transcription start site. A 5'-terminus of the predicted genes in the UCSC Genes track was considered a TSS. H3K27Ac peaks: The peak signals of K27-acetylated histone H3. A summary of the data appears in supplementary Table S8, and details of the compositions of the sites for each sequence appear in supplementary Table S6A–C.

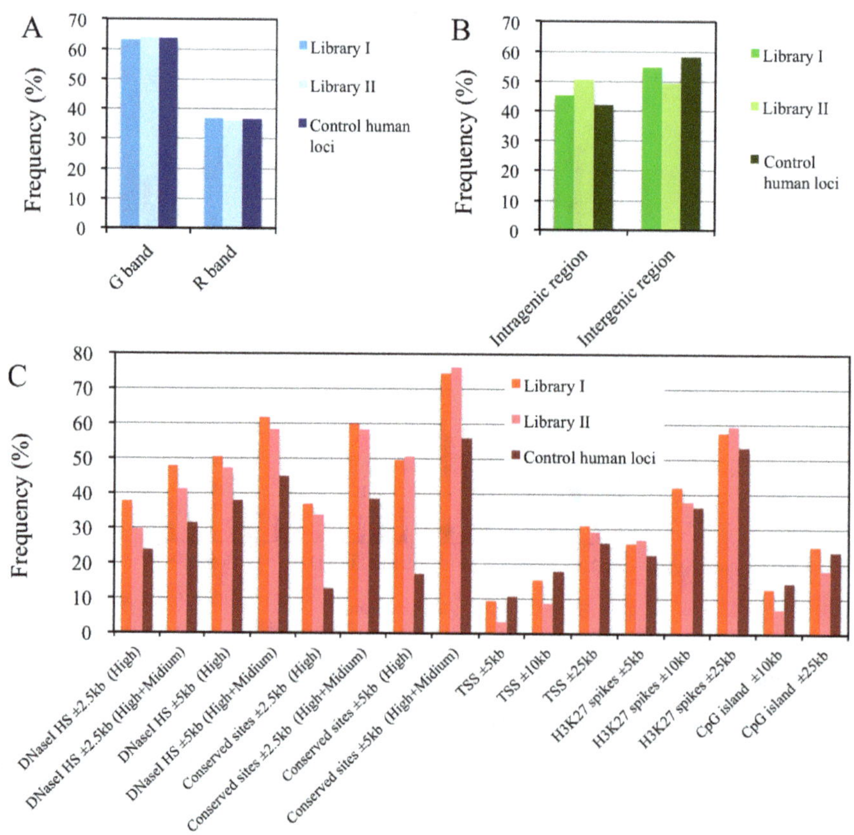

We examined the flanking regions of the mapped loci in the human genome using the project data from ENCODE [55] in the UCSC Genome Browser [54] to look for associations with the predicted genes (Figure 10B), transcription start sites (TSSs), DNase I hypersensitive sites (DNaseI HSs) suggestive of active chromatin domains, mammalian conserved sequences (indicating important genetic elements such as exons), CpG islands (implicated in regulation of gene expression), and the K27-acetylated histone H3 peaks associated with transcriptionally active chromatins (Figure 10C). To compare the compositions of these sites in the flanking sequences, 236 genomic regions from human chromosomes were selected as control loci. This is in keeping with the same R/G-band ratio of the libraries and where the number of control loci from each chromosome almost corresponds to a chromosome size (supplementary Table S6C). Approximately 50% of the putative origin loci from the libraries tended to be located in the intragenic regions, which compared with 42% in the corresponding control loci (Figure 10B). Increased numbers of conserved sites (mainly exons) were detected in the flanking regions of the loci from both libraries in 5 and 10 kb-windows compared with those from the control loci (Figure 10C). In a recent study on origin mapping (using human HeLa and normal lymphoblastoid cell lines), Mesner *et al.* [17] showed that significant fractions (ca. 50%) of the origins resided within the bodies of genes, as was also observed in this study. Valenzuela *et al.* [18] and Cadoret *et al.* [11] also observed preferential localization of replication origins at evolutionarily conserved sequences in about 1% of the human genome in their origin mapping studies. In the recent genome-wide origin mapping study by Besnard *et al.* [19], 57.3% of the origins were associated with genes (promoters, exons and introns), but the majority (45.9%) were found in introns. In the present study, DNase I HSs were also more frequently detected in the flanking sequences (in 5 and 10 kb-windows) of the putative origin loci than in the control loci (Figure 10C). These data strongly suggest a tendency for origin loci to be located in gene (exon)-rich intragenic regions containing open chromatin domains.

By contrast, TSSs, CpG islands, and K27-acetylated histone H3 peaks in the flanking regions of the 10, 20, and 50 kb windows, were similarly observed in both the putative origin loci and the control loci (Figure 10C), suggesting that the majority of the mapped loci are not located near the 5'-upstream regions of the genes. Interestingly, a recent study on genome-wide origin mapping indicated that less than 18% of the origins were associated with TSSs and CpG islands in the human genome [19], a value that is in keeping with our own observations. Although replication origins have been shown to be closely associated with CpG islands and/or TSSs (including promoters in the 5'-upstream regions of the genes) in previous studies with specific mammalian origins [56,57] and the comprehensively mapped origins in fly [15], mouse [12,15] and humans [11,14,18], we do not have any evidence that can account for the discrepancy between these findings and those of our own study. The potential cryptic origins identified in library II of the present study may have arisen from treatment with hydroxyurea. The majority of the origins in the libraries might represent subsets of the origins that are located in the intragenic regions of the genome, rather than in the 5'-upstream regions of the genes.

3. Experimental Section

3.1. Cell Culture

The human nucleotide excision repair-deficient cell line GM8207 (XP6BE[SV40]) from the XP-D female patient [35] was used in this study. Cells were maintained in Dulbecco's modified Eagle's medium (DMEM) supplemented with 10% dialyzed fetal calf serum (dFCS, Sigma, St. Louis, MO, USA or Gibco BRL, Gaithersburg, MD, USA) and 5 μg/mL gentamicin (Sigma) at 37 °C in a CO_2 incubator. The cells used in the experiments were free of *Mycoplasma* contamination.

3.2. UV Irradiation Experiments

Exponentially growing GM8207 cells were cultured in 15-cm-diameter culture dishes with 20 mL of DMEM containing 10% dFCS (*i.e.*, growth medium). The cells were synchronized by incubating at 37 °C for 10 h in the presence of 1 μg/mL aphidicolin (Wako, Osaka, Japan). After washing the cells with 20 mL of warm Ca^{2+} and Mg^{2+}-free phosphate-buffered saline (PBS(−)), the cells were cultured in the growth medium for a further 10 h in the presence of 0.03 μg/mL colcemide (Gibco BRL). Mitotic cells transferred into conical tubes by gentle pipetting were collected by centrifugation at 1,300 rpm for 5 min at room temperature. After washing the cells with 20 mL of warm DMEM (twice), they were suspended in growth medium. Approximately 1.5×10^6 cells were inoculated into a 6-cm culture dish with 5 mL of warm growth medium and cultured for a further 5 h at 37 °C. Cells synchronized in S phase were gently washed by soaking in 5 mL of warm PBS(−) with aspiration. Warm PBS(−) (0.4 mL) was then added to the dish and the cells were simultaneously irradiated using short wave ultraviolet light (254 nm, 500 J/m^2) under a UV lamp (UVGL-58, UVP Inc., Upland, CA, USA). After aspirating with PBS(−), 2 mL of warm growth medium containing 10 μM fluorodeoxyuridine (FUdR: Sigma) was added. After a 30-min culture at 37 °C, the cells were incubated further with 75 μM 5-BrdU (Sigma) for 15 min to label the newly synthesized DNA. In parallel, the cells in one plate were labeled with 10 μCi/mL of [^{3}H]TdR, (118 Ci/mmoL, Amersham, Piscataway, NJ, USA) to monitor the size of the labeled DNAs, as reported previously [58].

The labeling reaction was terminated by addition of sodium azide at a final concentration of 2 mM with chilling on ice, after which the cells were washed with ice-cold PBS(−) and fixed by addition of 0.7 mL of cold ethanol. Cells fixed in this way were scraped from the dish, transferred into Eppendorf tubes and collected by centrifugation. The cells were suspended in 0.7 mL of cold ethanol and stored at −2 0 °C in the dark for purification of the BrdU-incorporated nascent DNAs. For monitoring the influence of UV irradiation on DNA synthesis, the cells were irradiated with the UV doses indicated above, and pulse labeling of the DNA with [^{3}H]TdR was performed in FUdR-containing growth medium for 15 min then terminated as described above. For the pulse-chase experiments, after a 15-min pulse, the radioactive growth medium was replaced with 4 mL of warm growth medium containing 10 mM TdR, after which the cells were incubated at 37 °C for the time periods indicated above, and then harvested for alkaline sucrose density gradient centrifugation.

3.3. Chemical Crosslinking Experiments

DNA crosslinking with Trioxsalen and long-wave UV light was performed according to a previously reported procedure by Russev and Vassilev [33] with the modifications described below. DNA synthesizing GM8207 cells were used for analyzing the effects of crosslinking in DNA replication. Cells (1.5×10^6) synchronized in S phase were prepared by aphidicolin treatment for 15 h followed by a 2-h release. After removing the medium from a 6-cm dish by aspiration and washing the cells with PBS(−), 3 mL of PBS(−) with 15 μL of Trioxsalen (Sigma, 0.5 mg/mL in 50% ethanol) was added and the dish placed on ice for 2 min. The cells were irradiated under a high pressure mercury lamp (BHRF500WH, Iwasaki Electronics, Tokyo, Japan) at a distance of 10 cm (lid-to-lamp) using the conditions indicated above. The reagents were removed by aspiration and washing with 5 mL of PBS(−), after which the cells were incubated in 2 mL of the growth medium at 37 °C for 30 min. The experimental conditions for the pulse-label and chase are described in section 3.2. For nascent DNA preparation after DNA crosslinking, logarithmic growing GM8207 cells were used. Crosslinking was performed by 2-min irradiation (twice), followed by incubation in 2 mL of growth medium at 37 °C for 15 min, after which the cells were labeled with BrdU for 20 min. Labeling reactions were terminated and the cells were fixed as described in Section 3.2.

3.4. BrdU-Labeling of Nascent DNA Synthesized in Early S Phase

GM8207 cells were arrested in early S phase with replication inhibitors and labeled with BrdU to prepare nascent DNAs from the early-firing origins. Briefly, exponentially proliferating cells (≈50% confluent) in a 15-cm dish with 20 mL of growth medium were incubated with 1.25 μg/mL of aphidicolin at 37 °C for 12 h. The cells were released by removing the medium and successive washing with 20 mL of warm DMEM (twice), and cultured in 15 mL of warm growth medium containing 0.03 μg/mL colcemide for 12 h in the dark. Mitotic cells were collected into conical tubes and washed in 40 mL of warm DMEM twice as described in section 3.2. Cells (2×10^6) were inoculated and cultured in a collagen-coated 6-cm dish with 3 mL of warm growth medium containing 5 μg/mL aphidicolin and 1.5 mM hydroxyurea at 37 °C for 5 h. Thereafter, the cells were continuously labeled with 40 μM BrdU at 37 °C for 4 h in the dark and harvested as described in section 3.2.

3.5. Purification of BrdU-Labeled DNA

BrdU-labeled DNAs in the fixed cells were purified by fractionation using alkaline sucrose density gradient centrifugation and successive immunoprecipitation with an anti-BrdU antibody. Cells were handled under dim light or in the dark. Cells fixed in ethanol were pelleted, suspended and washed in 1 mL of ice cold PBS(−) by centrifugation at 5,000 rpm for 2 min (4 °C). The cells were gently resuspended in 0.3 mL of TNE (10 mM Tris-HCl, pH 7.5, 0.1 M NaCl, 1 mM EDTA, pH 8.0) and incubated at 37 °C overnight in the presence of 1% SDS and 0.3 mg/mL proteinase K (Merck, Darmstadt, Germany). After gentle extraction by rotation with phenol-chloroform for 20 min and successive centrifugation at 12,000 rpm for 5 min, the DNA in the aqueous layer was precipitated with 0.25 M KCl and 2.5 volume of ethanol. After washing with 1 mL of 80% ethanol by centrifugation at 12,000 rpm for 10 min, the pellet was air-dried and dissolved in 50 μL of autoclaved distilled water at

4 °C for 5 h. The DNA solution was loaded onto 0.2 mL of alkaline lysis solution (0.5 M NaOH, 0.7 M NaCl, 10 mM EDTA, and 2% sodium *N*-lauroylsarcosinate) overlaid on top of 35 mL of a linear sucrose gradient (5 to 25%) with 0.2 mL of a 60% sucrose cushion in the solution containing 0.3 M NaOH, 0.7 M NaCl, 10 mM EDTA, 0.1% sodium *N*-lauroylsarcosinate. The samples were allowed to stand at 4 °C for 7 h in the dark then centrifuged at 25,000 rpm for 16 h in a Beckman SW28 rotor. Fractions (1 mL each) were carefully collected from the top of the tube. Fractions containing newly synthesized DNA were pooled and each 8 mL of a fraction transferred into a silane-coated centrifugation tube and mixed with ammonium acetate, ethanol. Twenty micrograms of glycogen (Boehringer, Ingelheim, Germany) was used as a carrier. The tube was left at −80 °C for 1 h and the DNA precipitated by centrifugation at 23,000 rpm for 30 min in a SW28 rotor. The DNA was washed with 1 mL of 80% ethanol, air-dried and dissolved in 40 μL of TE buffer (10 mM Tris-HCl, 1 mM EDTA, pH 8.0). For DNA preparation from UV irradiated cells, the fixed cells were washed with PBS(−) and suspended in 50 μL of PBS(−). Samples were directly loaded onto the alkaline lysis solution and left at 4 °C for 8 h before the following centrifugation step. The distribution of nascent DNA in the fractions was monitored by a parallel centrifugation of [^{3}H]TdR-labeled cells as described in the above sections. [^{32}P]-labeled linear pBluescript DNA (*Xho*I cut) was used as a 3 kb size marker in a parallel tube. The distribution of the DNA molecules in a gradient was also confirmed by a parallel centrifugation with a [^{32}P]-labeled 1 kb DNA ladder (Gibco BRL) and successive alkaline gel electrophoresis. Approximately 1–2 mg of purified DNA was used for further immunopurifications. Purified DNA was gently mixed at 4 °C for 1 h in 0.3 mL of binding buffer (20 mM Tris-HCl, pH 8.0, 50 mM NaCl, 1 mM EDTA, 0.1% NP-40) containing 0.2 mL of anti-BrdU monoclonal antibody (clone B44, Becton and Dickinson, San Jose, CA, USA). Then, 20 μL of protein A/G-coated latex beads (Interfacial Dynamics Corp., Portland, OR, USA) were added and the solution incubated at 4 °C for a further 1.5 h in a rotator. Prior to the experiments, the binding capacity of the BrdU-labeled single-stranded DNA and the negligible nonspecific DNA binding of the latex beads were examined. The beads were collected by centrifugation, successively washed with 0.3 mL of binding buffer (once) followed by autoclaved distilled water (at least three times), then suspended in 0.1 mL of TE buffer containing 0.2% SDS and 0.5 mg/mL proteinase K. After incubation at 37 °C for 1.5 h and phenol-chloroform extraction, the DNA was purified by ethanol precipitation with 2 μg of glycogen.

3.6. Cloning of BrdU-Labeled DNA

Purified BrdU-labeled single-stranded DNA was modified through addition of a poly dC tail at the 3'-end with TdT (Gibco BRL) and dCTP according to the manufacturer's instructions. Reactions were terminated by phenol-chloroform extraction and the DNA collected by ethanol precipitation. The complementary strand of the poly dC-tailed single-stranded DNA was synthesized with DNA polymerase and an oligo dG primer. Synthesis of single-stranded DNA from UV irradiated cells was performed with Sequenase™ (version 2, USB, Cleveland, OH, USA). After incubation with DNA and 50 pmoL of the 5'-phosphorylated oligo dG/I primer (5'-GGIIGGGIIGGGIIGGGIIGG-3') at 37 °C for 30 min in a 50 μL reaction mixture, 40 units of Sequenase and the recommended concentrations of dithiothreitol and dNTPs were added and the reaction incubated for 60 min. For nascent DNA from early S phase cells, rTaq DNA polymerase (Toyobo, Tokyo, Japan) was used for DNA synthesis. The

50 μL reaction containing the DNA sample, 10 units of rTaq polymerase, 125 pmoL of the 5'-phosphorylated oligo dG_{12-18} primer (Pharmacia, Uppsala, Sweden), and other components were incubated at 65 °C for 4 min, according to the manufacturer's instructions. DNA synthesis in each reaction was monitored in parallel reactions containing [^{32}P]dCTP and successive 1.3% alkaline agarose gel electrophoresis and autoradiography. After phenol-chloroform extraction and ethanol precipitation, DNA ends were blunted using a DNA blunting kit (Takara Bio, Otsu, Shiga) and the 5' ends of the DNA were phosphorylated with T4 polynucleotide kinase (Toyobo or Gibco BRL), according to the manufacturer's instructions. DNA purified by phenol-chloroform extraction and ethanol precipitation was ligated to a DNA adaptor (UniAmp, Clontech, Mountain View, CA, USA) with T4 DNA ligase (Ligation high, Toyobo). Next, the adaptor-ligated DNAs were PCR amplified with an adaptor-specific primer. *Eco*RI and *Xho*I adaptors, and ΔTth DNA polymerase (Toyobo) and *Ex Taq* DNA polymerase (Takara Bio) were used for adaptor-ligation and PCR amplification of DNA from the UV irradiated and synchronized cells, respectively. After 40 cycles of amplification, the products were separated on a 2% agarose gel. Approximately 0.5–3 kb DNA molecules were purified from the gel by electro-elution followed by ethanol precipitation. The DNA ends were blunted and phosphorylated as described above. The resultant DNA was ligated to *Sma*I-digested and dephosphorylated pUC19 DNA and used to transform *E. coli* DH5α cells. Approximately 500–700 clones derived from the UV-irradiated and the synchronized cells were isolated and designated libraries I and II, respectively, and thereafter used for sequence analyses.

3.7. DNA Sequencing

Preparation and sequencing of plasmid DNA, followed by editing and assembly of the sequences were performed as described previously [45]. Briefly, nucleotide sequences from insert DNAs were read in both directions by an automated DNA sequencer (type 373A, ABI, Foster City, CA, USA). The large DNA sequences in some of the plasmids were sequenced using a SequiTherm™ long-read cycle sequencing kit (Epicentre Technologies, Madison, WI, USA) and analyzed using a long-read sequencer (type 4000L, LICOR, Lincoln, NE, USA). Sequence IDs that are preceded by "2D_" and "D1_–D5_" correspond to clones prepared from libraries I and II, respectively (supplementary Table S3). Non-human DNAs, mitochondrial DNAs, and human DNAs less than 200 bp were removed from the libraries. Nucleotide sequences were deposited in the DNA Data Bank of Japan (accession numbers: AB761625 to AB761978).

3.8. Competitive PCR

Competitive PCR was performed as a quantitative assay of the target sequences in the purified DNA fractions. Genomic DNA used as a control was prepared from confluent GM8207 cells in stationary phase. The procedure was conducted according to Diviacco *et al.* [36]. Briefly, a pair of outer PCR primers and two inner primers (with 20-bp tails at the 5'-ends), were prepared for each locus for amplification of the target DNA and the 20-bp larger competitor DNA. The details of the primers used in the study are shown in supplementary Table S1. Competitor DNA was amplified as described in reference [27] and purified from the gel for DNA quantification. Some competitor DNAs were connected by PCR and their products were cloned into pT7Blue DNA (Novagen, Darmstadt, Germany) to use as competitor DNA. The 10-μL reaction mixture contained 10 mM Tris-HCl, pH 8.3, 50 mM KCl, 1.5 mM $MgCl_2$, 0.2 mM each dNTP, 20 pmoL of each primer, 0.5 unit of rTaq DNA polymerase, and 2–5 ng of template DNA with either 40, 130, 400, 1,200, 3,600, 10,800, or 32,400 molecules of the competitor DNA. In some reactions, 0.1 unit of Perfect Match (Stratagene, La Jolla, CA, USA) was added. Competitive PCR was performed for 40–50 cycles at 94 °C for 20 s, 55–65 °C for 30 s, and 72 °C for 60 s in a GeneAmp PCR System 9600 thermal cycler (Perkin-Elmer, Foster City, CA, USA). Heat denaturation of the DNA at 94 °C as an initial PCR step was omitted. PCR products were separated by electrophoresis on a 10% acrylamide/15% glycerol gel, and then stained with ethidium bromide. Gel band images were acquired with a CCD camera and analyzed by NIH Image for quantitative analysis.

3.9. Sequence Data Analyses and In Silico Mapping to the Human Genome

G+C content determination and prediction of replication-related sequence motifs (*i.e.*, autonomously replicating sequence core sequence [ACS] [45], nuclear scaffold associated region [SAR] [47], nuclear matrix associated region [MAR] [46], topoisomerase II recognition consensus sequence [48], and a 36-bp human ARS [25]) were all performed using Genetyx MAC software (version 16, Genetyx Corp., Tokyo, Japan) as described previously [45]. The number of sequence motifs in a sequence was counted after removing redundant sites. Detection of repeat sequences, low complexity sequences and simple repeat elements including AT-rich elements was performed by the RepeatMasker program [59] under the default setting (AB-Blast engine) via the internet [60]. A single sequence that was assembled from all of the library sequences was used for G+C content and sequence motif content analyses. The sites of G-quadruplex sites with loop lengths of 1–7 nucleotides (G4L1-7) or with loop lengths of 1–15 nucleotides (G4L1-15) in their sequences were obtained using the QGRS Mapper program [61] (with the following setting: Max length, 30; Min-G-group, 2; indicated loop size) via the internet [62]. *In silico* mapping of the putative origin sequences to the human genome and assignment of the chromosome bands were performed by BLAST searches [63] against the *Homo sapiens* build 37.3 genome database (updated on August 2011) in the National Center for Biotechnology Information (NCBI) [64] and by BLAT searches [65] against the human genome assembly data (GRCh37/hg19) in the UCSC Genome Browser [66] and the Ensemble browser [67,68]. Information about the flanking genome sequences, including the predicted genes (the UCSC Genes track), DNase I hypersensitive sites, K27-acetylated histone H3 peaks, CpG islands, and conserved regions among vertebrates was obtained from the UCSC Genome Browser. For analysis of the nucleotide sequence features, 14 sequences from the human genome (each of the six sequences from R- and G-band regions and two from the boundary regions of the bands) were selected as controls; these had comparable G+C content (ca. 36%) to the sequences tested from the libraries (supplementary Table S5). Two hundred and thirty-six loci from the human genome were also selected for analysis of their flanking sequences; this was in keeping with the ratio of the length of each of the chromosomes, as well as with the comparable R/G-band ratio of the loci tested from the libraries (supplementary Table S7C).

4. Conclusions

The aim of this study was to isolate and clone nascent DNAs from replication forks that had arrested near origins of replication in the human genome. We characterized the effects of UV radiation and chemical crosslinking during DNA synthesis in human DNA repair-deficient cells and used UV radiation to clone the nascent DNAs from the origins in the exponentially proliferating cells (Figure 1A). In parallel, we cloned nascent DNAs from cells synchronized in early S phase in the presence of BrdU, aphidicolin and hydroxyurea. Under these conditions, replication forks that initiate in early S phase will arrest and BrdU-labeled nascent DNAs can accumulate around the origins (Figure 1B). Enrichment of the origin-derived DNAs by the UV-mediated origin-trapping method (Figure 6) and the potential origin activities of seven loci identified by the cell synchronization method (Figure 7) were confirmed by use of a quantitative competitive PCR assay, suggesting that successful concentration of the origin-derived DNAs was achieved by these origin-trapping methods. We also showed enrichment of the origin-derived fragments using Trioxsalen-mediated chemical crosslinking (Figures 5 and 6), which has previously been reported although the enriched DNAs from authentic origins were not identified in the mammalian cells in that study [34]. Thus, in addition to UV-radiation, chemical crosslinking may be a useful tool for trapping origins of replication. Sequence analyses of the isolated putative origins revealed distinct sequence compositions for each of the two libraries; this strongly suggests that different origins were isolated by each of the two origin-trapping methods. In the analysis of the flanking sequences, we showed that the putative origin loci from the libraries tended to be located within genes, including evolutionally conserved sites, as reported by others [11,17–19]; however, we failed to find any significant association of the loci with specific functional sites (e.g., CpG islands and TSSs) for transcriptional regulators. Although we do not have any evidence that could account for the discrepancy between our observations and those of others on specific functional sites, it is interesting that a recent genome-wide origin mapping study using human cells revealed that a limited fraction of the origins were associated with CpG islands and TSSs [19]. Several lines of evidence indicate that there is flexible usage of replication origins as well as the existence of efficient and inefficient origins (reviewed in [20]). In fact, in two independent studies on origin mapping using the same method (purification of λ exonuclease-resistant DNAs and microarray platforms) and the same cell lines (independent sub-clones of HeLa cells) [11,14], less than 14% of the origins identified were overlapping, suggesting that only a subset of the origins can be detected, even when the same methods are used. Taken together, these findings suggest the possibility that a subset of replication origins (including the putative origins in our study) associate poorly with TSSs and CpG islands.

We found that there are some potential issues to be resolved in our origin-trapping procedures. As shown in Figure 7, five out of 12 loci isolated from the early S phase cells exhibited poor origin activities in the competitive PCR experiment. In addition, although it has been suggested (in previous studies) that regions within R-bands preferentially replicate during early S phase, a relatively large fraction of G-band-derived sequences were identified in library II from the early S phase cells (Figure 10A). These sequences suggest the possibility of potential contamination from DNAs derived from non-origin regions in the libraries and/or from the firing of cryptic (or inefficient) origins by stringent treatments with replication inhibitors to induce cell cycle arrest, as observed in other studies [17,44]. Although genome-wide analyses of replication origins have been performed in mammalian cells whose genomes have already been sequenced, the origin-trapping methods with UV radiation or crosslinking described herein will be useful for obtaining structural information at the sequence level about the regions near the origins in eukaryotes for which no genomic sequences exist.

Acknowledgments

The authors thank Masanori Naitou, Fumihiro Fujimori, Makoto Abe, and Kazuo Iida for invaluable support with the experiments, especially the competitive PCR assays. We are also grateful to Hiroko Hagiwara for initial contributions to the *in vivo* and *in vitro* replication studies on UV-irradiated cells, Eiko Kohriki for sequence homology searches, Naoko Eki for data preparation, and all of the other laboratory members in RIKEN for technical support and discussion. This work was supported by grants to the Special Coordination Funds for Promoting Science and Technology from the Science and Technology Agency of Japan, and to the Life Science Research Project of RIKEN (F.H.). The authors declare no conflicts of interest.

References

1. Masai, H.; Matsumoto, S.; You, Z.; Yoshizawa-Sugata, N.; Oda, M. Eukaryotic chromosome DNA replication: Where, when, and how? *Annu. Rev. Biochem.* **2010**, *79*, 89–130.
2. Okuno, Y.; Satoh, H.; Sekiguchi, M.; Masukata, H. Clustered adenine/thymine stretches are essential for function of a fission yeast replication origin. *Mol. Cell Biol.* **1999**, *19*, 6699–6709.
3. Segurado, M.; de Luis, A.; Antequera, F. Genome-wide distribution of DNA replication origins at A+T-rich islands in *Schizosaccharomyces pombe*. *EMBO Rep.* **2003**, *4*, 1048–1053.
4. Dai, J.; Chuang, R.Y.; Kelly, T.J. DNA replication origins in the *Schizosaccharomyces pombe* genome. *Proc. Natl. Acad. Sci. USA* **2005**, *102*, 337–342.
5. Cotobal, C.; Segurado, M.; Antequera, F. Structural diversity and dynamics of genomic replication origins in *Schizosaccharomyces pombe*. *EMBO J.* **2010**, *29*, 934–942.
6. Dijkwel, P.A.; Hamlin, J.L. The Chinese hamster dihydrofolate reductase origin consists of multiple potential nascent-strand start sites. *Mol. Cell Biol.* **1995**, *15*, 3023–3031.
7. Hamlin, J.L.; Mesner, L.D.; Dijkwel, P.A. A winding road to origin discovery. *Chromosome Res.* **2010**, *18*, 45–61.
8. Vassilev, L.; Johnson, E.M. An initiation zone of chromosomal DNA replication located upstream of the *c-myc* gene in proliferating HeLa cells. *Mol. Cell Biol.* **1990**, *10*, 4899–4904.
9. Waltz, S.E.; Trivedi, A.A.; Leffak, M. DNA replication initiates non-randomly at multiple sites near the *c-myc* gene in HeLa cells. *Nucleic Acids Res.* **1996**, *24*, 1887–1894.
10. Giacca, M.; Zentilin, L.; Norio, P.; Diviacco, S.; Dimitrova, D.; Contreas, G.; Biamonti, G.; Perini, G.; Weighardt, F.; Riva, S.; *et al.* Fine mapping of a replication origin of human DNA. *Proc. Natl. Acad. Sci. USA* **1994**, *91*, 7119–7123.
11. Cadoret, J.C.; Meisch, F.; Hassan-Zadeh, V.; Luyten, I.; Guillet, C.; Duret, L.; Quesneville, H.; Prioleau, M.N. Genome-wide studies highlight indirect links between human replication origins and gene regulation. *Proc. Natl. Acad. Sci. USA* **2008**, *105*, 15837–15842.

12. Sequeira-Mendes, J.; Diaz-Uriarte, R.; Apedaile, A.; Huntley, D.; Brockdorff, N.; Gomez, M. Transcription initiation activity sets replication origin efficiency in mammalian cells. *PLoS Genet.* **2009**, *5*, e1000446.
13. Hansen, R.S.; Thomas, S.; Sandstrom, R.; Canfield, T.K.; Thurman, R.E.; Weaver, M.; Dorschner, M.O.; Gartler, S.M.; Stamatoyannopoulos, J.A. Sequencing newly replicated DNA reveals widespread plasticity in human replication timing. *Proc. Natl. Acad. Sci. USA* **2010**, *107*, 139–144.
14. Karnani, N.; Taylor, C.M.; Malhotra, A.; Dutta, A. Genomic study of replication initiation in human chromosomes reveals the influence of transcription regulation and chromatin structure on origin selection. *Mol. Biol. Cell* **2010**, *21*, 393–404.
15. Cayrou, C.; Coulombe, P.; Vigneron, A.; Stanojcic, S.; Ganier, O.; Peiffer, I.; Rivals, E.; Puy, A.; Laurent-Chabalier, S.; Desprat, R.; *et al.* Genome-scale analysis of metazoan replication origins reveals their organization in specific but flexible sites defined by conserved features. *Genome Res.* **2011**, *21*, 1438–1449.
16. Martin, M.M.; Ryan, M.; Kim, R.; Zakas, A.L.; Fu, H.; Lin, C.M.; Reinhold, W.C.; Davis, S.R.; Bilke, S.; Liu, H.; *et al.* Genome-wide depletion of replication initiation events in highly transcribed regions. *Genome Res.* **2011**, *21*, 1822–1832.
17. Mesner, L.D.; Valsakumar, V.; Karnani, N.; Dutta, A.; Hamlin, J.L.; Bekiranov, S. Bubble-chip analysis of human origin distributions demonstrates on a genomic scale significant clustering into zones and significant association with transcription. *Genome Res.* **2011**, *21*, 377–389.
18. Valenzuela, M.S.; Chen, Y.; Davis, S.; Yang, F.; Walker, R.L.; Bilke, S.; Lueders, J.; Martin, M.M.; Aladjem, M.I.; Massion, P.P.; *et al.* Preferential localization of human origins of DNA replication at the 5'-ends of expressed genes and at evolutionarily conserved DNA sequences. *PLoS One* **2011**, *6*, e17308.
19. Besnard, E.; Babled, A.; Lapasset, L.; Milhavet, O.; Parrinello, H.; Dantec, C.; Marin, J.M.; Lemaitre, J.M. Unraveling cell type-specific and reprogrammable human replication origin signatures associated with G-quadruplex consensus motifs. *Nat. Struct. Mol. Biol.* **2012**, *19*, 837–844.
20. Gilbert, D.M. Evaluating genome-scale approaches to eukaryotic DNA replication. *Nat. Rev. Genet.* **2010**, *11*, 673–684.
21. Mechali, M. Eukaryotic DNA replication origins: Many choices for appropriate answers. *Nat. Rev. Mol. Cell Biol.* **2010**, *11*, 728–738.
22. Dazy, S.; Gandrillon, O.; Hyrien, O.; Prioleau, M.N. Broadening of DNA replication origin usage during metazoan cell differentiation. *EMBO Rep.* **2006**, *7*, 806–811.
23. Hiratani, I.; Ryba, T.; Itoh, M.; Yokochi, T.; Schwaiger, M.; Chang, C.W.; Lyou, Y.; Townes, T.M.; Schubeler, D.; Gilbert, D.M. Global reorganization of replication domains during embryonic stem cell differentiation. *PLoS Biol.* **2008**, *6*, e245.
24. Nordman, J.; Li, S.; Eng, T.; Macalpine, D.; Orr-Weaver, T.L. Developmental control of the DNA replication and transcription programs. *Genome Res.* **2011**, *21*, 175–181.
25. Price, G.B.; Allarakhia, M.; Cossons, N.; Nielsen, T.; Diaz-Perez, M.; Friedlander, P.; Tao, L.; Zannis-Hadjopoulos, M. Identification of a *cis*-element that determines autonomous DNA replication in eukaryotic cells. *J. Biol. Chem.* **2003**, *278*, 19649–19659.

26. Berezney, R.; Coffey, D.S. Nuclear protein matrix: Association with newly synthesized DNA. *Science* **1975**, *189*, 291–293.
27. Valenzuela, M.S.; Mueller, G.C.; Dasgupta, S. Nuclear matrix-DNA complex resulting from *Eco*R1 digestion of HeLa nucleoids is enriched for DNA replicating forks. *Nucleic Acids Res.* **1983**, *11*, 2155–2164.
28. Nakamura, H.; Morita, T.; Sato, C. Structural organizations of replicon domains during DNA synthetic phase in the mammalian nucleus. *Exp. Cell Res.* **1986**, *165*, 291–297.
29. Razin, S.V.; Kekelidze, M.G.; Lukanidin, E.M.; Scherrer, K.; Georgiev, G.P. Replication origins are attached to the nuclear skeleton. *Nucleic Acids Res.* **1986**, *14*, 8189–8207.
30. Wilson, R.H.; Coverley, D. Relationship between DNA replication and the nuclear matrix. *Genes Cells* **2013**, *18*, 17–31.
31. Vassilev, L.T.; DePamphilis, M.L. Guide to identification of origins of DNA replication in eukaryotic cell chromosomes. *Crit. Rev. Biochem. Mol. Biol.* **1992**, *27*, 445–472.
32. Mesner, L.D.; Crawford, E.L.; Hamlin, J.L. Isolating apparently pure libraries of replication origins from complex genomes. *Mol. Cell* **2006**, *21*, 719–726.
33. Russev, G.; Vassilev, L. Isolation of a DNA fraction from Ehrlich ascites tumour cells containing the putative origin of replication. *J. Mol. Biol.* **1982**, *161*, 77–87.
34. Dimitrova, D.; Vassilev, L.; Anachkova, B.; Russev, G. Isolation and cloning of putative mouse DNA replication initiation sites: Binding to nuclear protein factors. *Nucleic Acids Res.* **1993**, *21*, 5554–5560.
35. Protic-Sabljic, M.; Seetharam, S.; Seidman, M.M.; Kraemer, K.H. An SV40-transformed xeroderma pigmentosum group D cell line: Establishment, ultraviolet sensitivity, transfection efficiency and plasmid mutation induction. *Mutat. Res.* **1986**, *166*, 287–294.
36. Diviacco, S.; Norio, P.; Zentilin, L.; Menzo, S.; Clementi, M.; Biamonti, G.; Riva, S.; Falaschi, A.; Giacca, M. A novel procedure for quantitative polymerase chain reaction by coamplification of competitive templates. *Gene* **1992**, *122*, 313–320.
37. Kitsberg, D.; Selig, S.; Keshet, I.; Cedar, H. Replication structure of the human β-globin gene domain. *Nature* **1993**, *366*, 588–590.
38. Aladjem, M.I.; Groudine, M.; Brody, L.L.; Dieken, E.S.; Fournier, R.E.; Wahl, G.M.; Epner, E.M. Participation of the human β-globin locus control region in initiation of DNA replication. *Science* **1995**, *270*, 815–819.
39. Hansen, R.S.; Canfield, T.K.; Fjeld, A.D.; Mumm, S.; Laird, C.D.; Gartler, S.M. A variable domain of delayed replication in *FRAXA* fragile X chromosomes: X inactivation-like spread of late replication. *Proc. Natl. Acad. Sci. USA* **1997**, *94*, 4587–4592.
40. Lin, H.B.; Dijkwel, P.A.; Hamlin, J.L. Promiscuous initiation on mammalian chromosomal DNA templates and its possible suppression by transcription. *Exp. Cell Res.* **2005**, *308*, 53–64.
41. Norio, P.; Kosiyatrakul, S.; Yang, Q.; Guan, Z.; Brown, N.M.; Thomas, S.; Riblet, R.; Schildkraut, C.L. Progressive activation of DNA replication initiation in large domains of the immunoglobulin heavy chain locus during B cell development. *Mol. Cell* **2005**, *20*, 575–587.
42. Lebofsky, R.; Heilig, R.; Sonnleitner, M.; Weissenbach, J.; Bensimon, A. DNA replication origin interference increases the spacing between initiation events in human cells. *Mol. Biol. Cell* **2006**, *17*, 5337–5345.

43. Guan, Z.; Hughes, C.M.; Kosiyatrakul, S.; Norio, P.; Sen, R.; Fiering, S.; Allis, C.D.; Bouhassira, E.E.; Schildkraut, C.L. Decreased replication origin activity in temporal transition regions. *J. Cell Biol.* **2009**, *187*, 623–635.
44. Anglana, M.; Apiou, F.; Bensimon, A.; Debatisse, M. Dynamics of DNA replication in mammalian somatic cells: Nucleotide pool modulates origin choice and interorigin spacing. *Cell* **2003**, *114*, 385–394.
45. Murakami, Y.; Naitou, M.; Hagiwara, H.; Shibata, T.; Ozawa, M.; Sasanuma, S.; Sasanuma, M.; Tsuchiya, Y.; Soeda, E.; Yokoyama, K.; *et al.* Analysis of the nucleotide sequence of chromosome VI from *Saccharomyces cerevisiae*. *Nat. Genet.* **1995**, *10*, 261–268.
46. Cockerill, P.N.; Garrard, W.T. Chromosomal loop anchorage of the kappa immunoglobulin gene occurs next to the enhancer in a region containing topoisomerase II sites. *Cell* **1986**, *44*, 273–282.
47. Gasser, S.M.; Laemmli, U.K. Cohabitation of scaffold binding regions with upstream/enhancer elements of three developmentally regulated genes of *D. melanogaster*. *Cell* **1986**, *46*, 521–530.
48. Spitzner, J.R.; Muller, M.T. A consensus sequence for cleavage by vertebrate DNA topoisomerase II. *Nucleic Acids Res.* **1988**, *16*, 5533–5556.
49. Mesner, L.D.; Hamlin, J.L.; Dijkwel, P.A. The matrix attachment region in the Chinese hamster dihydrofolate reductase origin of replication may be required for local chromatid separation. *Proc. Natl. Acad. Sci. USA* **2003**, *100*, 3281–3286.
50. Venter, J.C.; Adams, M.D.; Myers, E.W.; Li, P.W.; Mural, R.J.; Sutton, G.G.; Smith, H.O.; Yandell, M.; Evans, C.A.; Holt, R.A.; *et al.* The sequence of the human genome. *Science* **2001**, *291*, 1304–1351.
51. Drouin, R.; Holmquist, G.P.; Richer, C.L. High-resolution replication bands compared with morphologic G- and R-bands. *Adv. Hum. Genet.* **1994**, *22*, 47–115.
52. Watanabe, Y.; Fujiyama, A.; Ichiba, Y.; Hattori, M.; Yada, T.; Sakaki, Y.; Ikemura, T. Chromosome-wide assessment of replication timing for human chromosomes 11q and 21q: Disease-related genes in timing-switch regions. *Hum. Mol. Genet.* **2002**, *11*, 13–21.
53. Costantini, M.; Clay, O.; Auletta, F.; Bernardi, G. An isochore map of human chromosomes. *Genome Res.* **2006**, *16*, 536–541.
54. Meyer, L.R.; Zweig, A.S.; Hinrichs, A.S.; Karolchik, D.; Kuhn, R.M.; Wong, M.; Sloan, C.A.; Rosenbloom, K.R.; Roe, G.; Rhead, B.; *et al.* The UCSC Genome Browser database: Extensions and updates 2013. *Nucleic Acids Res.* **2013**, *41*, D64–D69.
55. Birney, E.; Stamatoyannopoulos, J.A.; Dutta, A.; Guigo, R.; Gingeras, T.R.; Margulies, E.H.; Weng, Z.; Snyder, M.; Dermitzakis, E.T.; Thurman, R.E.; *et al.* Identification and analysis of functional elements in 1% of the human genome by the ENCODE pilot project. *Nature* **2007**, *447*, 799–816.
56. Biamonti, G.; Giacca, M.; Perini, G.; Contreas, G.; Zentilin, L.; Weighardt, F.; Guerra, M.; Della Valle, G.; Saccone, S.; Riva, S.; *et al.* The gene for a novel human lamin maps at a highly transcribed locus of chromosome 19 which replicates at the onset of S-phase. *Mol. Cell Biol.* **1992**, *12*, 3499–3506.
57. Delgado, S.; Gomez, M.; Bird, A.; Antequera, F. Initiation of DNA replication at CpG islands in mammalian chromosomes. *EMBO J.* **1998**, *17*, 2426–2435.

58. Eki, T.; Murakami, Y.; Enomoto, T.; Hanaoka, F.; Yamada, M. Characterization of DNA replication at a restrictive temperature in a mouse DNA temperature-sensitive mutant, tsFT20 strain, containing heat-labile DNA polymerase α activity. *J. Biol. Chem.* **1986**, *261*, 8888–8893.
59. Smit, A.F.A.; Hubley, R.; Green, P. RepeatMasker Open-3.0 (1996–2010). Available online: http://www.repeatmasker.org/ (accessed on 10 April 2013).
60. RepeatMasker. Available online: http://www.repeatmasker.org/cgi-bin/WEBRepeatMasker/ (accessed on 10 April 2013).
61. Kikin, O.; D'Antonio, L.; Bagga, P.S. QGRS Mapper: A web-based server for predicting G-quadruplexes in nucleotide sequences. *Nucleic Acids Res.* **2006**, *34*, W676–W682.
62. QGRS Mapper. Available online: http://bioinformatics.ramapo.edu/QGRS/index.php/ (accessed on 10 April 2013).
63. Altschul, S.F.; Gish, W.; Miller, W.; Myers, E.W.; Lipman, D.J. Basic local alignment search tool. *J. Mol. Biol.* **1990**, *215*, 403–410.
64. The NCBI Blast. Available online: http://blast.ncbi.nlm.nih.gov/Blast.cgi/ (accessed on 10 April 2013).
65. Kent, W.J. BLAT—The BLAST-like alignment tool. *Genome Res.* **2002**, *12*, 656–664.
66. Human BLAT Search. Available online: http://genome.ucsc.edu/cgi-bin/hgBlat?hgsid=306482237&command=start/ (accessed on 10 April 2013).
67. Flicek, P.; Amode, M.R.; Barrell, D.; Beal, K.; Brent, S.; Carvalho-Silva, D.; Clapham, P.; Coates, G.; Fairley, S.; Fitzgerald, S.; *et al.* Ensembl 2012. *Nucleic Acids Res.* **2012**, *40*, D84–D90.
68. BLAST/BLAT. Available online: http://www.ensembl.org/Multi/blastview/ (accessed on 10 April 2013).

2

DNA Methylation Biomarkers: Cancer and Beyond

Thomas Mikeska [1] and Jeffrey M. Craig [2,3,*]

1 Genetic Technologies Ltd., Fitzroy, Victoria 3065, Australia; E-Mail: thomas.mikeska@gtglabs.com
2 Murdoch Childrens Research Institute, Royal Children's Hospital, Parkville, Victoria 3052, Australia
3 Department of Paediatrics, The University of Melbourne, Parkville, Victoria 3052, Australia

* Author to whom correspondence should be addressed; E-Mail: jeff.craig@mcri.edu.au;

Abstract: Biomarkers are naturally-occurring characteristics by which a particular pathological process or disease can be identified or monitored. They can reflect past environmental exposures, predict disease onset or course, or determine a patient's response to therapy. Epigenetic changes are such characteristics, with most epigenetic biomarkers discovered to date based on the epigenetic mark of DNA methylation. Many tissue types are suitable for the discovery of DNA methylation biomarkers including cell-based samples such as blood and tumor material and cell-free DNA samples such as plasma. DNA methylation biomarkers with diagnostic, prognostic and predictive power are already in clinical trials or in a clinical setting for cancer. Outside cancer, strong evidence that complex disease originates in early life is opening up exciting new avenues for the detection of DNA methylation biomarkers for adverse early life environment and for estimation of future disease risk. However, there are a number of limitations to overcome before such biomarkers reach the clinic. Nevertheless, DNA methylation biomarkers have great potential to contribute to personalized medicine throughout life. We review the current state of play for DNA methylation biomarkers, discuss the barriers that must be crossed on the way to implementation in a clinical setting, and predict their future use for human disease.

Keywords: cancer; diabetes; obesity; smoking; stress; autism; schizophrenia; bipolar disorder; depression; environmental factors

---

1. Introduction

A biomarker is any biological characteristic that can be objectively measured and evaluated as an indicator of normal biological process, pathogenic process, or pharmacological response to a therapeutic intervention [1]. Biomarkers can be used at any stage of a disease and can be associated with its cause or latency (risk biomarkers), onset (diagnostic biomarkers), clinical course (prognostic biomarkers), or response to treatment (predictive biomarkers) ([2–4] and references therein). Biomarkers can also be associated with specific environments (exposure biomarkers). As almost all complex human diseases are caused by a mixture of genetic and environmental variation, biomarkers, especially those antecedent to disease, can be influenced by either of these factors. Biomarkers can also reflect the mechanisms by which exposure and disease are related. They can stratify individuals according to risk or prognosis and they can be used as targets or surrogate endpoints in clinical trials. An ideal biomarker must be able to provide clinically-relevant information, be accurately measurable in multiple individuals, ideally across multiple populations [2,4]. In this review we focus on DNA methylation biomarkers, review the current state of the field, and discuss limitations and our expectations for the future.

2. Epigenetics and Disease Latency

Epigenetics refers to the molecular marks that influence gene function in a mitotically-heritable manner [5]. Epigenetic marks are themselves influenced by a mix of genetic and environmental variation [6]. A typical gene will be regulated by epigenetic marks present at one or more gene promoters, which are usually but not exclusively close to its transcriptional start site, and by one or more enhancers, which can be within the gene or a large distance away from the gene [7]. Such regions of transcriptional control exhibit molecular characteristics in the form of multiple, synergistic epigenetic marks.

Epigenetic marks include methylation of DNA at the cytosine residue of cytosine-phosphate-guanine (CpG) dinucleotides and covalent modifications of amino acid residues within histone proteins that are responsible for the primary packaging of DNA. Other cellular components, such as those involved in writing, reading, and erasing epigenetic marks, determine the local chromatin structure, which at two extremes can be open and active or closed and inactive [8].

In the human genome, DNA methylation occurs almost exclusively at CpG dinucleotides. The cytosine residue of a CpG dinucleotide can be covalently modified by adding a methyl group to its carbon-5 atom resulting in 5-methylcytosine. The methyl group is transferred from *S*-adenosyl-L-methionine to a cytosine residue via DNA methyltransferases (reviewed in [9,10]). CpG dinucleotides are unevenly distributed throughout the genome and are generally methylated [11]. Some CpG dinucleotides are clustered in regions known as CpG-islands, which can span hundreds to thousands of base pairs and are generally unmethylated [11].

The definition of a CpG island has been quite arbitrary and two algorithms have found widespread use throughout the scientific community to identify CpG-islands in genomic DNA sequences [12,13]. However, genome-wide studies have vastly increased our understanding of the human genome over the last few years, and more sophisticated algorithms for the identification of CpG-islands have been developed [14–16].

CpG islands are often, but not exclusively, located at gene promoters, where the methylation status is generally correlated with transcriptional gene activity [11]. DNA methylation can have other (regulatory) functions outside promoter regions, for example in intragenic regions [17,18], intergenic regions [19] and in regions of low CpG density [20]. DNA methylation performs a regulatory role at local and global levels. Global methylation is mainly determined by methylated CpG dinucleotides in highly repeated DNA sequences, such as satellite DNAs, which play an important function in maintaining genome stability [21]. DNA methylation level changes, namely local hypermethylation (gain of DNA methylation) and global hypomethylation (loss of DNA methylation), are often associated with a diseased state.

Most studies of the role of epigenetics in human disease have focused on investigating disease-associated DNA methylation changes and on determining the environmental influence on DNA methylation variations. Most of these have focused on cancer. It is now widely accepted that cancer results from a combination of genetic and epigenetic disruption or dysfunction (reviewed in [22]). Whereas the underlying causes of cancer remains largely elusive, it has also become clear that certain environmental factors such as the exposure to certain chemicals, toxins or heavy metals are capable of altering the epigenome and ultimately increase the risk of developing cancer [23–25].

Outside cancer, environmental influences on DNA methylation are the centre of the developmental origins of health and disease (DOHaD) phenomenon [26,27]. In this phenomenon, which grew out of the "fetal origins" hypothesis [28], adverse environment, *in utero* or in early postnatal life, programs the body for complex, non-communicable diseases including diabetes, cardiovascular disease (CVD) and neurodevelopmental disorders. Central to this phenomenon is the hypothesis that disease predisposition results when postnatal environment is mismatched to prenatal environment [29].

The DOHaD phenomenon involves a period of disease latency between the early origins and the later clinical manifestation. This latency may be in the form of a few years, for example with obesity and autism, or many decades, in the case of CVD. Non-epigenetic biomarkers of latent conditions such as CVD are already being developed and these include plasma high sensitivity C-reactive peptide, blood pressure, body mass index and artery wall thickness [30,31]. We discuss below how epigenetic biomarkers, in particular DNA methylation biomarkers, are being identified within the context of cancer and DOHaD.

3. Tissues and Bodily Fluids Suitable for Analysis of DNA Methylation Biomarkers

Almost any biological tissue sample or bodily fluid can be used for DNA methylation analysis. DNA methylation is the most robust epigenetic mark and will survive most sample storage conditions including, in the case of Guthrie neonatal blood spots, long-term drying [32]. DNA methylation can also be studied in histological specimens such as formalin-fixed paraffin-embedded (FFPE) samples [33] and microscopic preparations [34]. The robustness of DNA methylation marks makes DNA methylation analysis very attractive in a clinical environment as the analysis of gene expression pattern and histone modifications require more careful storage conditions, either with an RNA-preserving agent, by snap-freezing, or by cryopreservation of viable cells. In most cancers, (primary) tumor biopsies can be sampled but for the early detection of cancer and most other non-communicable diseases, only peripheral, easy-to-access tissues or bodily fluids can be collected. Such samples include venous peripheral blood, buccal epithelium or saliva, urine, stools, bronchial aspirates, and, even in some cases, muscle or adipose tissue [35–39] (Figure 1). At birth, placenta, umbilical cords and fetal membranes are also suitable tissues for analysis of DNA methylation [40–42].

Figure 1. Illustration of the variety of tissues that can be used to investigate DNA methylation biomarkers. Note that tumor tissue is not listed individually as a cancer can affect any part of the body.

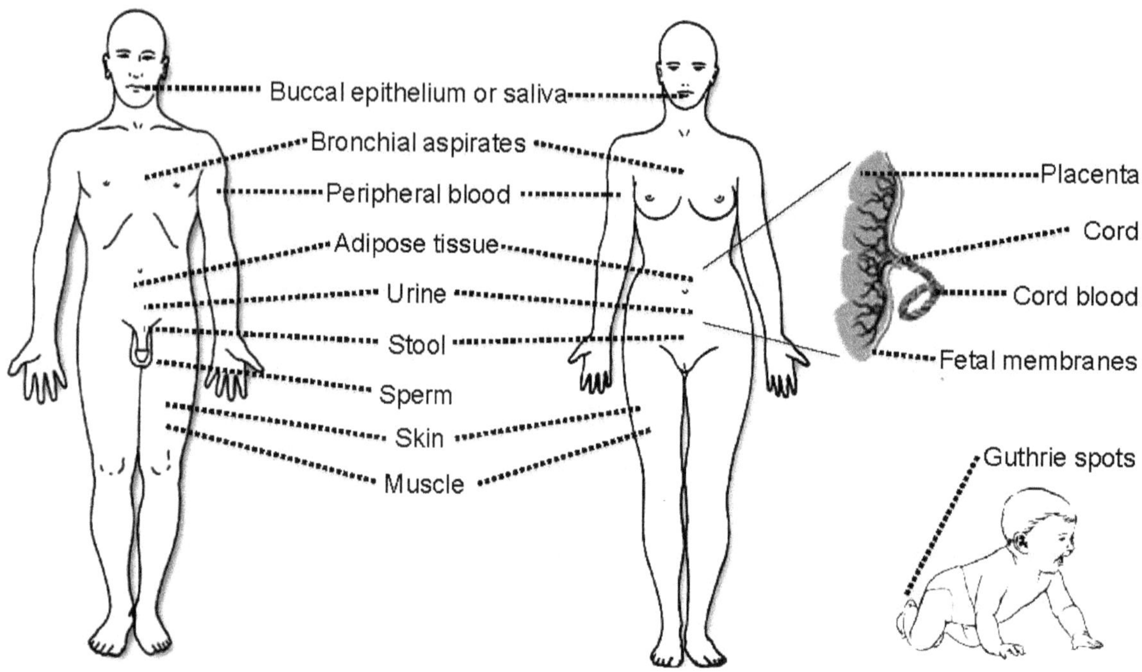

It is important to note that even though it is desirable to measure disease-associated methylation biomarkers in a disease-relevant tissue, this condition does not always need to be met if a methylation biomarker is tightly associated with disease state. This is especially the case for tissues such as the brain and heart that can only be sampled *post mortem*.

Cellular homogeneity within a tissue is also a desirable characteristic for a DNA methylation biomarker [43]. Tissues such as blood or even blood fractions such as mononuclear cells, exhibit cellular heterogeneity [44–46]. However, methods have been developed to control for such heterogeneity, using either differential cell counts [47] or *post hoc* in regression models [48–50].

4. Parameters for Developing DNA Methylation Biomarkers

Before we go into more detail about specific DNA methylation biomarkers, we will review the measures of particular importance for assay performance and the barriers that must be breached in developing DNA methylation biomarkers. The nomenclature we use in this review is generally already in use, although it has not been previously summarized in such a way. It is as follows: single studies provide *potential biomarkers*, which could be *validated* using an independent technique and *replicated* in an independent cohort, also known as external validation. Following the systematic review and/or meta-analysis of a large number of independent studies, they become *candidate clinical biomarkers* that can enter clinical trials. Once approved, they become *proven clinical biomarkers* (Table 1).

Table 1. Nomenclature used in this review for the stages of DNA methylation biomarker development.

Nomenclature	Description
Potential biomarker	Results of a single study
Validated biomarker	Same finding using an independent method
Replicated biomarker	Same finding in independent cohort(s)
Candidate clinical biomarker	Replicated in multiple cohorts and subjected to systematic review and meta-analysis; most likely undergoing clinical trials
Proven clinical biomarker	Used in clinical practice

4.1. Methods for DNA Methylation Biomarker Discovery

Genome-wide profiling of DNA methylation patterns of healthy and diseased individuals has enabled the identification of potential methylation biomarkers for many diseases, most prominently in cancer but also other diseases such as metabolic or neurodevelopmental disorders. Following initial studies using pre-selected candidate gene approaches [51–53], many different genome-wide methods have been developed and used in the scientific community for DNA methylation biomarker discovery and good overviews are provided elsewhere [54–58]. Other scientific publications review such methods in the context of methylome-wide association studies (MWAS), which utilize a variety of platforms [59,60]. Typically, MWAS, as a subset of epigenome-wide association studies (EWAS), involves regression of DNA methylation at thousands to millions of CpG dinucleotides or CpG-rich regions on disease phenotype, outcomes or interventions. Such analyses usually adjust for multiple testing to produce potential methylation biomarkers in the form of differentially-methylated probes (DMPs) or regions (DMRs). Often, DMPs or DMRs are validated using locus-specific methods. The next stages of discovery following replication involve longitudinal analysis to resolve the issue of cause *vs.* effect in MWAS, and importantly to show whether replicated biomarkers can be used to predict a disease before its clinical onset or predict clinical outcomes after onset or after therapeutic intervention. Following discovery of such replicated biomarkers, further replication followed by meta-analysis and/or systematic review are required, at which stage these candidate clinical methylation biomarkers are ready for clinical trials leading to clinical proven methylation biomarkers. In this review we will focus on single locus DNA methylation biomarkers at all stages of discovery.

4.2. DNA Methylation Assay Sensitivity and Specificity

Assay sensitivity describes the proportion of patients with disease who have a positive test result (true positive rate), whereas the assay specificity describes the proportion of patients without disease who have a negative test result (true negative result) [61]. The ideal assay would show 100% sensitivity and 100% specificity. In other words, the test is never positive for a disease-free patient and never negative for a patient with disease. However, this ideal scenario is rarely achieved. It is also important to note that an assay with a sensitivity of 50% and a specificity of 50% is no better than tossing a coin to decide if the patient is harboring the disease or is disease-free [61].

The receiver operating characteristic (ROC) curve is a fundamental tool for diagnostic test or biomarker evaluation and visually displays the interdependency of specificity and sensitivity [62,63].

In a ROC curve the true positive rate (sensitivity; y-axis) is plotted in function of the false positive rate (1-specificity; x-axis). The area under the curve is equal to the probability that a classifier will rank a randomly chosen positive instance higher than a randomly chosen negative one. In other words, for a well performing diagnostic test or biomarker the curve is located towards the upper left corner. On the other hand a less well-performing diagnostic test or biomarker is characterized by a curve close to a diagonal line, representing a state in which sensitivity and specificity are similar.

It is desirable to achieve values for sensitivity and specificity as high as possible. However, for some tests it might be acceptable to achieve a higher sensitivity by sacrificing assay specificity or *vice versa*. This could be the case in particular for diseases for which a misclassification would result in severe consequences for the patient [61]. Acceptable values for sensitivity and specificity of a testing procedure can be determined by comparing to existing values of a test currently considered as gold standard. It is also important to consider that a diagnostic test is providing information independent of the experience of a clinician, which sometimes varies dramatically among hospitals and countries. However, it remains to be determined how easily the new testing procedure can be implemented in a clinical environment.

4.3. Barriers to Developing, Testing and Using DNA Methylation Biomarkers

Despite the promise of epigenetic biomarkers, so far only a few DNA methylation-based candidate biomarkers have reached the potential for use in a clinical setting, and all these are mainly related to the field of cancer. As with disease phenotypes, each clinical DNA methylation biomarker would need to be measured accurately and reproducibly. Differences in DNA methylation between cases and controls may be large (e.g., more than 50%) in cancer but in other non-communicable diseases may often be less than 5%. Methods used to measure methylation must be accurate to well below this level of resolution. The analytical sensitivity of specific methods is discussed below. Next, variability within the population needs to be small to maximize assay sensitivity and specificity. Predictive power also needs to be high. Positive predictive power is the percentage of people with a positive test who actually get the disease. These hurdles are all similar to those for any clinical trial.

5. Methods Suitable for the Analysis of Locus-Specific DNA Methylation Biomarkers

Many different methods have been described for the investigation of locus-specific DNA methylation (reviewed in [58,64–67]). Whereas some methods use genomic DNA for methylation analysis, the majority of methods require bisulfite-treated DNA as starting material [68,69]. Bisulfite treatment converts unmethylated cytosines to uracil, whereas 5-methylcytosines are relatively inert under reaction conditions. Subsequent use of bisulfite-treated DNA in PCR replaces the uracils with thymines and 5-methylcytosines with cytosines. Therefore, the methylation status of a particular CpG dinucleotide is detected indirectly [70].

The use of bisulfite-treated DNA has three important consequences for downstream applications for DNA methylation detection. Firstly, a considerable loss of initial input DNA can occur, due to extensive DNA degradation during the preparation and purification of bisulfite-treated DNA [71–74]. Loss of amplifiable DNA can be critical in particular for those samples where only a limited amount of genomic DNA is available, such as those from very small biopsies. Secondly, a poor bisulfite conversion rate can lead to false-positive results. This is of particular importance for very sensitive DNA methylation detection methods, such as those based on methylation-specific PCR (MSP) [75]. However, the use of a commercially-available bisulfite conversion kit can help to improve DNA recovery and to control for a proper bisulfite conversion rate [67,72]. Thirdly, PCR amplification may sometimes be biased towards unmethylated or methylated templates due to differences in CG content [76]. However, different approaches have been described in the literature to overcome or at least to minimize a potential PCR amplification bias [77–80].

Another problem for most downstream applications is the presence of heterogeneous DNA methylation patterns at many gene loci [81–83]. Heterogeneous methylation patterns are characterized by the presence of multiple epialleles (alleles which differ in the pattern of methylated and unmethylated CpG dinucleotides across the analyzed region). As every sample has its own set of epialleles, it can complicate quantification of methylation (reviewed in [84]) and cut-off value settings for when to call a sample unmethylated or methylated. The need for cut-off values also demands the use of quantitative DNA methylation detection methods, in particular for those gene loci, which are hypomethylated (loss of DNA methylation), or where already variable background methylation is present in healthy individuals [81,85].

Despite the many methodologies available for DNA methylation analysis the methodological considerations and requirements of a molecular diagnostics laboratory renders only a fraction of these methods suitable for DNA methylation analysis in a clinical setting. Such methods would need to use small quantities of DNA of varying quality. The latter is of particular importance for formalin-fixed paraffin-embedded (FFPE) specimens where the DNA is often degraded and chemically modified [86]. Ideally, DNA methylation detection methods for clinical settings should be low cost, easy to use, automatable, and capable of processing many samples in parallel in order to minimize costs of future tests. In the following sections we will discuss methods for DNA methylation detection suitable for use in clinical settings or in a molecular diagnostic laboratory.

Bisulfite pyrosequencing (Qiagen, Hilden, Germany) is based on sequencing-by-synthesis methodology and uses bisulfite-treated DNA as starting material [87–89]. This method is relatively cost- and time-effective, and is suitable for DNA methylation analysis of single gene loci. DNA methylation can be determined at single CpG dinucleotide resolution but methylation levels are provided in a quantitative manner for each CpG site as an average across all epialleles amplified during PCR. The analytical sensitivity is about 5%–10% for individual CpG dinucleotides [90,91]. This approach has a high-throughput capacity and is well suited for the analysis of small PCR amplicons, such as those typically generated from FFPE specimens. Importantly, this approach allows to quality control for a sufficient bisulfite conversion rate. However, the downside of this approach is that the instrument required to perform DNA methylation analysis is rather costly.

The MassARRAY EpiTYPER (Sequenom Inc., San Diego, CA, USA) also requires bisulfite-treated DNA as starting material and uses matrix-assisted laser desorption ionization time-of-flight (MALDI-TOF) mass spectrometry to extract (semi-) quantitative DNA methylation information from shifts and intensities of fragment signals after base-specific cleavage of PCR amplified epialleles present at single gene loci [92]. DNA methylation levels are determined as an average for a single CpG dinucleotide, or for multiple CpG dinucleotides, clusters of CpGs on the same fragment or for multiple CpGs across all fragments of amplicons generated during PCR [93]. Nevertheless, this approach is suitable for providing an almost complete methylation profile across the region-of-interest [92]. The analytical sensitivity is similar to bisulfite pyrosequencing [93] and DNA methylation data obtained by both methods for the same set of CpG dinucleotides has been shown to be highly concordant [81]. Like bisulfite pyrosequencing, MassARRAY EpiTYPER is suitable for high sample throughput and also requires the purchase of an expensive instrument.

Methylation-sensitive high-resolution melting (MS-HRM) is an inexpensive, fast and medium to high throughput screening methodology for DNA methylation analysis at single gene loci [94,95]. This approach requires bisulfite-treated DNA as starting material and exploits the differential melting behavior of PCR products generated from unmethylated and methylated epialleles. The melting profile of an unknown sample is compared to melting profiles of a DNA methylation standard series. This allows the reliably detection of homogeneous methylation levels down to 1%–5%, and can detect the presence of heterogeneous methylation patterns. However, the presence of heterogeneous DNA methylation allows the estimation of methylation levels in a semi-quantitative or qualitative manner. This is because the presence of heterogeneous DNA methylation results in a complex melting profile that does not allow the ready estimation of the amount of methylated epialleles; the result is largely qualitative [84]. MS-HRM is quite attractive in a clinical environment as PCR amplification and subsequent DNA methylation analysis is performed in one tube, which minimizes the risk of sample mix-up and sample cross contamination [96]. However, MS-HRM is not suitable on its own for use in a clinical setting as this method is not capable to deliver quantitative methylation information. Nevertheless, MS-HRM PCR products can be further quantified for DNA methylation using bisulfite pyrosequencing [97].

Another group of important approaches for DNA methylation detection is based on methylation-specific PCR (MSP) [98]. The strength of MSP-based approaches comes from the high analytical sensitivity, which allows them to detect only few methylated epialleles in a large background of unmethylated epialleles. The high analytical sensitivity originates from PCR primers containing CpG dinucleotides that selectively amplify only methylated epialleles. However, conventional MSP is not suitable for use in clinical settings as this approach detects DNA methylation only in a qualitative manner [98]. This can result in an overestimation of methylation in particular for those samples where background methylation is already present in normal tissues [81]. Moreover, conventional MSP is difficult to standardize between different laboratories and is also well known to generate false-positive, as well false-negative results, especially when DNA of low quality is used as starting material, for example FFPE-derived DNA [99,100]. Nevertheless, quantitative offshoots of conventional MSP, such as MethyLight [101], ConLight-MSP [102], MS-FLAG [103], SMART-MSP [104] and HeavyMethyl [105] are potentially suitable approaches for use in a clinical environment. The latter approach has been already successfully applied for DNA methylation detection of *SHOX2* and *PITX2* (see below).

Methylation-sensitive multiplex ligation-dependent probe amplification (MS-MLPA) plays a key role in the diagnosis of genomic imprinting disorders (see below) [106]. Different to the methods described above, this method uses genomic DNA as starting material to produce semi-quantitative DNA methylation information for single CpG dinucleotides. MS-MLPA relies on CpG dinucleotide-specific probes and a digestion step using the methylation-sensitive restriction endonuclease *HhaI* prior to PCR amplification to distinguish unmethylated from methylated epialleles. DNA methylation levels are determined by comparing peak sizes of patient samples with control samples and the analytical sensitivity is approximately 5%–20% [107–110]. MS-MLPA is suitable for high-throughput screening, is relatively cost-effective and does not require non-standard laboratory instruments as the PCR amplification products are separated by capillary electrophoresis on a DNA analyzer instrument.

The use of genomic DNA for methylation analysis is quite attractive as it avoids problems associated with bisulfite treatment. However, as MS-MLPA is based on a digestion step with a methylation-sensitive restriction endonuclease, false-positive results can occur as a result of incomplete digestion, in particular with DNA of poor quality. The use of the restriction endonuclease *HhaI* also limits the investigation of DNA methylation to *HhaI* recognition sites and therefore provides only a limited view of the DNA methylation landscape of any region of interest. However, MS-MLPA is capable of analyzing up to 50 CpG dinucleotides at any one time and allows the determination of DNA methylation levels at different gene loci simultaneously. Moreover, MS-MLPA can be combined with gene copy number and point mutation detection, which makes it a quite flexible methodology [110].

Lessons Learned from the DNA Methylation Biomarker MGMT

The DNA repair gene O^6-methylguanine-DNA methyltransferase (*MGMT*) was first characterized in the early 1990s [111,112] and its key role in the resistance of malignant glioma to alkylating drugs was proposed repeatedly [113–115]. Approximately ten years later, a first link was established between *MGMT* methylation and improved patient outcome in response of malignant gliomas to the alkylating drug carmustine [116]. However, the relatively small number of patients investigated as well as some flaws in study design raised concerns of the validity of the results and warranted confirmation of the potential predictive biomarker *MGMT* (see comments in [116,117]).

Subsequently, *MGMT* methylation as a predictive biomarker of a patient's response to alkylating drug regimens was replicated on different sample cohorts with mixed success. Methylation of *MGMT* was shown to serves as a predictive biomarker for determining response of glioma and glioblastoma patients treated with the alkylating agent Temozolomide [117,118]. Nevertheless, another study was not able to replicate *MGMT* methylation as a predictive biomarker in glioblastoma patients treated with alkylating drug regimens [119].

However, the seminal findings of a clinical trial reported in 2005, conducted by Hegi and colleagues, clearly showed that glioblastoma patients treated with Temozolomide showed a survival benefit if the promoter-associated CpG-island of the *MGMT* gene was methylated [120]. Since then, several clinical trials have confirmed *MGMT* methylation as a candidate clinical biomarker for determining patient response to Temozolomide treatment and it is now a proven clinical biomarker (reviewed in [121]).

Since 2005, many research groups and commercial companies (Table 2) have spent much effort developing assays to investigate the methylation status of *MGMT* by using various methods and platforms [122–126]. However, these methods varied in analytical sensitivity and provided methylation information ranging from purely qualitative to quantitative. As consequence, the general lack of consensus for an agreed methodology and the widespread use of inappropriate methodologies slowed down the implementation of *MGMT* methylation analysis in molecular diagnostics [127].

Table 2. Commercially-available DNA methylation test kits for cancer. References are either systematic reviews/meta-analyses [1] or a set of corroborating references [2]. This table is an updated version of that shown in [127].

Gene(s)	Type of Biomarker	Type of Cancer	Diagnostic Test Kit: Brand Name (Manufacturer)	References
VIM	diagnostic	Colorectal	Cologuard (Exact Sciences)	[128] [1]
SEPT9	diagnostic	Colorectal	Epi proColon (Epigenomics) ColoVantage (Quest Diagnostics) RealTime mS9 (Abbott)	[129] [1]
SHOX2	diagnostic	Lung	Epi prolong (Epigenomics)	[130–135] [2]
GSTP1/APC/RASSF1A	diagnostic	Prostate	ConfirmMDx (MDx Health)	[136–138] [1]
MGMT	predictive	Glioblastoma	PredictMDx Glioblastoma (MDx Health) SALSA MS-MLPA probemix ME011 Mismatch Repair genes (MRC-Holland) PyroMark MGMT Kit (Qiagen)	[121,139,140] [1]

Several recent studies assessing the clinical utility of different methodologies for *MGMT* methylation detection favor quantitative approaches such as bisulfite pyrosequencing [141,142]. Quantitative approaches are necessary to determine cut-off values for methylation ranges related to clinical information such as prognosis [143,144]. However, methylation cut-off values are not universal for a particular gene and strongly depend on the method used for DNA methylation analysis. Even by using the same methodology for methylation analysis requires determination of cut-off values for each assay as these values also depend on the region of the gene investigated, PCR primers and PCR conditions used as well as minimal tumor content required [143–145]. It has also been recognized that careful studies of the entire *MGMT* promoter-associated CpG-island are required to determine those CpG dinucleotides or CpG clusters suitable as a surrogate biomarker for biological or clinical relevant information [83,146].

Keeping in mind that *MGMT* methylation was one of the first DNA methylation biomarkers to be identified, it is not surprising that it took a considerable amount of time until it found its way into the clinic. Advancements in study and clinical trial design will certainly help to speed up replication and clinical implementation of new DNA methylation biomarker. However, the current lack of an agreed methodology as the gold standard for DNA methylation analysis is still a roadblock to overcome. For a more detailed view on which milestones need to be achieved in bringing a DNA methylation biomarker into clinical practice we refer the interested reader elsewhere [127].

6. DNA Methylation Biomarkers

To date, the vast majority of replicated and candidate clinical DNA methylation biomarkers come from cancer research. Clinically-relevant DNA methylation biomarkers outside cancer exist for diseases originating from genomic imprinting disorders, such as Prader-Willi and Angelman syndrome (see below), and are currently being developed for a wide range of environmental agents and the chronic diseases to which they predispose. The following sections will give an overview of promising DNA methylation biomarkers for potential clinical use.

6.1. Candidate Clinical DNA Methylation Biomarkers for Cancer

A selection of candidate clinical DNA methylation biomarkers for cancer will be discussed below; many others have been described in greater detail elsewhere (e.g., [127,147–161]) or have been subject of systematic reviews and meta-analysis (e.g., [129,136–140,162]). Not surprisingly, much effort has been spent in identifying diagnostic DNA methylation biomarkers suitable for early detection and diagnosis of cancer. Early detection allows treatment of the cancer at a stage that is generally considered beneficial for disease outcome. Such tests could be blood-based or use other bodily fluids collected less invasively, which makes it very convenient to the patient. Prognostic biomarkers would provide information on a patient's overall survival if the disease is left untreated, whereas predictive biomarkers would be suitable for determining a patient's response to a certain drug regimen. The latter are of particular importance as they may help to minimize the health burden of patients, as well as to minimize costs for healthcare providers for unnecessary drug treatment.

DNA methylation-based candidate clinical biomarker genes for the early detection include vimentin (*VIM*) [128,163], septin 9 (*SEPT9*) [129,164], and syndecan 2 (*SDC2*) [165,166] for colorectal cancer, glutathione S-transferase pi 1 (*GSTP1*) for prostate cancer [136,167,168], and cyclin-dependent kinase inhibitor 2A (*CDKN2A*) [169,170] and short stature homeobox 2 (*SHOX2*) (see below) for lung cancer. These have already reached clinical potential and for some diagnostic test kits are commercially-available (Table 2). In the next sections we will provide an overview of *SHOX2*, *PITX2* and *MGMT* as good examples of diagnostic, prognostic and predictive biomarkers in cancer.

6.1.1. *SHOX2*

DNA methylation of the short stature homeobox 2 (*SHOX2*) gene was found to be a diagnostic clinical biomarker candidate for the detection of malignant lung disease even in patients where histology and cytology results are equivocal [135]. *SHOX2* methylation allowed the specific detection of malignant lung disease with a sensitivity of 60% and a specificity of 90% in blood plasma using HeavyMethyl, a quantitative methylation-specific PCR-based approach [134]. The highest assay sensitivity was achieved for small cell lung cancer (SCLC) cases with 80% and squamous cell carcinoma (SCC) with 63%, respectively, when compared to adenocarcinomas (AC) cases with a sensitivity of only 39%. However, the poor sensitivity for detecting adenocarcinomas could be improved by the addition of a second (or more) adenocarcinoma-specific biomarker. Not surprisingly, the sensitivity values obtained of the blood-plasma-based assay were lower compared to sensitivities seen from bronchial aspirates (SCLC: 97% (80%), SCC: 82% (63%), and AC: 47% (39%)); overall sensitivity and specificity were 68% (60%) and 95% (90%), respectively) as the tumor-derived amount of DNA is expected to be lower in blood than a lung-specific analyte [134,135]. However, a blood-based assay has the advantage of using specimens which have been collected with a far less invasive procedure compared to those obtained from bronchoscopy. Furthermore, a blood-based assay enables screening of asymptomatic patients whereas availability of bronchoscopy is limited to patients with suspected lung cancer. Noteworthy, elevated *SHOX2* methylation levels in pleural effusions do not only allow the detection of lung cancer but also the detection of other malignancies, such as breast cancer and gastrointestinal cancers [132,133]. However, assay sensitivity and specificity for these was not as good as for bronchial aspirates or blood. *SHOX2* methylation level in lymph node tissue obtained by endobrochial ultrasound with transbronchial needle aspiration (EBUS-TBNA) improved endoscopic lung cancer staging with an assay sensitivity and specificity of 94% and 99%, respectively [130].

DNA methylation of *SHOX2* not only provides diagnostic but also provides prognostic information for cancer patients [131,132]. Pleural effusion samples obtained from patients with different malignancies (including lung cancer) showed a shorter overall survival if elevated levels of *SHOX2* methylation were detected [132]. Contrarily, gain of *SHOX2* methylation in tumor tissues has been shown to be associated with good prognosis in lung cancer patients. The prognostic power of *SHOX2* methylation was further improved when combined with DNA methylation analysis of *PITX2* [131].

6.1.2. *PITX2*

The paired-like homeodomain 2 (*PITX2*) gene encodes the PITX2 transcription factor. DNA methylation status of the *PITX2* promoter has been identified as a candidate clinical biomarker in tumor tissues. This has provided prognostic information for breast cancer, prostate cancer, and lung cancer. *PITX2* methylation in estrogen receptor alpha positive breast cancer patients without lymph node metastasis has been associated with poor prognosis when treated without any systemic adjuvant therapy [171] as well as a higher risk of disease recurrence after surgery when treated with the antiestrogen Tamoxifen only [172,173]. Furthermore, methylation of the *PITX2* promoter was also associated with poor patient outcome in estrogen receptor alpha positive, HER-2/*neu*-negative breast cancer patients positive for lymph node metastasis when treated with an anthracycline-based adjuvant chemotherapy [174]. Methylation of *PITX2* in prostate cancer patients has also been shown to be a prognostic biomarker for an increased risk of biochemical recurrence after radical prostatectomy [175–177]. Importantly, the prognostic value of *PITX2* methylation was particularly high in tumor-enriched samples of patients at intermediate risk for whom further risk stratification is quite often challenging [176]. Interestingly, and different to breast and prostate cancer, increased DNA methylation levels of *PITX2* were associated with prolonged survival in lung cancer patients and requires further investigation [131].

6.1.3. *MGMT*

O^6-methylguanine DNA methyltransferase is a DNA repair protein that is encoded by the *MGMT* gene and is capable of removing alkyl residues directly from the O^6-position of guanines. However, if the DNA repair capacity of *MGMT* is impaired or inactivated, for example by DNA methylation, affected cells are less protected against mutagenic DNA adducts [178,179]. Therefore, tumor *MGMT* promoter methylation renders cancer cells susceptible to the cell damaging effects of drug regimens utilizing alkylating agents [116,180] (see also *Lessons learned from the DNA methylation biomarker MGMT*). *MGMT* was one of the first predictive DNA methylation biomarkers to determine a patient's response to alkylating chemotherapeutics and it was shown that glioblastoma patients with tumor *MGMT* promoter methylation have a survival benefit from Temozolomide chemotherapy [120,121].

The more frequent use of quantitative approaches such as bisulfite pyrosequencing to detect and measure *MGMT* methylation have revealed that the DNA methylation biomarker *MGMT* does not only have a predictive but also a prognostic clinical component (reviewed in [121,126]). Glioblastoma patients with more than 29% *MGMT* promoter methylation showed a longer progression-free and overall survival when treated with radiotherapy and Temozolomide [143]. A methylation cut-off value of 25% separated elderly glioblastoma patients into two groups with those having more than 25% of methylation had a better survival rate when treated with alkylating agents alone [144]. Tumor *MGMT* methylation status was also shown to have a prognostic value for progression-free survival of anaplastic glioma patients treated with radiotherapy alone [181,182].

7. DNA Methylation Biomarkers for Genomic Imprinting Disorders

Whereas most genes are expressed from both the maternal and paternal allele, imprinted genes are monoallelically expressed in a parent-of-origin-specific manner either from the maternal or the paternal allele. Only a small proportion of all human genes are imprinted and are often found clustered in imprinted domains and mono-allelic gene expression is controlled by differentially methylated regions (DMRs) (reviewed in [183]). Disrupted or altered imprinting patterns have been linked to pathological conditions termed genomic imprinting disorders (reviewed in [184]). Examples of imprinting disorders include Prader-Willi syndrome (PWS), Angelman syndrome (AS), Beckwith-Wiedemann syndrome (BWS) and Silver-Russell syndrome (SRS), which will be discussed briefly below.

PWS and AS are clinically distinct neurodevelopmental imprinting disorders, which have been linked to a region on the long arm of human chromosome 15 (15q11–q13; reviewed in [185]). This region consists of several imprinted genes and the absence of paternally expressed genes in this imprinting domain results in PWS, whereas the loss of maternally-expressed genes causes AS. Additionally, point mutations in the E6-AP ubiquitin-protein ligase (*UBE3A*) gene, which is also part of the imprinting domain account for approximately 10% of AS patients. In cases where PWS or AS is suspected, DNA methylation analysis of the PWS/AS critical region allows the reliable identification of more than 99% of PWS patients and about 80% of AS patients [186].

Two approaches are commonly used in molecular diagnostics for DNA methylation analysis of the PWS/AS critical region [186,187]. The first approach determines the methylation status at a single gene locus, the small nuclear ribonucleoprotein polypeptide N (*SNRPN*) gene, whereas the second approach determines the methylation status and gene copy number changes at several sites across the region [186]. DNA methylation analysis of the *SNRPN* gene is frequently determined by MSP [188,189] whereas the simultaneous investigation of methylation levels and gene copy numbers is determined by methylation-sensitive multiplex ligation-dependent probe amplification (MS-MLPA) [190]. Molecular diagnostics of PWS and AS is quite complex and challenging, and guidelines for molecular genetic testing and reporting PWS and AS have been developed [186]. Furthermore, a WHO international genetic reference panel for PWS and AS has been established and was successfully validated in an international multicentre study [187].

BWS and SRS are imprinting disorders, which have been associated with imprinted genes on chromosome region 11p15.5 [191–193]. This region is functionally divided into two domains: the first domain consists of the imprinted insulin-like growth factor gene 2 (*IGF2*) and the non-coding RNA *H19* and is controlled by DMR1 whereas the second region contains several imprinted genes, including cyclin-dependent kinase inhibitor 1C (*CDKN1C*), potassium voltage-gated channel, KQT-like subfamily, member 1 (*KCNQ1*) and KCNQ1 opposite strand/antisense transcript 1 (*KCNQ1OT1*), is controlled by DMR2. Loss of methylation at DMR2 (*KCNQ1OT1* hypomethylation), is the most frequent alteration, in around 50% of BWS patients [194] whereas loss of methylation at DMR1 (*H19* hypomethylation) is typically observed in SRS is found in around 40% of SRS patients [192,195]. As mentioned before, MS-MLPA allows the simultaneous investigation of methylation levels and gene copy numbers and has thus been considered well suited for detecting the majority of (epi-) genetic alterations associated with BWS and SRS in region 11p15.5 [196–198].

Most approaches for routine clinical DNA methylation analysis at single-gene loci in genomic imprinting disorders rely, most probably for historical reasons, on qualitative methylation detection methods. However, the diagnostic advantages of quantitative DNA methylation detection methodologies, such as bisulfite pyrosequencing [191,199,200], are being increasingly recognized and will be probably the preferred methods of choice for analyzing single gene loci in the near future.

8. DNA Methylation Biomarkers of Outcome in Chronic Diseases Other than Cancer

Given the likely early life origins for non-communicable disease, there are plenty of opportunities in which DNA methylation biomarkers could be used. Biomarkers for intrauterine environmental exposures such as maternal alcohol consumption or smoking could provide a way to measure exposures without the need for time-consuming, hard-to-administer questionnaires and where access to mothers is not possible. DNA methylation risk biomarkers could be used to stratify risk for latent non-communicable disease before the onset of disease. They could also be used to monitor progression from first symptoms to full disease. After disease onset, they could be used for predicting survival and response to therapy as they are beginning to do with cancer. Below, we review data from the most promising studies of environmental, risk, diagnostic, prognostic, and predictive DNA methylation biomarkers.

8.1. DNA Methylation Biomarkers for Adverse Environments

There have been a large number of environmental agents linked to epigenetic change, including toxins, stress and nutrition, and these have been reviewed elsewhere [201–203]. Below, we focus on two that have yielded replicated DNA methylation biomarkers: smoking and stress.

8.1.1. *AHRR* Methylation and Smoking

Exposure to adverse environments at all stages of life have been shown to influence the epigenome (reviewed in [39,42,204,205]). However, a replicated association has been found for only one: the effect of DNA methylation on the aryl hydrocarbon receptor repressor (*AHRR*) gene involved in the detoxification of chemicals found in tobacco smoke. As of June 2014, ten independent methylome-wide studies using Illumina Infinium HM450 arrays (containing probes for about 480,000 CpG dinucleotides located in functionally-relevant regions of the genome [206]) had all identified the same smoking-associated probe, cg05575921, located in a region of intermediate CpG density (CpG-island shore) 450 bp upstream of a CpG island in the third intron of the *AHRR* gene [207–216] (Table 3).

Two studies focused on the effect of maternal smoking in umbilical cord blood [209,215], which they and others [217] replicated in independent sample cohorts. Others found an association of adult smoking with *AHRR* methylation in blood [207,208,210–212,214], lung tissue [211] or blood lymphoblastoid cell lines [213]. No effects were seen at birth in placenta or buccal epithelium [217] and effects were seen elsewhere in the *AHRR* gene in lung alveolar macrophage DNA but not at the cg05575921 probe [213]. Three studies performed within-cohort validation using locus-specific DNA methylation analysis [207,211,212] and six studies replicated their findings in adults in independent cohorts [208,209,211,212,214,215]. Two studies showed evidence of a role for the region surrounding probe cg05575921 in regulation of *AHRR* expression [211,213]. All found an inverse relationship between smoking and DNA methylation with an effect size ranging from −4% in neonates of mothers who smoked throughout pregnancy [215] to −24.4% in adult current smokers [212].

Similar effects were seen in Europeans, African Americans [207], and South Asians [208]. The latter study found that current smokers were identified with 100% sensitivity and 97% specificity in Europeans and with 80% sensitivity and 95% specificity in South Asians. Timing-specific effects were also found; prenatal smoking only exerted an effect when mothers smoked during a significant part of gestation [217,218]. Furthermore, associations found at birth were also present at 18 months of age [217] but in adulthood, DNA methylation levels were similar in never smokers and in former smokers [212]. Clearly, loss of methylation at and around the *AHRR* cg05575921 probe is strongly associated with first or second hand exposure to smoking. Importantly, one study found an association in adults with smoking, but not tobacco snuff consumption, implicating that a product(s) of tobacco combustion is responsible for the loss of DNA methylation rather than tobacco itself. Further work is needed to link this loss to the timing of prenatal smoking, and postnatal passive and active smoking, and its relationship with downstream health outcomes previously associated with *AHRR* polymorphisms such as cancers [219–221] and endometriosis [222].

In addition to probe cg05575921, a number of CpG dinucleotides have been significantly associated with prior smoking. Table 4 lists these probes, using a cut-off of those that have been identified by four or more studies. These include two further CpG dinucleotides from *AHRR* [207,209–213], one from the thrombin receptor-like 3 (*F2RL3*) gene [208,211,214,216], one from the growth factor independent 1 transcription repressor (*GFI1*) gene and two from the myosin 1G (*MYO1G*) gene. In addition, two intergenic smoking-associated CpG sites have been replicated across multiple studies [207–209,211–214], all coinciding with regions of DNAse hypersensitivity, suggesting functional significance. Potentially, one or more of these CpG dinucleotides could be used in combination with probe cg05575921 as DNA methylation biomarkers for smoking.

Table 3. Summary of findings for the relationship between smoking and DNA methylation within the *AHRR* gene. Data refer to *AHRR* HM450 probe cg05575921 unless otherwise stated. Summary includes details of assay platform, age of subjects, details of exposure, tissues examined, number of subjects, whether *AHRR* expression was also measured, whether findings were validated or replicated and effect size (methylation levels in smokers compared to non-smokers).

Reference	Platform	Age, Median	Exposure	Tissue	N	Effects Elsewhere in AHRR	*AHRR* Expression	Vali-dation	Repli-cation	Effect Size	Notes
[213]	HM450	Adults, 45	Current smoking	LCLs & alveolar MP	119/19 [1]	yes	Yes [2]	No	No	−15%/NS	
[209]	HM450	Birth	Maternal smoking [3]	Whole CB	1062/36 [4]	Yes	No	No	Yes [4]	−7.5%/−7.7% [4]	Multiple hits in the aryl hydrocarbon signaling pathway. Authors have since shown that effects are specific for maternal smoking through at least gestational week 18 [218]
[207]	HM450	Adults, 49	Current smoking	PBMC	111	Yes	No	Yes	No	−15%	African Americans
[208]	HM450	Adults, 48	Current smoking	Whole PB	81/84 [5]	No	No	No	Yes [6]	−22%	Former smokers same as never smokers; changed only slightly after adjusting for cell composition
[210]	HM450	Adults, 22	Current serum cotinine	PBMC	107	yes	No	No	No	−20% [7]	
[211]	HM450	Adults, 51/55/49/? [8]	Current smoking	Whole PB, lung tissue	184/190/180/27	yes	Yes [9]	Yes	Yes	−17%/−14%/NS/NS [10]	Replicated in a mouse model of smoking exposure

Table 3. *Cont.*

Reference	Platform	Age, Median	Exposure	Tissue	N	Effects Elsewhere in AHRR	*AHRR* Expression	Vali-dation	Repli-cation	Effect Size	Notes
[212]	HM450	Adults, 60/53 [11]	Current smoking	Whole PB	749/232 [11]	yes	No	Yes	Yes [11]	−24/−23% [11]	methylation-specific protein binding patterns were observed for cg05575921; levels in former smokers revert to levels similar to never smokers over time
[215]	HM450	Birth	Maternal smoking	Whole CB	889	yes	No	No	Yes	−4%	Replicated a previous study [209]
[214]	HM450	Adults, 43	Current smoking	Whole PB	432	yes	No [13]	No	Yes	−7.4%	Replicated a previous study [212]; no effect with tobacco snuff
[216]	HM450	Female adults, 57	Current smoking	Whole PB	200	No	No	No	Yes	−8%	Former and never smokers had similar methylation levels
[217]	Sequenom EpiTyper	Birth & 18 months	Maternal smoking	CBMC, buccal epithelium, placenta	46/15/24 [12]	yes	Yes [14]	n/a	Y	−10%/NS/NS [12]	No effect if mother smoked early pregnancy only; effects of smoking stable to 18 months of age

[1] refers to the two different cell types tested; [2] *AHRR* expression in alveolar macrophages was inversely correlated with methylation of probe cg05575921; [3] measured using plasma cotinine at gestational week 18; [4] replicated using data from maternal smoking in pregnancy in an independent cohort; [5] data on Europeans replicated in South East Asians; [6] replicated across two ethnic groups; [7] effect size calculated from the regression line, highest to lowest plasma cotinine; [8] discovery, replication and validation groups are subsets of the same cohort and were analyzed along with lung tissue samples from a separate cohort; [9] *AHRR* expression in lung tissue was inversely correlated with methylation of probe cg05575921; [10] no difference with probe cg05575921, differences found for *AHRR* probes cg21161138 and cg23576855 (magnitudes similar to those seen in blood); [11] discovery and replication subsets of the same cohort; [12] significant associations between methylation and expression seen at six genes other than *AHRR*; [13] CBMC/buccal epithelium/placenta; [14] *AHRR* expression non-significantly higher in CBMCs in newborns exposed to smoking in pregnancy than those not exposed. Abbreviations: LCL, lymphoblastoid cell lines; MP, macrophages; PB, peripheral blood; CB, cord blood; MC, mononuclear cells; NS, not significant.

Table 4. Other HM450 probes with significant correlations with smoking in at least four studies. Probes are included if found to be significantly associated with smoking in at least four independent studies. DHS, DNAse hypersensitive site, indicative of regulatory potential.

Probe	Gene	References
cg03991871	*AHRR*	[209,212,213,215]
cg21161138	*AHRR*	[207,209–212,215]
cg03636183	*F2RL3*	[208,211,214,216]
cg09935388	*GFI1*	[208,209,212,214,215]
cg22132788	*MYO1G*	[208–210,214]
cg12803068	*MYO1G*	[210,212,215,218]
cg21566642	Intergenic (CpG island, DHS)	[207,208,211,212]
cg06126421	Intergenic (enhancer, DHS)	[207,208,211,212,214]

8.1.2. *NR3C1* Methylation and Stress

Stress triggers the activation of the hypothalamus-pituitary-adrenal axis, resulting in the production of glucocorticoids by the adrenal glands. By binding to receptors in the brain, glucocorticoids induce changes in gene expression and in turn, health and behavior [223]. Landmark studies with rats have shown that lack of maternal licking and grooming at birth resulted in an increased level of DNA methylation within the exon 1_7 promoter of the glucocorticoid receptor gene *Nr3c1* in rat hippocampus, in particular at a region that binds nerve growth factor-inducible protein-A (NGFI-A) [224,225]. Since then, studies of the equivalent region in humans (exon 1F of the *NR3C1* gene) have found decreased DNA methylation in cord blood [226,227] and placenta [228] associated with maternal anxiety during pregnancy. Others have shown that violence towards women during pregnancy can have a similar effect [229,230]. Even extremes of stress experienced prior to conception, in the form of the holocaust, were also found to correlate with *NR3C1* exon 1F methylation, albeit in opposite directions depending on the sex of the parent [231]. Methylation analysis of various tissues from adults, either alive or *post mortem*, have found long-lasting effects of abuse [232–235] or death of a parent [235,236] during childhood on *NR3C1* exon 1F. In addition, adults with post-traumatic stress disorder had decreased DNA methylation at the same [237] or alternate [238] *NR3C1* promoters. Of further interest, three studies have shown that methylation of *NR3C1* exon 1F can predict health outcomes, whether predicting quality of movement and attention at birth [239], response to psychotherapy in adults with posttraumatic stress disorder [240] or response to threat-associated stress in adult females [241]. In the latter study, DNA methylation levels at *NR3C1*, the estrogen receptor alpha gene *ESR1* and the serotonin transporter gene *SL6A4* each had independent predictive power. Furthermore, a model containing data from all those genes accounted for half of the variance in total cortisol output. Rat studies showing that the adverse effects and DNA methylation changes associated with early neglect could be reversed in adulthood by methyl-donor rich diet [242] or the histone deacetylase inhibitor Trichostatin A [243], suggesting that *NR3C1* methylation could be use to monitor response to future interventions in humans.

Clearly, methylation at *NR3C1* promoters has the potential to be developed into a variety of candidate biomarkers. In addition, despite yielding no replicated stress biomarkers to date, the small (typically 1%–2%) effect sizes for *NR3C1* methylation would suggest that there may be better DNA methylation-based stress biomarkers out there, discoverable using epigenome-wide approaches [244–248].

8.2. DNA Methylation Risk Biomarkers at Birth

Measuring DNA methylation in five candidate genes in DNA from umbilical cords, Godfrey and colleagues found that methylation of two genes correlated with childhood adiposity as measured by fat mass and trunk/limb fat ratio in 78 nine-year-olds [42]. Methylation of the retinoic acid X receptor alpha (*RXRA*) gene and the endothelial nitric oxide synthase (*ENOS*) gene, together with sex, explained 25% of the variance in adiposity at age nine. Data for *RXRA* were replicated in a second cohort of 239 six-year-olds [42]. Other studies have identified associations between *RXRA* methylation in cord blood at birth and bone mineral density at age four [249] and between methylation of the alkaline phosphatase *ALPL* and body mass index at nine years of age [250]. However, the first association could not be replicated in another sample cohort whereas for the second association no replication study was performed.

8.3. DNA Methylation Biomarkers during Childhood

Rakyan and colleagues identified 132 CpG dinucleotides whose methylation levels differed significantly in twin pairs discordant for type 1 diabetes and which were subsequently validated with an independent method and replicated in a further set of twin pairs [251]. Two-thirds of these CpG dinucleotides were also present in singletons prior to the onset of overt symptoms of type 1 diabetes but positive for diabetes-associated autoantibodies. If those findings can be further replicated, this could provide single or panels of DNA methylation candidate clinical biomarkers predicting the onset of type 1 diabetes. A potential biomarker study found that DNA methylation within the promoter of the peroxisomal proliferator activated receptor gamma (*PPARG*) gene in blood at age five to seven years predicted obesity risk from nine to 14 years [252]. However, these results have yet to be replicated.

Autism spectrum disorder (ASD) describes a related set of neurodevelopmental disorders of childhood characterized by social deficits and communication difficulties, stereotyped or repetitive behaviors and interests, and in some cases, cognitive delays. To date, a small number of ASD MWAS have been performed, using a variety of platforms, on lymphoblastoid cell lines [253], peripheral blood [254,255], buccal epithelium [256], *post mortem* occipital cortex and cerebellum [257], and dorsolateral prefrontal cortex, temporal cortex and cerebellum [258]. ASD-specific DNA methylation was found in all but one study [257] and in the rest, although ASD-specific methylation was often validated within the study, only one study attempted to replicate across cohorts and tissues [258]. In this study, three significant ASD-associated array probes discovered in temporal cortex were replicated in such a manner. ASD-specific DNA methylation found within the proline-rich transmembrane protein 1 (*PRRT1*) gene was replicated in prefrontal *post mortem* cortex and cerebellum, methylation of *c11orf21* was replicated in prefrontal cortex and methylation at an intergenic site near the zinc finger gene *ZFP37* was replicated in a sex-specific manner in cerebellum. The only differentially methylated gene replicated in two separate studies is the olfactory receptor gene *OR2L13* found in buccal epithelium [256] and peripheral blood [254]. Further replication will be required to develop this potential biomarker for ASD.

8.4. DNA Methylation Biomarkers in Adults

Cardiovascular disease (CVD) and its precursors are receiving arguably the greatest attention in MWAS outside cancer [259–261]. DNA methylation biomarkers could help ascertain risk early in life, help with diagnosis and predict response to interventions. Below, we report some of the more advanced such studies.

Levels of fasting glucose and insulin and measures of insulin resistance are used to test for early signs of diabetes and they have been subject to a recent MWAS [262]. This study divided up a cohort of 837 non-diabetic individuals at a median age of 48 years into discovery and replication subsets. Using HM450 arrays on DNA from CD4+ T cells, the investigators found significant associations between methylation of two CpG sites with the ATP-binding cassette gene *ABCG1*, involved in macrophage cholesterol and phospholipids transport, with insulin resistance, with one associated with insulin "of borderline significance". The CpG site with the strongest association with insulin and insulin resistance was also strongly associated with nearby single-nucleotide polymorphisms, implying that differences in genetic sequence can alter the epigenetic functionality of a genomic region. Another recent study replicated across two cohorts a DNA methylation biomarker for triglyceride levels at the carnitine palmitoyltransferase gene *CPT1A* in the same cell type [263]. In this study, *CPT1A* methylation explained 11.6% and 5.5% of the variation in triglyceride levels in the discovery and replication cohorts, respectively.

Although several studies have discovered associations between DNA methylation and obesity [264], few studies have searched for risk or predictive DNA methylation biomarkers in adulthood. In one study that did, males with a history of CVD had higher global DNA methylation than those without [57]. Those who went on to develop symptoms of CVD six years later had intermediate levels of global DNA methylation. In other study, a type 2 diabetes-specific CpG dinucleotide in the first intron of the fat mass and obesity-associated gene *FTO* predicted the onset of symptoms between ages 30 and 43 in a cohort of initially asymptomatic adults [265]. Replication is required for both studies.

Two unreplicated studies resulted in potential predictive DNA methylation biomarkers for response to weight loss programs in adults. In the first, obese women with better response to dietary intervention showed significantly lower levels of DNA methylation at promoters of the leptin (*LEP*) and TNF-alpha (*TNF*) genes than the non-responder group [266]. Although no differences were found between responder and non-responder groups in *LEP* and *TNF* gene expression, if replicated, the potential predictive methylation biomarker would still be valid on its own. In a similar study of obese men, DNA methylation levels in several CpG dinucleotides located in the ATPase *ATP10A* and the CD44 antigen (*CD44*) genes showed statistical baseline differences depending on the weight-loss outcome [266]. Again, these finding have not yet been replicated.

In a search for potential DNA methylation biomarkers of postpartum depression using MWAS and a parallel study in mice, Guintivano and colleagues found that DNA methylation at the heterochromatin protein 1 binding protein 3 (*HP1BP3*), and tetratricopeptide repeat domain 9B (*TTC9B*) genes predicted postpartum depression in the original and replication cohorts [267]. Adjustment for blood cell heterogeneity resulted in a higher specificity (96%) in both cohorts compared to unadjusted values.

Schizophrenia is a psychotic disorder, and bipolar disorder is a mood disorder but both have similar symptoms and they are often studied together. Many potential DNA methylation biomarker studies and MWAS have been conducted for these disorders ([268–274] and references therein). Despite the heterogeneity of platforms and tissues used in these studies, a small number of potential diagnostic schizophrenia- and/or bipolar disorder-associated biomarkers have been identified. The serotonin receptor 2A (*HTR2A*) gene was differentially methylated in both disorders in two brain regions (frontal cortex and the anterior cingulate) [270], replicating the findings of a previous study [275,276]. Similar results were also found in saliva of patients with these disorders [277]. Another gene differentially methylated in two brain regions in both disorders was the dystrobrevin binding protein gene *DTNBP1*, also found in an MWAS of frontal cortex of females with both disorders [278] and in all individuals with schizophrenia [268]. The reelin (*RLN*) gene was differentially methylated in individuals with schizophrenia using an MWAS [271], as it was for schizophrenia and bipolar disorder in a MWAS of brain regions [270], replicating previous findings [279,280]. Other potential DNA methylation biomarkers for psychoses include the human leukocyte antigen (HLA) gene *HCG9* and the serotonin transported gene *SCL6A4* (*5HTT*). *HCG9* was identified in patients with schizophrenia or bipolar disorder in an MWAS of frontal cortex [278] and in brain, blood and sperm in an MWAS for bipolar disorder [281]. *SLC6A4* was differentially methylated in an MWAS of saliva and frontal cortex in individuals with schizophrenia [272], similar to previous findings in lymphoblastoid cell lines and brain tissue of individuals with bipolar disorder in a study that included cross-cohort replication [282]. No studies have investigated the possibility of using above associative biomarkers as potential risk biomarkers in early life. However, a subset of studies has found associations between DNA methylation and medication for schizophrenia or bipolar disorder [273,276]. Clearly, there is much promise for future potential biomarkers of risk, diagnosis and prognosis in schizophrenia and bipolar disorder.

More longitudinal studies at stages of life are required to generate DNA methylation biomarkers for exposure and outcome in chronic diseases other than cancer. Birth cohorts and the retrospective utility of birth dried blood spot Guthrie cards [283] will be essential for this search.

8.5. DNA Methylation Biomarkers of Aging

A number of individual MWAS have looked at the relationship of DNA methylation and aging, with the intention of developing age-specific biomarkers for forensic applications and for investigating premature cellular aging. Three independent meta-analyses have been performed on such datasets [284–286]. The first [284] reviewed six MWAS datasets from Infinium HM27 arrays containing probes for about 27,000 CpG sites [287] on a variety of cell types. None of the 1,093 age-associated probe CpG dinucleotides replicated across all six studies. However, probes at two genes, neuronal pentraxin II (*NPTX2*) and phosphodiesterase 4C (*PDE4C*), did overlap in five of the six studies. The second study [285] performed an analysis of DNA methylation from whole blood from 575 individuals ranging from newborns to age 78 from published HM27 datasets and replicated with a further group of four similar datasets. This yielded 99 significantly age-associated probes including the same *PDE4C* CpG probe cg17861230 as the first study. An even more extensive study of 39 "training" and 32 "test" HM27 and HM450 datasets of more than 7,000 samples from multiple tissues yielded 353 "age predictor" CpG dinucleotides, which included one (cg13899108) in

PDE4C [286] just 420 bp from the CpG site identified in the first two studies. Although this locus is the most validated age-related CpG dinucleotide, these analyses show that sometimes, a combination of several CpG dinucleotides may be more accurate than a single CpG site. A recent large single analysis measured age-associated DNA methylation in whole blood DNA from 656 individuals using HM450 arrays [288]. In this tissue, investigators identified 70,387 significant age-associated CpG dinucleotides, of which, 53,670 were replicated in an independent dataset. Data was not available to identify whether the *PDE4C* locus mentioned above was among this dataset. The study went on to develop a predictive model of aging that included methylation data and clinical parameters such as gender and body mass index. We predict that this is how most DNA methylation biomarkers will be used in the future. The model selected a set of 71 age-associated methylation biomarkers that were highly predictive of age. Although *PDE4C* was not among this subset, another probe within the subset, cg09809672 associated with the EDAR-associated death domain (*EDARADD*) gene was also identified as age-associated in two of the other studies [285,286]. Importantly, this study also found evidence of an accelerated epigenetic aging in tumor tissue [288] and a further study has since identified epigenetic age acceleration as a risk factor for mortality [289]. Clearly, age-associated DNA methylation biomarkers have more applications than forensic medicine.

9. Integrating Epigenetic Data into Disease Risk Models

Although DNA methylation biomarkers can be used by themselves, the emerging field of molecular pathological epidemiology proposes that they can be integrated into models of disease risk together with other factors [4,290,291]. Such factors include transcriptomic, proteomic, metabolomic, microbiome, and neuroimaging data. The logic is that combinations of risk biomarkers will provide more accurate estimation of disease risk, particularly when dealing with individuals, due to inter- and intra-individual biological variation. Based on principles similar to systems and network biology and a variety of modeling methods, this field is in its infancy but is the next logical step for DNA methylation biomarkers and is already yielding promising results for genetic biomarkers [292].

10. Future Prospects

An increasing tendency to harmonize appropriate methods for DNA methylation detection and reference standards will accelerate the development of DNA methylation biomarkers for cancer and for other diseases. This tendency will be synergistically enhanced by next generation sequencing methodology, which has unlocked a new area of possibilities. This relatively new methodology opens the avenue for routine testing of DNA methylation biomarker panels rather than the selective choice of individual biomarkers. The use of appropriate DNA methylation biomarker panels will prove beneficial where the disease phenotype is quite heterogeneous. It is also expected that the genetic component of disease will be further revealed, which will subsequently allows the strengthening of biomarker panels by combining genetic and DNA methylation biomarker panels [293].

It is not only important to have appropriate epi(genetic) biomarker panels available for certain diseases or risk stratification but also to translate them into clinical actionable information. If no clinical action is available there is a risk of adverse psychological impacts among patients and a risk of those patients being disadvantaged by healthcare providers. However, there is also an enormous potential that affected patients can use the knowledge to their benefit allowing them to actively prevent or delay the early onset of certain diseases.

11. Conclusions

DNA methylation biomarkers are promising and valuable biomarkers which are heading for the molecular diagnostic laboratory. This is particular true for methylation biomarkers in cancer where the biomarkers are currently being used for early detection. However, the uptake of DNA methylation biomarkers is quite slow and will still require a considerable amount of time until the field reaches its full potential. The development of DNA methylation biomarkers for cancer and other diseases has also been slowed down by the lack of standardized methodologies and reference standards for use in DNA methylation detection. The still widespread use of inappropriate methods in combination with inappropriate controls still produces potential DNA methylation biomarkers, which may not be replicated. The need for methods of quantitative DNA methylation detection is becoming more and more obvious and is critical where only small differences in methylation values determine a diseased or disease-free state. Finally, the availability of DNA methylation biomarkers in diseases other than cancer is still in its very early steps but in time, their transition to a clinical setting will follow as it has for cancer.

Acknowledgments

Jeffrey M. Craig is supported by the Murdoch Childrens Research Institute (MCRI), the National Health and Medical Research Council, the National Institutes of Health, and the Australian Twin Registry. This work was supported by the Victorian Government's Operational Infrastructure Support Program.

Author Contributions

Jeffrey M. Craig conceived the idea; Jeffrey M. Craig and Thomas Mikeska were involved in planning, writing and editing the manuscript.

References

1. Naylor, S. Biomarkers: Current perspectives and future prospects. *Expert Rev. Mol. Diagn.* **2003**, *3*, 525–529.
2. Mayeux, R. Biomarkers: Potential uses and limitations. *NeuroRx* **2004**, *1*, 182–188.
3. Strimbu, K.; Tavel, J.A. What are biomarkers? *Curr. Opin. HIV AIDS* **2010**, *5*, 463–466.
4. Ogino, S.; Lochhead, P.; Chan, A.T.; Nishihara, R.; Cho, E.; Wolpin, B.M.; Meyerhardt, J.A.; Meissner, A.; Schernhammer, E.S.; Fuchs, C.S.; *et al.* Molecular pathological epidemiology of epigenetics: Emerging integrative science to analyze environment, host, and disease. *Mod. Pathol.* **2013**, *26*, 465–484.
5. Bird, A. Perceptions of epigenetics. *Nature* **2007**, *447*, 396–398.
6. Teh, A.L.; Pan, H.; Chen, L.; Ong, M.L.; Dogra, S.; Wong, J.; Macisaac, J.L.; Mah, S.M.; McEwen, L.M.; Saw, S.M.; *et al.* The effect of genotype and *in utero* environment on

inter-individual variation in neonate DNA methylomes. *Genome Res.* **2014**, doi:10.1101/gr.171439.113.

7. De Laat, W.; Duboule, D. Topology of mammalian developmental enhancers and their regulatory landscapes. *Nature* **2013**, *502*, 499–506.
8. Quina, A.S.; Buschbeck, M.; di Croce, L. Chromatin structure and epigenetics. *Biochem. Pharmacol.* **2006**, *72*, 1563–1569.
9. Jurkowska, R.Z.; Jurkowski, T.P.; Jeltsch, A. Structure and function of mammalian DNA methyltransferases. *Chembiochem* **2011**, *12*, 206–222.
10. Jeltsch, A.; Jurkowska, R.Z. New concepts in DNA methylation. *Trends Biochem. Sci.* **2014**, *39*, 310–318.
11. Deaton, A.M.; Bird, A. CpG islands and the regulation of transcription. *Genes Dev.* **2011**, *25*, 1010–1022.
12. Gardiner-Garden, M.; Frommer, M. CpG islands in vertebrate genomes. *J. Mol. Biol.* **1987**, *196*, 261–282.
13. Takai, D.; Jones, P.A. Comprehensive analysis of CpG islands in human chromosomes 21 and 22. *Proc. Natl. Acad. Sci. USA* **2002**, *99*, 3740–3745.
14. Zhao, Z.; Han, L. CpG islands: Algorithms and applications in methylation studies. *Biochem. Biophys. Res. Commun.* **2009**, *382*, 643–645.
15. Wu, H.; Caffo, B.; Jaffee, H.A.; Irizarry, R.A.; Feinberg, A.P. Redefining CpG islands using hidden Markov models. *Biostatistics* **2010**, *11*, 499–514.
16. Chuang, L.Y.; Huang, H.C.; Lin, M.C.; Yang, C.H. Particle swarm optimization with reinforcement learning for the prediction of CpG islands in the human genome. *PLoS One* **2011**, *6*, e21036.
17. Kulis, M.; Queiros, A.C.; Beekman, R.; Martin-Subero, J.I. Intragenic DNA methylation in transcriptional regulation, normal differentiation and cancer. *Biochim. Biophys. Acta* **2013**, *1829*, 1161–1174.
18. Oberdoerffer, S. A conserved role for intragenic DNA methylation in alternative pre-mRNA splicing. *Transcription* **2012**, *3*, 106–109.
19. Illingworth, R.S.; Gruenewald-Schneider, U.; Webb, S.; Kerr, A.R.; James, K.D.; Turner, D.J.; Smith, C.; Harrison, D.J.; Andrews, R.; Bird, A.P. Orphan CpG islands identify numerous conserved promoters in the mammalian genome. *PLoS Genet.* **2010**, *6*, e1001134.
20. Irizarry, R.A.; Ladd-Acosta, C.; Wen, B.; Wu, Z.; Montano, C.; Onyango, P.; Cui, H.; Gabo, K.; Rongione, M.; Webster, M.; *et al.* The human colon cancer methylome shows similar hypo- and hypermethylation at conserved tissue-specific CpG island shores. *Nat. Genet.* **2009**, *41*, 178–186.
21. Ehrlich, M. DNA methylation and cancer-associated genetic instability. *Adv. Exp. Med. Biol.* **2005**, *570*, 363–392.
22. Choi, J.D.; Lee, J.S. Interplay between Epigenetics and Genetics in Cancer. *Genomics Inform.* **2013**, *11*, 164–173.
23. Heng, H.H.; Liu, G.; Stevens, J.B.; Bremer, S.W.; Ye, K.J.; Ye, C.J. Genetic and epigenetic heterogeneity in cancer: The ultimate challenge for drug therapy. *Curr. Drug Targets* **2010**, *11*, 1304–1316.

24. Huang, S. Genetic and non-genetic instability in tumor progression: Link between the fitness landscape and the epigenetic landscape of cancer cells. *Cancer Metastasis Rev.* **2013**, *32*, 423–448.
25. Stecklein, S.R.; Jensen, R.A.; Pal, A. Genetic and epigenetic signatures of breast cancer subtypes. *Front. Biosci.* **2012**, *4*, 934–949.
26. Gluckman, P.D.; Hanson, M.A.; Buklijas, T. A conceptual framework for the developmental origins of health and disease. *J. Dev. Orig. Health Dis.* **2010**, *1*, 6–18.
27. Gluckman, P.D.; Hanson, M.A.; Pinal, C. The developmental origins of adult disease. *Matern. Child Nutr.* **2005**, *1*, 130–141.
28. Barker, D.J. The fetal and infant origins of adult disease. *BMJ* **1990**, *301*, 1111.
29. Godfrey, K.M.; Lillycrop, K.A.; Burdge, G.C.; Gluckman, P.D.; Hanson, M.A. Epigenetic mechanisms and the mismatch concept of the developmental origins of health and disease. *Pediatr. Res.* **2007**, *61*, R5–R10.
30. McNeal, C.J.; Wilson, D.P.; Christou, D.; Bush, R.L.; Shepherd, L.G.; Santiago, J.; Wu, G.Y. The use of surrogate vascular markers in youth at risk for premature cardiovascular disease. *J. Pediatr. Endocrinol. Metab.* **2009**, *22*, 195–211.
31. Urbina, E.M.; Williams, R.V.; Alpert, B.S.; Collins, R.T.; Daniels, S.R.; Hayman, L.; Jacobson, M.; Mahoney, L.; Mietus-Snyder, M.; Rocchini, A.; *et al.* Noninvasive assessment of subclinical atherosclerosis in children and adolescents: Recommendations for standard assessment for clinical research: A scientific statement from the American Heart Association. *Hypertension* **2009**, *54*, 919–950.
32. Wong, N.C.; Morley, R.; Saffery, R.; Craig, J.M. Archived guthrie blood spots as a novel source for quantitative DNA methylation analysis. *Biotechniques* **2008**, *45*, 423–428.
33. Thirlwell, C.; Eymard, M.; Feber, A.; Teschendorff, A.; Pearce, K.; Lechner, M.; Widschwendter, M.; Beck, S. Genome-wide DNA methylation analysis of archival formalin-fixed paraffin-embedded tissue using the Illumina Infinium HumanMethylation27 BeadChip. *Methods* **2010**, *52*, 248–254.
34. Wong, N.C.; Ashley, D.; Chatterton, Z.; Parkinson-Bates, M.; Ng, H.K.; Halemba, M.S.; Kowalczyk, A.; Bedo, J.; Wang, Q.; Bell, K.; *et al.* A distinct DNA methylation signature defines pediatric pre-B cell acute lymphoblastic leukemia. *Epigenetics* **2012**, *7*, 535–541.
35. Ribel-Madsen, R.; Fraga, M.F.; Jacobsen, S.; Bork-Jensen, J.; Lara, E.; Calvanese, V.; Fernandez, A.F.; Friedrichsen, M.; Vind, B.F.; Hojlund, K.; *et al.* Genome-wide analysis of DNA methylation differences in muscle and fat from monozygotic twins discordant for type 2 diabetes. *PLoS One* **2012**, *7*, e51302.
36. Talens, R.P.; Boomsma, D.I.; Tobi, E.W.; Kremer, D.; Jukema, J.W.; Willemsen, G.; Putter, H.; Slagboom, P.E.; Heijmans, B.T. Variation, patterns, and temporal stability of DNA methylation: Considerations for epigenetic epidemiology. *FASEB J.* **2010**, *24*, 3135–3144.
37. Souren, N.Y.; Tierling, S.; Fryns, J.P.; Derom, C.; Walter, J.; Zeegers, M.P. DNA methylation variability at growth-related imprints does not contribute to overweight in monozygotic twins discordant for BMI. *Obesity* **2011**, *19*, 1519–1522.
38. Belinsky, S.A.; Palmisano, W.A.; Gilliland, F.D.; Crooks, L.A.; Divine, K.K.; Winters, S.A.; Grimes, M.J.; Harms, H.J.; Tellez, C.S.; Smith, T.M.; *et al.* Aberrant promoter methylation in bronchial epithelium and sputum from current and former smokers. *Cancer Res.* **2002**, *62*, 2370–2377.

39. Barnes, S.K.; Ozanne, S.E. Pathways linking the early environment to long-term health and lifespan. *Prog. Biophys. Mol. Biol.* **2011**, *106*, 323–336.
40. Ollikainen, M.; Smith, K.R.; Joo, E.J.; Ng, H.K.; Andronikos, R.; Novakovic, B.; Abdul Aziz, N.K.; Carlin, J.B.; Morley, R.; Saffery, R.; *et al.* DNA methylation analysis of multiple tissues from newborn twins reveals both genetic and intrauterine components to variation in the human neonatal epigenome. *Hum. Mol. Genet.* **2010**, *19*, 4176–4188.
41. Dittrich, B.; Buiting, K.; Gross, S.; Horsthemke, B. Characterization of a methylation imprint in the Prader-Willi syndrome chromosome region. *Hum. Mol. Genet.* **1993**, *2*, 1995–1999.
42. Godfrey, K.M.; Sheppard, A.; Gluckman, P.D.; Lillycrop, K.A.; Burdge, G.C.; McLean, C.; Rodford, J.; Slater-Jefferies, J.L.; Garratt, E.; Crozier, S.R.; *et al.* Epigenetic gene promoter methylation at birth is associated with child's later adiposity. *Diabetes* **2011**, *60*, 1528–1534.
43. Lowe, R.; Rakyan, V.K. Correcting for cell-type composition bias in epigenome-wide association studies. *Genome Med.* **2014**, *6*, 23.
44. Adalsteinsson, B.T.; Gudnason, H.; Aspelund, T.; Harris, T.B.; Launer, L.J.; Eiriksdottir, G.; Smith, A.V.; Gudnason, V. Heterogeneity in white blood cells has potential to confound DNA methylation measurements. *PLoS One* **2012**, *7*, e46705.
45. Loh, M.; Liem, N.; Lim, P.L.; Vaithilingam, A.; Cheng, C.L.; Salto-Tellez, M.; Yong, W.P.; Soong, R. Impact of sample heterogeneity on methylation analysis. *Diagn. Mol. Pathol.* **2010**, *19*, 243–247.
46. Montano, C.M.; Irizarry, R.A.; Kaufmann, W.E.; Talbot, K.; Gur, R.E.; Feinberg, A.P.; Taub, M.A. Measuring cell-type specific differential methylation in human brain tissue. *Genome Biol.* **2013**, *14*, R94.
47. Moverare-Skrtic, S.; Mellstrom, D.; Vandenput, L.; Ehrich, M.; Ohlsson, C. Peripheral blood leukocyte distribution and body mass index are associated with the methylation pattern of the androgen receptor promoter. *Endocrine* **2009**, *35*, 204–210.
48. Zou, J.; Lippert, C.; Heckerman, D.; Aryee, M.; Listgarten, J. Epigenome-wide association studies without the need for cell-type composition. *Nat. Methods* **2014**, *11*, 309–311.
49. Accomando, W.P.; Wiencke, J.K.; Houseman, E.A.; Nelson, H.H.; Kelsey, K.T. Quantitative reconstruction of leukocyte subsets using DNA methylation. *Genome Biol.* **2014**, *15*, R50.
50. Houseman, E.A.; Accomando, W.P.; Koestler, D.C.; Christensen, B.C.; Marsit, C.J.; Nelson, H.H.; Wiencke, J.K.; Kelsey, K.T. DNA methylation arrays as surrogate measures of cell mixture distribution. *BMC Bioinform.* **2012**, *13*, 86.
51. Herceg, Z.; Hainaut, P. Genetic and epigenetic alterations as biomarkers for cancer detection, diagnosis and prognosis. *Mol. Oncol.* **2007**, *1*, 26–41.
52. Iliopoulos, D.; Guler, G.; Han, S.Y.; Johnston, D.; Druck, T.; McCorkell, K.A.; Palazzo, J.; McCue, P.A.; Baffa, R.; Huebner, K. Fragile genes as biomarkers: Epigenetic control of WWOX and FHIT in lung, breast and bladder cancer. *Oncogene* **2005**, *24*, 1625–1633.
53. Verma, M.; Manne, U. Genetic and epigenetic biomarkers in cancer diagnosis and identifying high risk populations. *Crit. Rev. Oncol./Hematol.* **2006**, *60*, 9–18.
54. Zilberman, D.; Henikoff, S. Genome-wide analysis of DNA methylation patterns. *Development* **2007**, *134*, 3959–3965.

55. Zuo, T.; Tycko, B.; Liu, T.M.; Lin, J.J.; Huang, T.H. Methods in DNA methylation profiling. *Epigenomics* **2009**, *1*, 331–345.
56. Kalari, S.; Pfeifer, G.P. Identification of driver and passenger DNA methylation in cancer by epigenomic analysis. *Adv. Genet.* **2010**, *70*, 277–308.
57. Kim, M.; Long, T.I.; Arakawa, K.; Wang, R.; Yu, M.C.; Laird, P.W. DNA methylation as a biomarker for cardiovascular disease risk. *PLoS One* **2010**, *5*, e9692.
58. How Kit, A.; Nielsen, H.M.; Tost, J. DNA methylation based biomarkers: Practical considerations and applications. *Biochimie* **2012**, *94*, 2314–2337.
59. Michels, K.B.; Binder, A.M.; Dedeurwaerder, S.; Epstein, C.B.; Greally, J.M.; Gut, I.; Houseman, E.A.; Izzi, B.; Kelsey, K.T.; Meissner, A.; *et al.* Recommendations for the design and analysis of epigenome-wide association studies. *Nat. Methods* **2013**, *10*, 949–955.
60. Rakyan, V.K.; Down, T.A.; Balding, D.J.; Beck, S. Epigenome-wide association studies for common human diseases. *Nat. Rev. Genet.* **2011**, *12*, 529–541.
61. Wians, F.H. Clinical laboratory tests: Which, why, and what do the results mean? *Lab Med.* **2009**, *40*, 105–113.
62. Linnet, K.; Bossuyt, P.M.; Moons, K.G.; Reitsma, J.B. Quantifying the accuracy of a diagnostic test or marker. *Clin. Chem.* **2012**, *58*, 1292–1301.
63. Wentzensen, N.; Wacholder, S. From differences in means between cases and controls to risk stratification: A business plan for biomarker development. *Cancer Discov.* **2013**, *3*, 148–157.
64. Fraga, M.F.; Esteller, M. DNA methylation: A profile of methods and applications. *Biotechniques* **2002**, *33*, 632, 634, 636–649.
65. Dahl, C.; Guldberg, P. DNA methylation analysis techniques. *Biogerontology* **2003**, *4*, 233–250.
66. Kristensen, L.S.; Hansen, L.L. PCR-based methods for detecting single-locus DNA methylation biomarkers in cancer diagnostics, prognostics, and response to treatment. *Clin. Chem.* **2009**, *55*, 1471–1483.
67. Hernandez, H.G.; Tse, M.Y.; Pang, S.C.; Arboleda, H.; Forero, D.A. Optimizing methodologies for PCR-based DNA methylation analysis. *Biotechniques* **2013**, *55*, 181–197.
68. Jorda, M.; Peinado, M.A. Methods for DNA methylation analysis and applications in colon cancer. *Mutat. Res.* **2010**, *693*, 84–93.
69. Shen, L.; Waterland, R.A. Methods of DNA methylation analysis. *Curr. Opin. Clin. Nutr. Metab. Care* **2007**, *10*, 576–581.
70. Frommer, M.; McDonald, L.E.; Millar, D.S.; Collis, C.M.; Watt, F.; Grigg, G.W.; Molloy, P.L.; Paul, C.L. A genomic sequencing protocol that yields a positive display of 5-methylcytosine residues in individual DNA strands. *Proc. Natl. Acad. Sci. USA* **1992**, *89*, 1827–1831.
71. Grunau, C.; Clark, S.J.; Rosenthal, A. Bisulfite genomic sequencing: Systematic investigation of critical experimental parameters. *Nucleic Acids Res.* **2001**, *29*, e65.
72. Holmes, E.E.; Jung, M.; Meller, S.; Leisse, A.; Sailer, V.; Zech, J.; Mengdehl, M.; Garbe, L.A.; Uhl, B.; Kristiansen, G.; *et al.* Performance evaluation of kits for bisulfite-conversion of DNA from tissues, cell lines, FFPE tissues, aspirates, lavages, effusions, plasma, serum, and urine. *PLoS One* **2014**, *9*, e93933.
73. Tanaka, K.; Okamoto, A. Degradation of DNA by bisulfite treatment. *Bioorg. Med. Chem. Lett.* **2007**, *17*, 1912–1915.

74. Munson, K.; Clark, J.; Lamparska-Kupsik, K.; Smith, S.S. Recovery of bisulfite-converted genomic sequences in the methylation-sensitive QPCR. *Nucleic Acids Res.* **2007**, *35*, 2893–2903.
75. Brandes, J.C.; Carraway, H.; Herman, J.G. Optimal primer design using the novel primer design program: MSPprimer provides accurate methylation analysis of the ATM promoter. *Oncogene* **2007**, *26*, 6229–6237.
76. Warnecke, P.M.; Stirzaker, C.; Melki, J.R.; Millar, D.S.; Paul, C.L.; Clark, S.J. Detection and measurement of PCR bias in quantitative methylation analysis of bisulphite-treated DNA. *Nucleic Acids Res.* **1997**, *25*, 4422–4426.
77. Chhibber, A.; Schroeder, B.G. Single-molecule polymerase chain reaction reduces bias: Application to DNA methylation analysis by bisulfite sequencing. *Anal. Biochem.* **2008**, *377*, 46–54.
78. Moskalev, E.A.; Zavgorodnij, M.G.; Majorova, S.P.; Vorobjev, I.A.; Jandaghi, P.; Bure, I.V.; Hoheisel, J.D. Correction of PCR-bias in quantitative DNA methylation studies by means of cubic polynomial regression. *Nucleic Acids Res.* **2011**, *39*, e77.
79. Shen, L.; Guo, Y.; Chen, X.; Ahmed, S.; Issa, J.P. Optimizing annealing temperature overcomes bias in bisulfite pcr methylation analysis. *Biotechniques* **2007**, *42*, 48–58.
80. Wojdacz, T.K.; Hansen, L.L.; Dobrovic, A. A new approach to primer design for the control of PCR bias in methylation studies. *BMC Res. Notes* **2008**, *1*, 54.
81. Claus, R.; Wilop, S.; Hielscher, T.; Sonnet, M.; Dahl, E.; Galm, O.; Jost, E.; Plass, C. A systematic comparison of quantitative high-resolution DNA methylation analysis and methylation-specific PCR. *Epigenetics* **2012**, *7*, 772–780.
82. Raval, A.; Tanner, S.M.; Byrd, J.C.; Angerman, E.B.; Perko, J.D.; Chen, S.S.; Hackanson, B.; Grever, M.R.; Lucas, D.M.; Matkovic, J.J.; *et al.* Downregulation of death-associated protein kinase 1 (DAPK1) in chronic lymphocytic leukemia. *Cell* **2007**, *129*, 879–890.
83. Shah, N.; Lin, B.; Sibenaller, Z.; Ryken, T.; Lee, H.; Yoon, J.G.; Rostad, S.; Foltz, G. Comprehensive analysis of MGMT promoter methylation: Correlation with MGMT expression and clinical response in GBM. *PLoS One* **2011**, *6*, e16146.
84. Mikeska, T.; Candiloro, I.L.; Dobrovic, A. The implications of heterogeneous DNA methylation for the accurate quantification of methylation. *Epigenomics* **2010**, *2*, 561–573.
85. Reddy, A.N.; Jiang, W.W.; Kim, M.; Benoit, N.; Taylor, R.; Clinger, J.; Sidransky, D.; Califano, J.A. Death-associated protein kinase promoter hypermethylation in normal human lymphocytes. *Cancer Res.* **2003**, *63*, 7694–7698.
86. Srinivasan, M.; Sedmak, D.; Jewell, S. Effect of fixatives and tissue processing on the content and integrity of nucleic acids. *Am. J. Pathol.* **2002**, *161*, 1961–1971.
87. Colella, S.; Shen, L.; Baggerly, K.A.; Issa, J.P.; Krahe, R. Sensitive and quantitative universal Pyrosequencing methylation analysis of CpG sites. *Biotechniques* **2003**, *35*, 146–150.
88. Tost, J.; Dunker, J.; Gut, I.G. Analysis and quantification of multiple methylation variable positions in CpG islands by Pyrosequencing. *Biotechniques* **2003**, *35*, 152–156.
89. Mikeska, T.; Felsberg, J.; Hewitt, C.A.; Dobrovic, A. Analysing DNA methylation using bisulphite pyrosequencing. *Methods Mol. Biol.* **2011**, *791*, 33–53.
90. Dejeux, E.; Audard, V.; Cavard, C.; Gut, I.G.; Terris, B.; Tost, J. Rapid identification of

promoter hypermethylation in hepatocellular carcinoma by pyrosequencing of etiologically homogeneous sample pools. *J. Mol. Diagn.* **2007**, *9*, 510–520.

91. Lillycrop, K.A.; Phillips, E.S.; Torrens, C.; Hanson, M.A.; Jackson, A.A.; Burdge, G.C. Feeding pregnant rats a protein-restricted diet persistently alters the methylation of specific cytosines in the hepatic PPAR alpha promoter of the offspring. *Br. J. Nutr.* **2008**, *100*, 278–282.
92. Ehrich, M.; Nelson, M.R.; Stanssens, P.; Zabeau, M.; Liloglou, T.; Xinarianos, G.; Cantor, C.R.; Field, J.K.; van den Boom, D. Quantitative high-throughput analysis of DNA methylation patterns by base-specific cleavage and mass spectrometry. *Proc. Natl. Acad. Sci. USA* **2005**, *102*, 15785–15790.
93. Coolen, M.W.; Statham, A.L.; Gardiner-Garden, M.; Clark, S.J. Genomic profiling of CpG methylation and allelic specificity using quantitative high-throughput mass spectrometry: Critical evaluation and improvements. *Nucleic Acids Res.* **2007**, *35*, e119.
94. Wojdacz, T.K.; Dobrovic, A. Methylation-sensitive high resolution melting (MS-HRM): A new approach for sensitive and high-throughput assessment of methylation. *Nucleic Acids Res.* **2007**, *35*, e41.
95. Mikeska, T.; Dobrovic, A. Methylation-sensitive high resolution melting for the rapid analysis of DNA methylation. In *Epigenetics: A Reference Manual*; Craig, J.M., Wong, N.C., Eds.; Caister Academic Press: Norwich, UK, 2011; pp. 325–335.
96. Candiloro, I.L.; Mikeska, T.; Dobrovic, A. Closed-tube PCR methods for locus-specific DNA methylation analysis. *Methods Mol. Biol.* **2011**, *791*, 55–71.
97. Candiloro, I.L.; Mikeska, T.; Dobrovic, A. Assessing combined methylation-sensitive high resolution melting and pyrosequencing for the analysis of heterogeneous DNA methylation. *Epigenetics* **2011**, *6*, 500–507.
98. Herman, J.G.; Graff, J.R.; Myohanen, S.; Nelkin, B.D.; Baylin, S.B. Methylation-specific PCR: A novel PCR assay for methylation status of CpG islands. *Proc. Natl. Acad. Sci. USA* **1996**, *93*, 9821–9826.
99. Preusser, M.; Elezi, L.; Hainfellner, J.A. Reliability and reproducibility of PCR-based testing of O^6-methylguanine-DNA methyltransferase gene (MGMT) promoter methylation status in formalin-fixed and paraffin-embedded neurosurgical biopsy specimens. *Clin. Neuropathol.* **2008**, *27*, 388–390.
100. Hamilton, M.G.; Roldan, G.; Magliocco, A.; McIntyre, J.B.; Parney, I.; Easaw, J.C. Determination of the methylation status of MGMT in different regions within glioblastoma multiforme. *J. Neurooncol.* **2011**, *102*, 255–260.
101. Eads, C.A.; Danenberg, K.D.; Kawakami, K.; Saltz, L.B.; Blake, C.; Shibata, D.; Danenberg, P.V.; Laird, P.W. MethyLight: A high-throughput assay to measure DNA methylation. *Nucleic Acids Res.* **2000**, *28*, e32.
102. Rand, K.; Qu, W.; Ho, T.; Clark, S.J.; Molloy, P. Conversion-specific detection of DNA methylation using real-time polymerase chain reaction (ConLight-MSP) to avoid false positives. *Methods* **2002**, *27*, 114–120.
103. Bonanno, C.; Shehi, E.; Adlerstein, D.; Makrigiorgos, G.M. MS-FLAG, a novel real-time signal generation method for methylation-specific PCR. *Clin. Chem.* **2007**, *53*, 2119–2127.

104. Kristensen, L.S.; Mikeska, T.; Krypuy, M.; Dobrovic, A. Sensitive Melting Analysis after Real Time- Methylation Specific PCR (SMART-MSP): High-throughput and probe-free quantitative DNA methylation detection. *Nucleic Acids Res.* **2008**, *36*, e42.
105. Cottrell, S.E.; Distler, J.; Goodman, N.S.; Mooney, S.H.; Kluth, A.; Olek, A.; Schwope, I.; Tetzner, R.; Ziebarth, H.; Berlin, K. A real-time PCR assay for DNA-methylation using methylation-specific blockers. *Nucleic Acids Res.* **2004**, *32*, e10.
106. Nygren, A.O.; Ameziane, N.; Duarte, H.M.; Vijzelaar, R.N.; Waisfisz, Q.; Hess, C.J.; Schouten, J.P.; Errami, A. Methylation-specific MLPA (MS-MLPA): Simultaneous detection of CpG methylation and copy number changes of up to 40 sequences. *Nucleic Acids Res.* **2005**, *33*, e128.
107. Serizawa, R.R.; Ralfkiaer, U.; Dahl, C.; Lam, G.W.; Hansen, A.B.; Steven, K.; Horn, T.; Guldberg, P. Custom-designed MLPA using multiple short synthetic probes: Application to methylation analysis of five promoter CpG islands in tumor and urine specimens from patients with bladder cancer. *J. Mol. Diagn.* **2010**, *12*, 402–408.
108. Gatta, V.; Gennaro, E.; Franchi, S.; Cecconi, M.; Antonucci, I.; Tommasi, M.; Palka, G.; Coviello, D.; Stuppia, L.; Grasso, M. MS-MLPA analysis for FMR1 gene: Evaluation in a routine diagnostic setting. *BMC Med. Genet.* **2013**, *14*, 79.
109. Jeuken, J.W.; Cornelissen, S.J.; Vriezen, M.; Dekkers, M.M.; Errami, A.; Sijben, A.; Boots-Sprenger, S.H.; Wesseling, P. MS-MLPA: An attractive alternative laboratory assay for robust, reliable, and semiquantitative detection of MGMT promoter hypermethylation in gliomas. *Lab. Investig.* **2007**, *87*, 1055–1065.
110. Homig-Holzel, C.; Savola, S. Multiplex ligation-dependent probe amplification (MLPA) in tumor diagnostics and prognostics. *Diagn. Mol. Pathol.* **2012**, *21*, 189–206.
111. Tano, K.; Shiota, S.; Collier, J.; Foote, R.S.; Mitra, S. Isolation and structural characterization of a cDNA clone encoding the human DNA repair protein for O^6-alkylguanine. *Proc. Natl. Acad. Sci. USA* **1990**, *87*, 686–690.
112. Natarajan, A.T.; Vermeulen, S.; Darroudi, F.; Valentine, M.B.; Brent, T.P.; Mitra, S.; Tano, K. Chromosomal localization of human O^6-methylguanine-DNA methyltransferase (MGMT) gene by in situ hybridization. *Mutagenesis* **1992**, *7*, 83–85.
113. Nutt, C.L.; Costello, J.F.; Bambrick, L.L.; Yarosh, D.B.; Swinnen, L.J.; Chambers, A.F.; Cairncross, J.G. O^6-methylguanine-DNA methyltransferase in tumors and cells of the oligodendrocyte lineage. *Can. J. Neurol. Sci.* **1995**, *22*, 111–115.
114. Mineura, K.; Yanagisawa, T.; Watanabe, K.; Kowada, M.; Yasui, N. Human brain tumor O(6)-methylguanine-DNA methyltransferase mRNA and its significance as an indicator of selective chloroethylnitrosourea chemotherapy. *Int. J. Cancer* **1996**, *69*, 420–425.
115. Silber, J.R.; Bobola, M.S.; Ghatan, S.; Blank, A.; Kolstoe, D.D.; Berger, M.S. O^6-methylguanine-DNA methyltransferase activity in adult gliomas: Relation to patient and tumor characteristics. *Cancer Res.* **1998**, *58*, 1068–1073.
116. Esteller, M.; Garcia-Foncillas, J.; Andion, E.; Goodman, S.N.; Hidalgo, O.F.; Vanaclocha, V.; Baylin, S.B.; Herman, J.G. Inactivation of the DNA-repair gene MGMT and the clinical response of gliomas to alkylating agents. *N. Engl. J. Med.* **2000**, *343*, 1350–1354.
117. Hegi, M.E.; Diserens, A.C.; Godard, S.; Dietrich, P.Y.; Regli, L.; Ostermann, S.; Otten, P.; van Melle, G.; de Tribolet, N.; Stupp, R. Clinical trial substantiates the predictive value of

O-6-methylguanine-DNA methyltransferase promoter methylation in glioblastoma patients treated with temozolomide. *Clin. Cancer Res.* **2004**, *10*, 1871–1874.

118. Paz, M.F.; Yaya-Tur, R.; Rojas-Marcos, I.; Reynes, G.; Pollan, M.; Aguirre-Cruz, L.; Garcia-Lopez, J.L.; Piquer, J.; Safont, M.J.; Balana, C.; *et al.* CpG island hypermethylation of the DNA repair enzyme methyltransferase predicts response to temozolomide in primary gliomas. *Clin. Cancer Res.* **2004**, *10*, 4933–4938.
119. Blanc, J.L.; Wager, M.; Guilhot, J.; Kusy, S.; Bataille, B.; Chantereau, T.; Lapierre, F.; Larsen, C.J.; Karayan-Tapon, L. Correlation of clinical features and methylation status of MGMT gene promoter in glioblastomas. *J. Neurooncol.* **2004**, *68*, 275–283.
120. Hegi, M.E.; Diserens, A.C.; Gorlia, T.; Hamou, M.F.; de Tribolet, N.; Weller, M.; Kros, J.M.; Hainfellner, J.A.; Mason, W.; Mariani, L.; *et al.* MGMT gene silencing and benefit from temozolomide in glioblastoma. *N. Engl. J. Med.* **2005**, *352*, 997–1003.
121. Wick, W.; Weller, M.; van den Bent, M.; Sanson, M.; Weiler, M.; von Deimling, A.; Plass, C.; Hegi, M.; Platten, M.; Reifenberger, G. MGMT testing-the challenges for biomarker-based glioma treatment. *Nat. Rev. Neurol.* **2014**, *10*, 372–385.
122. Christians, A.; Hartmann, C.; Benner, A.; Meyer, J.; von Deimling, A.; Weller, M.; Wick, W.; Weiler, M. Prognostic value of three different methods of MGMT promoter methylation analysis in a prospective trial on newly diagnosed glioblastoma. *PLoS One* **2012**, *7*, e33449.
123. Karayan-Tapon, L.; Quillien, V.; Guilhot, J.; Wager, M.; Fromont, G.; Saikali, S.; Etcheverry, A.; Hamlat, A.; Loussouarn, D.; Campion, L.; *et al.* Prognostic value of O^6-methylguanine-DNA methyltransferase status in glioblastoma patients, assessed by five different methods. *J. Neurooncol.* **2010**, *97*, 311–322.
124. Mikeska, T.; Bock, C.; el-Maarri, O.; Hubner, A.; Ehrentraut, D.; Schramm, J.; Felsberg, J.; Kahl, P.; Buttner, R.; Pietsch, T.; *et al.* Optimization of quantitative MGMT promoter methylation analysis using pyrosequencing and combined bisulfite restriction analysis. *J. Mol. Diagn.* **2007**, *9*, 368–381.
125. Quillien, V.; Lavenu, A.; Karayan-Tapon, L.; Carpentier, C.; Labussiere, M.; Lesimple, T.; Chinot, O.; Wager, M.; Honnorat, J.; Saikali, S.; *et al.* Comparative assessment of 5 methods (methylation-specific polymerase chain reaction, MethyLight, pyrosequencing, methylation-sensitive high-resolution melting, and immunohistochemistry) to analyze O^6-methylguanine-DNA-methyltranferase in a series of 100 glioblastoma patients. *Cancer* **2012**, *118*, 4201–4211.
126. Weller, M.; Stupp, R.; Reifenberger, G.; Brandes, A.A.; van den Bent, M.J.; Wick, W.; Hegi, M.E. MGMT promoter methylation in malignant gliomas: Ready for personalized medicine? *Nat. Rev. Neurol.* **2010**, *6*, 39–51.
127. Mikeska, T.; Bock, C.; Do, H.; Dobrovic, A. DNA methylation biomarkers in cancer: Progress towards clinical implementation. *Expert Rev. Mol. Diagn.* **2012**, *12*, 473–487.
128. Li, Y.W.; Kong, F.M.; Zhou, J.P.; Dong, M. Aberrant promoter methylation of the vimentin gene may contribute to colorectal carcinogenesis: A meta-analysis. *Tumour Biol.* **2014**, *35*, 6783–6790.
129. Payne, S.R. From discovery to the clinic: The novel DNA methylation biomarker (m)sept9 for the detection of colorectal cancer in blood. *Epigenomics* **2010**, *2*, 575–585.

130. Darwiche, K.; Zarogoulidis, P.; Baehner, K.; Welter, S.; Tetzner, R.; Wohlschlaeger, J.; Theegarten, D.; Nakajima, T.; Freitag, L. Assessment of shox2 methylation in ebus-tbna specimen improves accuracy in lung cancer staging. *Ann. Oncol.* **2013**, *24*, 2866–2870.
131. Dietrich, D.; Hasinger, O.; Liebenberg, V.; Field, J.K.; Kristiansen, G.; Soltermann, A. DNA methylation of the homeobox genes pitx2 and shox2 predicts outcome in non-small-cell lung cancer patients. *Diagn. Mol. Pathol.* **2012**, *21*, 93–104.
132. Dietrich, D.; Jung, M.; Puetzer, S.; Leisse, A.; Holmes, E.E.; Meller, S.; Uhl, B.; Schatz, P.; Ivascu, C.; Kristiansen, G. Diagnostic and prognostic value of shox2 and sept9 DNA methylation and cytology in benign, paramalignant and malignant pleural effusions. *PloS One* **2013**, *8*, e84225.
133. Ilse, P.; Biesterfeld, S.; Pomjanski, N.; Fink, C.; Schramm, M. Shox2 DNA methylation is a tumour marker in pleural effusions. *Cancer genomics proteomics* **2013**, *10*, 217–223.
134. Kneip, C.; Schmidt, B.; Seegebarth, A.; Weickmann, S.; Fleischhacker, M.; Liebenberg, V.; Field, J.K.; Dietrich, D. Shox2 DNA methylation is a biomarker for the diagnosis of lung cancer in plasma. *J. Thorac. Oncol.* **2011**, *6*, 1632–1638.
135. Schmidt, B.; Liebenberg, V.; Dietrich, D.; Schlegel, T.; Kneip, C.; Seegebarth, A.; Flemming, N.; Seemann, S.; Distler, J.; Lewin, J.; *et al.* Shox2 DNA methylation is a biomarker for the diagnosis of lung cancer based on bronchial aspirates. *BMC Cancer* **2010**, *10*, 600.
136. Wu, T.; Giovannucci, E.; Welge, J.; Mallick, P.; Tang, W.Y.; Ho, S.M. Measurement of gstp1 promoter methylation in body fluids may complement psa screening: A meta-analysis. *Br. J. Cancer* **2011**, *105*, 65–73.
137. Chen, Y.; Li, J.; Yu, X.; Li, S.; Zhang, X.; Mo, Z.; Hu, Y. Apc gene hypermethylation and prostate cancer: A systematic review and meta-analysis. *Eur. J. Hum. Genet.* **2013**, *21*, 929–935.
138. Pan, J.; Chen, J.; Zhang, B.; Chen, X.; Huang, B.; Zhuang, J.; Mo, C.; Qiu, S. Association between rassf1a promoter methylation and prostate cancer: A systematic review and meta-analysis. *PloS One* **2013**, *8*, e75283.
139. Yin, A.A.; Zhang, L.H.; Cheng, J.X.; Dong, Y.; Liu, B.L.; Han, N.; Zhang, X. The predictive but not prognostic value of mgmt promoter methylation status in elderly glioblastoma patients: A meta-analysis. *PloS One* **2014**, *9*, e85102.
140. Zhang, K.; Wang, X.Q.; Zhou, B.; Zhang, L. The prognostic value of mgmt promoter methylation in glioblastoma multiforme: A meta-analysis. *Fam. cancer* **2013**, *12*, 449–458.
141. Preusser, M.; Berghoff, A.S.; Manzl, C.; Filipits, M.; Weinhausel, A.; Pulverer, W.; Dieckmann, K.; Widhalm, G.; Wohrer, A.; Knosp, E.; *et al.* Clinical neuropathology practice news 1–2014: Pyrosequencing meets clinical and analytical performance criteria for routine testing of mgmt promoter methylation status in glioblastoma. *Clin. Neuropathol.* **2014**, *33*, 6–14.
142. Quillien, V.; Lavenu, A.; Sanson, M.; Legrain, M.; Dubus, P.; Karayan-Tapon, L.; Mosser, J.; Ichimura, K.; Figarella-Branger, D. Outcome-based determination of optimal pyrosequencing assay for mgmt methylation detection in glioblastoma patients. *J. Neurooncol.* **2014**, *116*, 487–496.
143. Dunn, J.; Baborie, A.; Alam, F.; Joyce, K.; Moxham, M.; Sibson, R.; Crooks, D.; Husband, D.; Shenoy, A.; Brodbelt, A.; *et al.* Extent of mgmt promoter methylation correlates with outcome in glioblastomas given temozolomide and radiotherapy. *Br. J. Cancer* **2009**, *101*, 124–131.
144. Reifenberger, G.; Hentschel, B.; Felsberg, J.; Schackert, G.; Simon, M.; Schnell, O.; Westphal, M.; Wick, W.; Pietsch, T.; Loeffler, M.; *et al.* Predictive impact of mgmt promoter methylation in

glioblastoma of the elderly. *Int. J. Cancer.* **2012**, *131*, 1342–1350.

145. Oberstadt, M.C.; Bien-Moller, S.; Weitmann, K.; Herzog, S.; Hentschel, K.; Rimmbach, C.; Vogelgesang, S.; Balz, E.; Fink, M.; Michael, H.; *et al.* Epigenetic modulation of the drug resistance genes mgmt, abcb1 and abcg2 in glioblastoma multiforme. *BMC Cancer* **2013**, *13*, 617.
146. Everhard, S.; Tost, J.; el Abdalaoui, H.; Criniere, E.; Busato, F.; Marie, Y.; Gut, I.G.; Sanson, M.; Mokhtari, K.; Laigle-Donadey, F.; *et al.* Identification of regions correlating mgmt promoter methylation and gene expression in glioblastomas. *Neuro-oncology* **2009**, *11*, 348–356.
147. Heyn, H.; Esteller, M. DNA methylation profiling in the clinic: Applications and challenges. *Nat. Rev. Genet.* **2012**, *13*, 679–692.
148. Delpu, Y.; Cordelier, P.; Cho, W.C.; Torrisani, J. DNA methylation and cancer diagnosis. *Int. J. Mol. Sci.* **2013**, *14*, 15029–15058.
149. Heichman, K.A.; Warren, J.D. DNA methylation biomarkers and their utility for solid cancer diagnostics. *Clin. Chem. Lab. Med.* **2012**, *50*, 1707–1721.
150. Bardhan, K.; Liu, K. Epigenetics and colorectal cancer pathogenesis. *Cancers* **2013**, *5*, 676–713.
151. Colussi, D.; Brandi, G.; Bazzoli, F.; Ricciardiello, L. Molecular pathways involved in colorectal cancer: Implications for disease behavior and prevention. *Int. J. Mol. Sci.* **2013**, *14*, 16365–16385.
152. Gyparaki, M.T.; Basdra, E.K.; Papavassiliou, A.G. DNA methylation biomarkers as diagnostic and prognostic tools in colorectal cancer. *J. Mol. Med. (Berlin, Germany)* **2013**, *91*, 1249–1256.
153. Balgkouranidou, I.; Liloglou, T.; Lianidou, E.S. Lung cancer epigenetics: Emerging biomarkers. *Biomark. Med.* **2013**, *7*, 49–58.
154. Jones, A.; Lechner, M.; Fourkala, E.O.; Kristeleit, R.; Widschwendter, M. Emerging promise of epigenetics and DNA methylation for the diagnosis and management of women's cancers. *Epigenomics* **2010**, *2*, 9–38.
155. Day, T.K.; Bianco-Miotto, T. Common gene pathways and families altered by DNA methylation in breast and prostate cancers. *Endocr. Relat. Cancer* **2013**, *20*, R215–R232.
156. Kandimalla, R.; van Tilborg, A.A.; Zwarthoff, E.C. DNA methylation-based biomarkers in bladder cancer. *Nat. Rev. Urol.* **2013**, *10*, 327–335.
157. Fukushige, S.; Horii, A. Road to early detection of pancreatic cancer: Attempts to utilize epigenetic biomarkers. *Cancer Lett.* **2014**, *342*, 231–237.
158. Greenberg, E.S.; Chong, K.K.; Huynh, K.T.; Tanaka, R.; Hoon, D.S. Epigenetic biomarkers in skin cancer. *Cancer Lett.* **2014**, *342*, 170–177.
159. Zoratto, F.; Rossi, L.; Verrico, M.; Papa, A.; Basso, E.; Zullo, A.; Tomao, L.; Romiti, A.; Lo Russo, G.; Tomao, S. Focus on genetic and epigenetic events of colorectal cancer pathogenesis: Implications for molecular diagnosis. *Tumour Biol.* **2014**, *35*, 6195–6206.
160. McDevitt, M.A. Clinical applications of epigenetic markers and epigenetic profiling in myeloid malignancies. *Semin. Oncol.* **2012**, *39*, 109–122.
161. Rodriguez-Vicente, A.E.; Diaz, M.G.; Hernandez-Rivas, J.M. Chronic lymphocytic leukemia: A clinical and molecular heterogenous disease. *Cancer Genet.* **2013**, *206*, 49–62.
162. Li, Y.S.; Xie, Q.; Yang, D.Y.; Zheng, Y. Role of rassf1a promoter methylation in the pathogenesis of hepatocellular carcinoma: A meta-analysis of 21 cohort studies. *Mol. Biol. Rep.* **2014**, *41*, 3925–3933.

163. Chen, W.D.; Han, Z.J.; Skoletsky, J.; Olson, J.; Sah, J.; Myeroff, L.; Platzer, P.; Lu, S.; Dawson, D.; Willis, J.; *et al.* Detection in fecal DNA of colon cancer-specific methylation of the nonexpressed vimentin gene. *J. Natl. Cancer Inst.* **2005**, *97*, 1124–1132.
164. deVos, T.; Tetzner, R.; Model, F.; Weiss, G.; Schuster, M.; Distler, J.; Steiger, K.V.; Grutzmann, R.; Pilarsky, C.; Habermann, J.K.; *et al.* Circulating methylated sept9 DNA in plasma is a biomarker for colorectal cancer. *Clin. Chem.* **2009**, *55*, 1337–1346.
165. Oh, T.; Kim, N.; Moon, Y.; Kim, M.S.; Hoehn, B.D.; Park, C.H.; Kim, T.S.; Kim, N.K.; Chung, H.C.; An, S. Genome-wide identification and validation of a novel methylation biomarker, sdc2, for blood-based detection of colorectal cancer. *J. Mol. Diagn.* **2013**, *15*, 498–507.
166. Mitchell, S.M.; Ross, J.P.; Drew, H.R.; Ho, T.; Brown, G.S.; Saunders, N.F.; Duesing, K.R.; Buckley, M.J.; Dunne, R.; Beetson, I.; *et al.* A panel of genes methylated with high frequency in colorectal cancer. *BMC Cancer* **2014**, *14*, 54.
167. Goessl, C.; Krause, H.; Muller, M.; Heicappell, R.; Schrader, M.; Sachsinger, J.; Miller, K. Fluorescent methylation-specific polymerase chain reaction for DNA-based detection of prostate cancer in bodily fluids. *Cancer Res.* **2000**, *60*, 5941–5945.
168. Goessl, C.; Muller, M.; Heicappell, R.; Krause, H.; Straub, B.; Schrader, M.; Miller, K. DNA-based detection of prostate cancer in urine after prostatic massage. *Urology* **2001**, *58*, 335–338.
169. Merlo, A.; Herman, J.G.; Mao, L.; Lee, D.J.; Gabrielson, E.; Burger, P.C.; Baylin, S.B.; Sidransky, D. 5' cpg island methylation is associated with transcriptional silencing of the tumour suppressor p16/cdkn2/mts1 in human cancers. *Nat. Med.* **1995**, *1*, 686–692.
170. Sterlacci, W.; Tzankov, A.; Veits, L.; Zelger, B.; Bihl, M.P.; Foerster, A.; Augustin, F.; Fiegl, M.; Savic, S. A comprehensive analysis of p16 expression, gene status, and promoter hypermethylation in surgically resected non-small cell lung carcinomas. *J. Thorac. Oncol.* **2011**, *6*, 1649–1657.
171. Nimmrich, I.; Sieuwerts, A.M.; Meijer-van Gelder, M.E.; Schwope, I.; Bolt-de Vries, J.; Harbeck, N.; Koenig, T.; Hartmann, O.; Kluth, A.; Dietrich, D.; *et al.* DNA hypermethylation of pitx2 is a marker of poor prognosis in untreated lymph node-negative hormone receptor-positive breast cancer patients. *Breast Cancer Res. Treat.* **2008**, *111*, 429–437.
172. Harbeck, N.; Nimmrich, I.; Hartmann, A.; Ross, J.S.; Cufer, T.; Grutzmann, R.; Kristiansen, G.; Paradiso, A.; Hartmann, O.; Margossian, A.; *et al.* Multicenter study using paraffin-embedded tumor tissue testing pitx2 DNA methylation as a marker for outcome prediction in tamoxifen-treated, node-negative breast cancer patients. *J. Clin. Oncol.* **2008**, *26*, 5036–5042.
173. Maier, S.; Nimmrich, I.; Koenig, T.; Eppenberger-Castori, S.; Bohlmann, I.; Paradiso, A.; Spyratos, F.; Thomssen, C.; Mueller, V.; Nahrig, J.; *et al.* DNA-methylation of the homeodomain transcription factor PITX2 reliably predicts risk of distant disease recurrence in tamoxifen-treated, node-negative breast cancer patients—Technical and clinical validation in a multi-centre setting in collaboration with the European Organisation for Research and Treatment of Cancer (EORTC) PathoBiology group. *Eur. J. Cancer* **2007**, *43*, 1679–1686.
174. Hartmann, O.; Spyratos, F.; Harbeck, N.; Dietrich, D.; Fassbender, A.; Schmitt, M.; Eppenberger-Castori, S.; Vuaroqueaux, V.; Lerebours, F.; Welzel, K.; *et al.* DNA methylation markers predict outcome in node-positive, estrogen receptor-positive breast cancer with adjuvant anthracycline-based chemotherapy. *Clin. Cancer Res.* **2009**, *15*, 315–323.

175. Banez, L.L.; Sun, L.; van Leenders, G.J.; Wheeler, T.M.; Bangma, C.H.; Freedland, S.J.; Ittmann, M.M.; Lark, A.L.; Madden, J.F.; Hartman, A.; *et al.* Multicenter clinical validation of PITX2 methylation as a prostate specific antigen recurrence predictor in patients with post-radical prostatectomy prostate cancer. *J. Urol.* **2010**, *184*, 149–156.
176. Dietrich, D.; Hasinger, O.; Banez, L.L.; Sun, L.; van Leenders, G.J.; Wheeler, T.M.; Bangma, C.H.; Wernert, N.; Perner, S.; Freedland, S.J.; *et al.* Development and clinical validation of a real-time PCR assay for PITX2 DNA methylation to predict prostate-specific antigen recurrence in prostate cancer patients following radical prostatectomy. *J. Mol. Diagn.* **2013**, *15*, 270–279.
177. Weiss, G.; Cottrell, S.; Distler, J.; Schatz, P.; Kristiansen, G.; Ittmann, M.; Haefliger, C.; Lesche, R.; Hartmann, A.; Corman, J.; *et al.* DNA methylation of the PITX2 gene promoter region is a strong independent prognostic marker of biochemical recurrence in patients with prostate cancer after radical prostatectomy. *J. Urol.* **2009**, *181*, 1678–1685.
178. Kaina, B.; Christmann, M.; Naumann, S.; Roos, W.P. MGMT: Key node in the battle against genotoxicity, carcinogenicity and apoptosis induced by alkylating agents. *DNA Repair* **2007**, *6*, 1079–1099.
179. Sharma, S.; Salehi, F.; Scheithauer, B.W.; Rotondo, F.; Syro, L.V.; Kovacs, K. Role of MGMT in tumor development, progression, diagnosis, treatment and prognosis. *Anticancer Res.* **2009**, *29*, 3759–3768.
180. Esteller, M.; Gaidano, G.; Goodman, S.N.; Zagonel, V.; Capello, D.; Botto, B.; Rossi, D.; Gloghini, A.; Vitolo, U.; Carbone, A.; *et al.* Hypermethylation of the DNA repair gene O(6)-methylguanine DNA methyltransferase and survival of patients with diffuse large B-cell lymphoma. *J. Natl. Cancer Inst.* **2002**, *94*, 26–32.
181. Van den Bent, M.J.; Dubbink, H.J.; Sanson, M.; van der Lee-Haarloo, C.R.; Hegi, M.; Jeuken, J.W.; Ibdaih, A.; Brandes, A.A.; Taphoorn, M.J.; Frenay, M.; *et al.* MGMT promoter methylation is prognostic but not predictive for outcome to adjuvant PCV chemotherapy in anaplastic oligodendroglial tumors: A report from EORTC Brain Tumor Group Study 26951. *J. Clin. Oncol.* **2009**, *27*, 5881–5886.
182. Wick, W.; Hartmann, C.; Engel, C.; Stoffels, M.; Felsberg, J.; Stockhammer, F.; Sabel, M.C.; Koeppen, S.; Ketter, R.; Meyermann, R.; *et al.* NOA-04 randomized phase III trial of sequential radiochemotherapy of anaplastic glioma with procarbazine, lomustine, and vincristine or temozolomide. *J. Clin. Oncol.* **2009**, *27*, 5874–5880.
183. Skaar, D.A.; Li, Y.; Bernal, A.J.; Hoyo, C.; Murphy, S.K.; Jirtle, R.L. The human imprintome: Regulatory mechanisms, methods of ascertainment, and roles in disease susceptibility. *ILAR J.* **2012**, *53*, 341–358.
184. Peters, J. The role of genomic imprinting in biology and disease: An expanding view. *Nat. Rev. Genet.* **2014**, *15*, 517–530.
185. Butler, M.G. Genomic imprinting disorders in humans: A mini-review. *J. Assist. Reprod. Genet.* **2009**, *26*, 477–486.
186. Ramsden, S.C.; Clayton-Smith, J.; Birch, R.; Buiting, K. Practice guidelines for the molecular analysis of Prader-Willi and Angelman syndromes. *BMC Med. Genet.* **2010**, *11*, 70.
187. Boyle, J.; Hawkins, M.; Barton, D.E.; Meaney, K.; Guitart, M.; O'Grady, A.; Tobi, S.;

Ramsden, S.C.; Elles, R.; Gray, E.; *et al.* Establishment of the first WHO international genetic reference panel for Prader Willi and Angelman syndromes. *Eur. J. Hum. Genet.* **2011**, *19*, 857–864.

188. Kubota, T.; Das, S.; Christian, S.L.; Baylin, S.B.; Herman, J.G.; Ledbetter, D.H. Methylation-specific PCR simplifies imprinting analysis. *Nat. Genet.* **1997**, *16*, 16–17.
189. Zeschnigk, M.; Lich, C.; Buiting, K.; Doerfler, W.; Horsthemke, B. A single-tube PCR test for the diagnosis of Angelman and Prader-Willi syndrome based on allelic methylation differences at the SNRPN locus. *Eur. J. Hum. Genet.* **1997**, *5*, 94–98.
190. Henkhaus, R.S.; Kim, S.J.; Kimonis, V.E.; Gold, J.A.; Dykens, E.M.; Driscoll, D.J.; Butler, M.G. Methylation-specific multiplex ligation-dependent probe amplification and identification of deletion genetic subtypes in Prader-Willi syndrome. *Genet. Test. Mol. Biomark.* **2012**, *16*, 178–186.
191. Bourque, D.K.; Penaherrera, M.S.; Yuen, R.K.; van Allen, M.I.; McFadden, D.E.; Robinson, W.P. The utility of quantitative methylation assays at imprinted genes for the diagnosis of fetal and placental disorders. *Clin. Genet.* **2011**, *79*, 169–175.
192. Eggermann, T.; Begemann, M.; Binder, G.; Spengler, S. Silver-Russell syndrome: Genetic basis and molecular genetic testing. *Orphanet J. Rare Dis.* **2010**, *5*, 19.
193. Weksberg, R.; Shuman, C.; Beckwith, J.B. Beckwith-Wiedemann syndrome. *Eur. J. Hum. Genet.* **2010**, *18*, 8–14.
194. Smilinich, N.J.; Day, C.D.; Fitzpatrick, G.V.; Caldwell, G.M.; Lossie, A.C.; Cooper, P.R.; Smallwood, A.C.; Joyce, J.A.; Schofield, P.N.; Reik, W.; *et al.* A maternally methylated CpG island in KvLQT1 is associated with an antisense paternal transcript and loss of imprinting in Beckwith-Wiedemann syndrome. *Proc. Natl. Acad. Sci. USA* **1999**, *96*, 8064–8069.
195. Horike, S.; Ferreira, J.C.; Meguro-Horike, M.; Choufani, S.; Smith, A.C.; Shuman, C.; Meschino, W.; Chitayat, D.; Zackai, E.; Scherer, S.W.; *et al.* Screening of DNA methylation at the H19 promoter or the distal region of its ICR1 ensures efficient detection of chromosome 11p15 epimutations in Russell-Silver syndrome. *Am. J. Med. Genet. A* **2009**, *149A*, 2415–2423.
196. Lukova, M.; Todorova, A.; Todorov, T.; Mitev, V. Different methylation patterns in BWS/SRS cases clarified by MS-MLPA. *Mol. Biol. Rep.* **2013**, *40*, 263–268.
197. Priolo, M.; Sparago, A.; Mammi, C.; Cerrato, F.; Lagana, C.; Riccio, A. MS-MLPA is a specific and sensitive technique for detecting all chromosome 11p15.5 imprinting defects of BWS and SRS in a single-tube experiment. *Eur. J. Hum. Genet.* **2008**, *16*, 565–571.
198. Scott, R.H.; Douglas, J.; Baskcomb, L.; Nygren, A.O.; Birch, J.M.; Cole, T.R.; Cormier-Daire, V.; Eastwood, D.M.; Garcia-Minaur, S.; Lupunzina, P.; *et al.* Methylation-specific multiplex ligation-dependent probe amplification (MS-MLPA) robustly detects and distinguishes 11p15 abnormalities associated with overgrowth and growth retardation. *J. Med. Genet.* **2008**, *45*, 106–113.
199. Calvello, M.; Tabano, S.; Colapietro, P.; Maitz, S.; Pansa, A.; Augello, C.; Lalatta, F.; Gentilin, B.; Spreafico, F.; Calzari, L.; *et al.* Quantitative DNA methylation analysis improves epigenotype-phenotype correlations in Beckwith-Wiedemann syndrome. *Epigenetics* **2013**, *8*, 1053–1060.
200. White, H.E.; Durston, V.J.; Harvey, J.F.; Cross, N.C. Quantitative analysis of SNRPN(correction

of SRNPN) gene methylation by pyrosequencing as a diagnostic test for Prader-Willi syndrome and Angelman syndrome. *Clin. Chem.* **2006**, *52*, 1005–1013.

201. Feil, R.; Fraga, M.F. Epigenetics and the environment: Emerging patterns and implications. *Nat. Rev. Genet.* **2011**, *13*, 97–109.
202. Mazzio, E.A.; Soliman, K.F. Basic concepts of epigenetics: Impact of environmental signals on gene expression. *Epigenetics* **2012**, *7*, 119–130.
203. Cortessis, V.K.; Thomas, D.C.; Levine, A.J.; Breton, C.V.; Mack, T.M.; Siegmund, K.D.; Haile, R.W.; Laird, P.W. Environmental epigenetics: Prospects for studying epigenetic mediation of exposure-response relationships. *Hum. Genet.* **2012**, *131*, 1565–1589.
204. Hogg, K.; Price, E.M.; Hanna, C.W.; Robinson, W.P. Prenatal and perinatal environmental influences on the human fetal and placental epigenome. *Clin. Pharmacol. Ther.* **2012**, *92*, 716–726.
205. Mirbahai, L.; Chipman, J.K. Epigenetic memory of environmental organisms: A reflection of lifetime stressor exposures. *Mutat. Res. Genet. Toxicol. Environ. Mutagen.* **2014**, *764–765*, 10–17.
206. Bibikova, M.; Barnes, B.; Tsan, C.; Ho, V.; Klotzle, B.; Le, J.M.; Delano, D.; Zhang, L.; Schroth, G.P.; Gunderson, K.L.; *et al.* High density DNA methylation array with single CpG site resolution. *Genomics* **2011**, *98*, 288–295.
207. Dogan, M.V.; Shields, B.; Cutrona, C.; Gao, L.; Gibbons, F.X.; Simons, R.; Monick, M.; Brody, G.H.; Tan, K.; Beach, S.R.; *et al.* The effect of smoking on DNA methylation of peripheral blood mononuclear cells from African American women. *BMC Genomics* **2014**, *15*, 151.
208. Elliott, H.R.; Tillin, T.; McArdle, W.L.; Ho, K.; Duggirala, A.; Frayling, T.M.; Davey Smith, G.; Hughes, A.D.; Chaturvedi, N.; Relton, C.L. Differences in smoking associated DNA methylation patterns in South Asians and Europeans. *Clin. Epigenetics* **2014**, *6*, 4.
209. Joubert, B.R.; Haberg, S.E.; Nilsen, R.M.; Wang, X.; Vollset, S.E.; Murphy, S.K.; Huang, Z.; Hoyo, C.; Midttun, O.; Cupul-Uicab, L.A.; *et al.* 450K epigenome-wide scan identifies differential DNA methylation in newborns related to maternal smoking during pregnancy. *Environ. Health Perspect.* **2012**, *120*, 1425–1431.
210. Philibert, R.A.; Beach, S.R.; Lei, M.K.; Brody, G.H. Changes in DNA methylation at the aryl hydrocarbon receptor repressor may be a new biomarker for smoking. *Clin. Epigenetics* **2013**, *5*, 19.
211. Shenker, N.S.; Polidoro, S.; van Veldhoven, K.; Sacerdote, C.; Ricceri, F.; Birrell, M.A.; Belvisi, M.G.; Brown, R.; Vineis, P.; Flanagan, J.M. Epigenome-wide association study in the European Prospective Investigation into Cancer and Nutrition (EPIC-Turin) identifies novel genetic loci associated with smoking. *Hum. Mol. Genet.* **2013**, *22*, 843–851.
212. Zeilinger, S.; Kuhnel, B.; Klopp, N.; Baurecht, H.; Kleinschmidt, A.; Gieger, C.; Weidinger, S.; Lattka, E.; Adamski, J.; Peters, A.; *et al.* Tobacco smoking leads to extensive genome-wide changes in DNA methylation. *PLoS One* **2013**, *8*, e63812.
213. Monick, M.M.; Beach, S.R.; Plume, J.; Sears, R.; Gerrard, M.; Brody, G.H.; Philibert, R.A. Coordinated changes in AHRR methylation in lymphoblasts and pulmonary macrophages from smokers. *Am. J. Med. Genet. B Neuropsychiatr. Genet.* **2012**, *159B*, 141–151.
214. Besingi, W.; Johansson, A. Smoke-related DNA methylation changes in the etiology of human disease. *Hum. Mol. Genet.* **2014**, *23*, 2290–2297.
215. Markunas, C.A.; Xu, Z.; Harlid, S.; Wade, P.A.; Lie, R.T.; Taylor, J.A.; Wilcox, A.J. Identification of DNA Methylation Changes in Newborns Related to Maternal Smoking during

Pregnancy. *Environ. Health Perspect.* **2014**, doi:10.1289/ehp.1307892.

216. Harlid, S.; Xu, Z.; Panduri, V.; Sandler, D.P.; Taylor, J.A. CpG Sites Associated with Cigarette Smoking: Analysis of Epigenome-Wide Data from the Sister Study. *Environ. Health Perspect.* **2014**, *122*, 673–678.
217. Novakovic, B.; Ryan, J.; Pereira, N.; Boughton, B.; Craig, J.M.; Saffery, R. Postnatal stability, tissue, and time specific effects of AHRR methylation change in response to maternal smoking in pregnancy. *Epigenetics* **2014**, *9*, 377–386.
218. Joubert, B.R.; Haberg, S.E.; Bell, D.A.; Nilsen, R.M.; Vollset, S.E.; Midttun, O.; Ueland, P.M.; Wu, M.C.; Nystad, W.; Peddada, S.D.; *et al.* Maternal Smoking and DNA Methylation in Newborns: *In Utero* Effect or Epigenetic Inheritance? *Cancer Epidemiol. Biomark. Prev.* **2014**, doi:10.1158/1055-9965.EPI-13-1256.
219. Brokken, L.J.; Lundberg-Giwercman, Y.; Meyts, E.R.; Eberhard, J.; Stahl, O.; Cohn-Cedermark, G.; Daugaard, G.; Arver, S.; Giwercman, A. Association between polymorphisms in the aryl hydrocarbon receptor repressor gene and disseminated testicular germ cell cancer. *Front. Endocrinol.* **2013**, *4*, 4.
220. Cauchi, S.; Stucker, I.; Cenee, S.; Kremers, P.; Beaune, P.; Massaad-Massade, L. Structure and polymorphisms of human aryl hydrocarbon receptor repressor (AhRR) gene in a French population: Relationship with CYP1A1 inducibility and lung cancer. *Pharmacogenetics* **2003**, *13*, 339–347.
221. Liang, Y.; Li, W.W.; Yang, B.W.; Tao, Z.H.; Sun, H.C.; Wang, L.; Xia, J.L.; Qin, L.X.; Tang, Z.Y.; Fan, J.; *et al.* Aryl hydrocarbon receptor nuclear translocator is associated with tumor growth and progression of hepatocellular carcinoma. *Int. J. Cancer* **2012**, *130*, 1745–1754.
222. Tsuchiya, M.; Katoh, T.; Motoyama, H.; Sasaki, H.; Tsugane, S.; Ikenoue, T. Analysis of the AhR, ARNT, and AhRR gene polymorphisms: Genetic contribution to endometriosis susceptibility and severity. *Fertil. Steril.* **2005**, *84*, 454–458.
223. Lupien, S.J.; McEwen, B.S.; Gunnar, M.R.; Heim, C. Effects of stress throughout the lifespan on the brain, behaviour and cognition. *Nat. Rev. Neurosci.* **2009**, *10*, 434–445.
224. Meaney, M.J.; Szyf, M. Environmental programming of stress responses through DNA methylation: Life at the interface between a dynamic environment and a fixed genome. *Dialogues Clin. Neurosci.* **2005**, *7*, 103–123.
225. Weaver, I.C.; Cervoni, N.; Champagne, F.A.; D'Alessio, A.C.; Sharma, S.; Seckl, J.R.; Dymov, S.; Szyf, M.; Meaney, M.J. Epigenetic programming by maternal behavior. *Nat. Neurosci.* **2004**, *7*, 847–854.
226. Oberlander, T.F.; Weinberg, J.; Papsdorf, M.; Grunau, R.; Misri, S.; Devlin, A.M. Prenatal exposure to maternal depression, neonatal methylation of human glucocorticoid receptor gene (NR3C1) and infant cortisol stress responses. *Epigenetics* **2008**, *3*, 97–106.
227. Hompes, T.; Izzi, B.; Gellens, E.; Morreels, M.; Fieuws, S.; Pexsters, A.; Schops, G.; Dom, M.; van Bree, R.; Freson, K.; *et al.* Investigating the influence of maternal cortisol and emotional state during pregnancy on the DNA methylation status of the glucocorticoid receptor gene (NR3C1) promoter region in cord blood. *J. Psychiatr. Res.* **2013**, *47*, 880–891.
228. Conradt, E.; Lester, B.M.; Appleton, A.A.; Armstrong, D.A.; Marsit, C.J. The roles of DNA

methylation of NR3C1 and 11beta-HSD2 and exposure to maternal mood disorder *in utero* on newborn neurobehavior. *Epigenetics* **2013**, *8*, 1321–1329.

229. Radtke, K.M.; Ruf, M.; Gunter, H.M.; Dohrmann, K.; Schauer, M.; Meyer, A.; Elbert, T. Transgenerational impact of intimate partner violence on methylation in the promoter of the glucocorticoid receptor. *Transl. Psychiatry* **2011**, *1*, e21.
230. Mulligan, C.J.; D'Errico, N.C.; Stees, J.; Hughes, D.A. Methylation changes at NR3C1 in newborns associate with maternal prenatal stress exposure and newborn birth weight. *Epigenetics* **2012**, *7*, 853–857.
231. Yehuda, R.; Daskalakis, N.P.; Lehrner, A.; Desarnaud, F.; Bader, H.N.; Makotkine, I.; Flory, J.D.; Bierer, L.M.; Meaney, M.J. Influences of Maternal and Paternal PTSD on Epigenetic Regulation of the Glucocorticoid Receptor Gene in Holocaust Survivor Offspring. *Am. J. Psychiatry* **2014**, *171*, 872–880.
232. McGowan, P.O.; Sasaki, A.; D'Alessio, A.C.; Dymov, S.; Labonte, B.; Szyf, M.; Turecki, G.; Meaney, M.J. Epigenetic regulation of the glucocorticoid receptor in human brain associates with childhood abuse. *Nat. Neurosci.* **2009**, *12*, 342–348.
233. Perroud, N.; Paoloni-Giacobino, A.; Prada, P.; Olie, E.; Salzmann, A.; Nicastro, R.; Guillaume, S.; Mouthon, D.; Stouder, C.; Dieben, K.; *et al.* Increased methylation of glucocorticoid receptor gene (NR3C1) in adults with a history of childhood maltreatment: A link with the severity and type of trauma. *Transl. Psychiatry* **2011**, *1*, e59.
234. Labonte, B.; Yerko, V.; Gross, J.; Mechawar, N.; Meaney, M.J.; Szyf, M.; Turecki, G. Differential glucocorticoid receptor exon 1(B), 1(C), and 1(H) expression and methylation in suicide completers with a history of childhood abuse. *Biol. Psychiatry* **2012**, *72*, 41–48.
235. Tyrka, A.R.; Price, L.H.; Marsit, C.; Walters, O.C.; Carpenter, L.L. Childhood adversity and epigenetic modulation of the leukocyte glucocorticoid receptor: Preliminary findings in healthy adults. *PLoS One* **2012**, *7*, e30148.
236. Melas, P.A.; Wei, Y.; Wong, C.C.; Sjoholm, L.K.; Aberg, E.; Mill, J.; Schalling, M.; Forsell, Y.; Lavebratt, C. Genetic and epigenetic associations of MAOA and NR3C1 with depression and childhood adversities. *Int. J. Neuropsychopharmacol.* **2013**, *16*, 1513–1528.
237. Yehuda, R.; Flory, J.D.; Bierer, L.M.; Henn-Haase, C.; Lehrner, A.; Desarnaud, F.; Makotkine, I.; Daskalakis, N.P.; Marmar, C.R.; Meaney, M.J. Lower Methylation of Glucocorticoid Receptor Gene Promoter 1 in Peripheral Blood of Veterans with Posttraumatic Stress Disorder. *Biol. Psychiatry* **2014**, doi:10.1016/j.biopsych.2014.02.006.
238. Labonte, B.; Azoulay, N.; Yerko, V.; Turecki, G.; Brunet, A. Epigenetic modulation of glucocorticoid receptors in posttraumatic stress disorder. *Transl. Psychiatry* **2014**, *4*, e368.
239. Bromer, C.; Marsit, C.J.; Armstrong, D.A.; Padbury, J.F.; Lester, B. Genetic and epigenetic variation of the glucocorticoid receptor (NR3C1) in placenta and infant neurobehavior. *Dev. Psychobiol.* **2013**, *55*, 673–683.
240. Yehuda, R.; Daskalakis, N.P.; Desarnaud, F.; Makotkine, I.; Lehrner, A.L.; Koch, E.; Flory, J.D.; Buxbaum, J.D.; Meaney, M.J.; Bierer, L.M. Epigenetic Biomarkers as Predictors and Correlates of Symptom Improvement Following Psychotherapy in Combat Veterans with PTSD. *Front. Psychiatry* **2013**, *4*, 118.

241. Edelman, S.; Shalev, I.; Uzefovsky, F.; Israel, S.; Knafo, A.; Kremer, I.; Mankuta, D.; Kaitz, M.; Ebstein, R.P. Epigenetic and genetic factors predict women's salivary cortisol following a threat to the social self. *PLoS One* **2012**, *7*, e48597.

242. Weaver, I.C.; Champagne, F.A.; Brown, S.E.; Dymov, S.; Sharma, S.; Meaney, M.J.; Szyf, M. Reversal of maternal programming of stress responses in adult offspring through methyl supplementation: Altering epigenetic marking later in life. *J. Neurosci.* **2005**, *25*, 11045–11054.

243. Weaver, I.C.; Meaney, M.J.; Szyf, M. Maternal care effects on the hippocampal transcriptome and anxiety-mediated behaviors in the offspring that are reversible in adulthood. *Proc. Natl. Acad. Sci. USA* **2006**, *103*, 3480–3485.

244. Essex, M.J.; Thomas Boyce, W.; Hertzman, C.; Lam, L.L.; Armstrong, J.M.; Neumann, S.M.; Kobor, M.S. Epigenetic Vestiges of Early Developmental Adversity: Childhood Stress Exposure and DNA Methylation in Adolescence. *Child Dev.* **2011**, doi:10.1111/j.1467-8624.2011.01641.x.

245. Uddin, M.; Aiello, A.E.; Wildman, D.E.; Koenen, K.C.; Pawelec, G.; de Los Santos, R.; Goldmann, E.; Galea, S. Epigenetic and immune function profiles associated with posttraumatic stress disorder. *Proc. Natl. Acad. Sci. USA* **2010**, *107*, 9470–9475.

246. Borghol, N.; Suderman, M.; McArdle, W.; Racine, A.; Hallett, M.; Pembrey, M.; Hertzman, C.; Power, C.; Szyf, M. Associations with early-life socio-economic position in adult DNA methylation. *Int. J. Epidemiol.* **2012**, *41*, 62–74.

247. Mehta, D.; Klengel, T.; Conneely, K.N.; Smith, A.K.; Altmann, A.; Pace, T.W.; Rex-Haffner, M.; Loeschner, A.; Gonik, M.; Mercer, K.B.; *et al.* Childhood maltreatment is associated with distinct genomic and epigenetic profiles in posttraumatic stress disorder. *Proc. Natl. Acad. Sci. USA* **2013**, *110*, 8302–8307.

248. Naumova, O.Y.; Lee, M.; Koposov, R.; Szyf, M.; Dozier, M.; Grigorenko, E.L. Differential patterns of whole-genome DNA methylation in institutionalized children and children raised by their biological parents. *Dev. Psychopathol.* **2012**, *24*, 143–155.

249. Harvey, N.C.; Sheppard, A.; Godfrey, K.M.; McLean, C.; Garratt, E.; Ntani, G.; Davies, L.; Murray, R.; Inskip, H.M.; Gluckman, P.D.; *et al.* Childhood Bone Mineral Content Is Associated With Methylation Status of the RXRA Promoter at Birth. *J. Bone Miner. Res.* **2014**, *29*, 600–607.

250. Relton, C.L.; Groom, A.; St Pourcain, B.; Sayers, A.E.; Swan, D.C.; Embleton, N.D.; Pearce, M.S.; Ring, S.M.; Northstone, K.; Tobias, J.H.; *et al.* DNA methylation patterns in cord blood DNA and body size in childhood. *PLoS One* **2012**, *7*, e31821.

251. Rakyan, V.K.; Beyan, H.; Down, T.A.; Hawa, M.I.; Maslau, S.; Aden, D.; Daunay, A.; Busato, F.; Mein, C.A.; Manfras, B.; *et al.* Identification of type 1 diabetes-associated DNA methylation variable positions that precede disease diagnosis. *PLoS Genet.* **2011**, *7*, e1002300.

252. Clarke-Harris, R.; Wilkin, T.J.; Hosking, J.; Pinkney, J.; Jeffery, A.N.; Metcalf, B.S.; Godfrey, K.M.; Voss, L.D.; Lillycrop, K.A.; Burdge, G.C. Peroxisomal proliferator activated receptor-gamma-co-activator-1alpha promoter methylation in blood at 5–7 years predicts adiposity from 9 to 14 years (earlybird 50). *Diabetes* **2014**, *63*, 2528–2537.

253. Nguyen, A.; Rauch, T.A.; Pfeifer, G.P.; Hu, V.W. Global methylation profiling of lymphoblastoid cell lines reveals epigenetic contributions to autism spectrum disorders and a novel autism candidate gene, RORA, whose protein product is reduced in autistic brain. *FASEB J.* **2010**, *24*,

3036–3051.
254. Wong, C.C.; Meaburn, E.L.; Ronald, A.; Price, T.S.; Jeffries, A.R.; Schalkwyk, L.C.; Plomin, R.; Mill, J. Methylomic analysis of monozygotic twins discordant for autism spectrum disorder and related behavioural traits. *Mol. Psychiatry* **2014**, *19*, 495–503.
255. Wang, Y.; Fang, Y.; Zhang, F.; Xu, M.; Zhang, J.; Yan, J.; Ju, W.; Brown, W.T.; Zhong, N. Hypermethylation of the enolase gene (ENO2) in autism. *Eur. J. Pediatr.* **2014**, *173*, 1233–1244.
256. Berko, E.R.; Suzuki, M.; Beren, F.; Lemetre, C.; Alaimo, C.M.; Calder, R.B.; Ballaban-Gil, K.; Gounder, B.; Kampf, K.; Kirschen, J.; *et al.* Mosaic epigenetic dysregulation of ectodermal cells in autism spectrum disorder. *PLoS Genet.* **2014**, *10*, e1004402.
257. Ginsberg, M.R.; Rubin, R.A.; Falcone, T.; Ting, A.H.; Natowicz, M.R. Brain transcriptional and epigenetic associations with autism. *PLoS One* **2012**, *7*, e44736.
258. Ladd-Acosta, C.; Hansen, K.D.; Briem, E.; Fallin, M.D.; Kaufmann, W.E.; Feinberg, A.P. Common DNA methylation alterations in multiple brain regions in autism. *Mol. Psychiatry* **2013**, doi:10.1038/mp.2013.114.
259. Sun, C.; Burgner, D.P.; Ponsonby, A.L.; Saffery, R.; Huang, R.C.; Vuillermin, P.J.; Cheung, M.; Craig, J.M. Effects of early-life environment and epigenetics on cardiovascular disease risk in children: Highlighting the role of twin studies. *Pediatr. Res.* **2013**, *73*, 523–530.
260. Chang, C.P.; Bruneau, B.G. Epigenetics and cardiovascular development. *Annu. Rev. Physiol.* **2012**, *74*, 41–68.
261. Shirodkar, A.V.; Marsden, P.A. Epigenetics in cardiovascular disease. *Curr. Opin. Cardiol.* **2011**, *26*, 209–215.
262. Hidalgo, B.; Irvin, M.R.; Sha, J.; Zhi, D.; Aslibekyan, S.; Absher, D.; Tiwari, H.K.; Kabagambe, E.K.; Ordovas, J.M.; Arnett, D.K. Epigenome-wide association study of fasting measures of glucose, insulin, and HOMA-IR in the Genetics of Lipid Lowering Drugs and Diet Network study. *Diabetes* **2014**, *63*, 801–807.
263. Irvin, M.R.; Zhi, D.; Joehanes, R.; Mendelson, M.; Aslibekyan, S.; Claas, S.A.; Thibeault, K.S.; Patel, N.; Day, K.; Waite Jones, L.; *et al.* Epigenome-Wide Association Study of Fasting Blood Lipids in the Genetics of Lipid Lowering Drugs and Diet Network Study. *Circulation* **2014**, *130*, 565–572.
264. Campion, J.; Milagro, F.I.; Martinez, J.A. Individuality and epigenetics in obesity. *Obes. Rev.* **2009**, *10*, 383–392.
265. Toperoff, G.; Aran, D.; Kark, J.D.; Rosenberg, M.; Dubnikov, T.; Nissan, B.; Wainstein, J.; Friedlander, Y.; Levy-Lahad, E.; Glaser, B.; *et al.* Genome-wide survey reveals predisposing diabetes type 2-related DNA methylation variations in human peripheral blood. *Hum. Mol. Genet.* **2012**, *21*, 371–383.
266. Cordero, P.; Campion, J.; Milagro, F.I.; Goyenechea, E.; Steemburgo, T.; Javierre, B.M.; Martinez, J.A. Leptin and TNF-alpha promoter methylation levels measured by MSP could predict the response to a low-calorie diet. *J. Physiol. Biochem.* **2011**, *67*, 463–470.
267. Guintivano, J.; Arad, M.; Gould, T.D.; Payne, J.L.; Kaminsky, Z.A. Antenatal prediction of postpartum depression with blood DNA methylation biomarkers. *Mol. Psychiatry* **2014**, *19*, 633.
268. Wockner, L.F.; Noble, E.P.; Lawford, B.R.; Young, R.M.; Morris, C.P.; Whitehall, V.L.; Voisey, J. Genome-wide DNA methylation analysis of human brain tissue from schizophrenia patients.

Transl. Psychiatry **2014**, *4*, e339.

269. Nishioka, M.; Bundo, M.; Koike, S.; Takizawa, R.; Kakiuchi, C.; Araki, T.; Kasai, K.; Iwamoto, K. Comprehensive DNA methylation analysis of peripheral blood cells derived from patients with first-episode schizophrenia. *J. Hum. Genet.* **2013**, *58*, 91–97.
270. Xiao, Y.; Camarillo, C.; Ping, Y.; Arana, T.B.; Zhao, H.; Thompson, P.M.; Xu, C.; Su, B.B.; Fan, H.; Ordonez, J.; *et al.* The DNA methylome and transcriptome of different brain regions in schizophrenia and bipolar disorder. *PLoS One* **2014**, *9*, e95875.
271. Aberg, K.A.; McClay, J.L.; Nerella, S.; Clark, S.; Kumar, G.; Chen, W.; Khachane, A.N.; Xie, L.; Hudson, A.; Gao, G.; *et al.* Methylome-wide association study of schizophrenia: Identifying blood biomarker signatures of environmental insults. *JAMA Psychiatry* **2014**, *71*, 255–264.
272. Abdolmaleky, H.M.; Nohesara, S.; Ghadirivasfi, M.; Lambert, A.W.; Ahmadkhaniha, H.; Ozturk, S.; Wong, C.K.; Shafa, R.; Mostafavi, A.; Thiagalingam, S. DNA hypermethylation of serotonin transporter gene promoter in drug naive patients with schizophrenia. *Schizophr. Res.* **2014**, *152*, 373–380.
273. Nishioka, M.; Bundo, M.; Kasai, K.; Iwamoto, K. DNA methylation in schizophrenia: Progress and challenges of epigenetic studies. *Genome Med.* **2012**, *4*, 96.
274. Kato, T.; Iwamoto, K. Comprehensive DNA methylation and hydroxymethylation analysis in the human brain and its implication in mental disorders. *Neuropharmacology* **2014**, *80*, 133–139.
275. Abdolmaleky, H.M.; Smith, C.L.; Faraone, S.V.; Shafa, R.; Stone, W.; Glatt, S.J.; Tsuang, M.T. Methylomics in psychiatry: Modulation of gene-environment interactions may be through DNA methylation. *Am. J. Med. Genet. B Neuropsychiatr. Genet.* **2004**, *127B*, 51–59.
276. Abdolmaleky, H.M.; Yaqubi, S.; Papageorgis, P.; Lambert, A.W.; Ozturk, S.; Sivaraman, V.; Thiagalingam, S. Epigenetic dysregulation of HTR2A in the brain of patients with schizophrenia and bipolar disorder. *Schizophr. Res.* **2011**, *129*, 183–190.
277. Ghadirivasfi, M.; Nohesara, S.; Ahmadkhaniha, H.R.; Eskandari, M.R.; Mostafavi, S.; Thiagalingam, S.; Abdolmaleky, H.M. Hypomethylation of the serotonin receptor type-2A gene (HTR2A) at T102C polymorphic site in DNA derived from the saliva of patients with schizophrenia and bipolar disorder. *Am. J. Med. Genet. B Neuropsychiatr. Genet.* **2011**, *156* 536–545.
278. Mill, J.; Tang, T.; Kaminsky, Z.; Khare, T.; Yazdanpanah, S.; Bouchard, L.; Jia, P.; Assadzadeh, A.; Flanagan, J.; Schumacher, A.; *et al.* Epigenomic profiling reveals DNA-methylation changes associated with major psychosis. *Am. J. Hum. Genet.* **2008**, *82*, 696–711.
279. Abdolmaleky, H.M.; Cheng, K.H.; Russo, A.; Smith, C.L.; Faraone, S.V.; Wilcox, M.; Shafa, R.; Glatt, S.J.; Nguyen, G.; Ponte, J.F.; *et al.* Hypermethylation of the reelin (RELN) promoter in the brain of schizophrenic patients: A preliminary report. *Am. J. Med. Genet. B Neuropsychiatr. Genet.* **2005**, *134B*, 60–66.
280. Grayson, D.R.; Jia, X.; Chen, Y.; Sharma, R.P.; Mitchell, C.P.; Guidotti, A.; Costa, E. Reelin promoter hypermethylation in schizophrenia. *Proc. Natl. Acad. Sci. USA* **2005**, *102*, 9341–9346.
281. Kaminsky, Z.; Tochigi, M.; Jia, P.; Pal, M.; Mill, J.; Kwan, A.; Ioshikhes, I.; Vincent, J.B.; Kennedy, J.L.; Strauss, J.; *et al.* A multi-tissue analysis identifies HLA complex group 9 gene methylation differences in bipolar disorder. *Mol. Psychiatry* **2012**, *17*, 728–740.

282. Sugawara, H.; Iwamoto, K.; Bundo, M.; Ueda, J.; Miyauchi, T.; Komori, A.; Kazuno, A.; Adati, N.; Kusumi, I.; Okazaki, Y.; *et al.* Hypermethylation of serotonin transporter gene in bipolar disorder detected by epigenome analysis of discordant monozygotic twins. *Transl. Psychiatry* **2011**, *1*, e24.
283. Cruickshank, M.N.; Pitt, J.; Craig, J.M. Going back to the future with Guthrie-powered epigenome-wide association studies. *Genome Med.* **2012**, *4*, 83.
284. Tsai, P.C.; Spector, T.D.; Bell, J.T. Using epigenome-wide association scans of DNA methylation in age-related complex human traits. *Epigenomics* **2012**, *4*, 511–526.
285. Weidner, C.I.; Lin, Q.; Koch, C.M.; Eisele, L.; Beier, F.; Ziegler, P.; Bauerschlag, D.O.; Jockel, K.H.; Erbel, R.; Muhleisen, T.W.; *et al.* Aging of blood can be tracked by DNA methylation changes at just three CpG sites. *Genome Biol.* **2014**, *15*, R24.
286. Horvath, S. DNA methylation age of human tissues and cell types. *Genome Biol.* **2013**, *14*, R115.
287. Bibikova, M.; Le, J.; Barnes, B.; Saedinia-Melnyk, S.; Zhou, L.; Shen, R.; Gunderson, K.L. Geneom-wide methylation profiling using Infinium assay. *Epigenomics* **2009**, *1*, 177–200.
288. Hannum, G.; Guinney, J.; Zhao, L.; Zhang, L.; Hughes, G.; Sadda, S.; Klotzle, B.; Bibikova, M.; Fan, J.B.; Gao, Y.; *et al.* Genome-wide methylation profiles reveal quantitative views of human aging rates. *Mol. Cell* **2013**, *49*, 359–367.
289. Gibbs, W.W. Biomarkers and ageing: The clock-watcher. *Nature* **2014**, *508*, 168–170.
290. Ogino, S.; Stampfer, M. Lifestyle factors and microsatellite instability in colorectal cancer: The evolving field of molecular pathological epidemiology. *J. Natl. Cancer Inst.* **2010**, *102*, 365–367.
291. Bishehsari, F.; Mahdavinia, M.; Vacca, M.; Malekzadeh, R.; Mariani-Costantini, R. Epidemiological transition of colorectal cancer in developing countries: Environmental factors, molecular pathways, and opportunities for prevention. *World J. Gastroenterol.* **2014**, *20*, 6055–6072.
292. Whelan, R.; Watts, R.; Orr, C.A.; Althoff, R.R.; Artiges, E.; Banaschewski, T.; Barker, G.J.; Bokde, A.L.; Buchel, C.; Carvalho, F.M.; *et al.* Neuropsychosocial profiles of current and future adolescent alcohol misusers. *Nature* **2014**, *512*, 185–189.
293. Kim, D.S.; Lee, J.Y.; Lee, S.M.; Choi, J.E.; Cho, S.; Park, J.Y. Promoter methylation of the RGC32 gene in nonsmall cell lung cancer. *Cancer* **2011**, *117*, 590–596.

3

Whole Exome Sequencing of Extreme Morbid Obesity Patients: Translational Implications for Obesity and Related Disorders

Gilberto Paz-Filho [1,†], Margaret C.S. Boguszewski [2,†], Claudio A. Mastronardi [1], Hardip R. Patel [1], Angad S. Johar [1], Aaron Chuah [1], Gavin A. Huttley [1], Cesar L. Boguszewski [2], Ma-Li Wong [3], Mauricio Arcos-Burgos [1] and Julio Licinio [3,*]

[1] Genome Biology Department, The John Curtin School of Medical Research, The Australian National University, Garran Rd, building 131, Acton, Canberra, ACT 0200, Australia;
E-Mails: gilberto.pazfilho@anu.edu.au (G.P.-F.); Claudio.mastronardi@anu.edu.au (C.A.M.); Hardip.Patel@anu.edu.au (H.R.P); u4842377@anu.edu.au (A.S.J.); Aaron.Chuah@anu.edu.au (A.C.); gavin.huttley@anu.edu.au (G.A.H.); Mauricio.Arcos-Burgos@anu.edu.au (M.A.-B.)

[2] Endocrine Division (SEMPR), Department of Internal Medicine, Federal University of Parana, Avenida Agostinho Leão Junior, 285-Alto da Glória. CEP 80030-110, Curitiba-PR, Brazil;
E-Mails: margabogus@uol.com.br (M.C.S.B.); clbogus@uol.com.br (C.L.B.)

[3] Mind and Brain Theme, South Australian Health and Medical Research Institute, and Department of Psychiatry, School of Medicine, Flinders University, PO Box 11060 Adelaide SA 5001, Adelaide, Australia; E-Mail: maliwong7@gmail.com (M.-L.W.)

† These authors contributed equally to this work.

* Author to whom correspondence should be addressed; E-Mail: julio.licinio@sahmri.com;

Abstract: Whole-exome sequencing (WES) is a new tool that allows the rapid, inexpensive and accurate exploration of Mendelian and complex diseases, such as obesity. To identify sequence variants associated with obesity, we performed WES of family trios of one male teenager and one female child with severe early-onset obesity. Additionally, the teenager patient had hypopituitarism and hyperprolactinaemia. A comprehensive bioinformatics analysis found *de novo* and compound heterozygote sequence variants with a damaging effect on genes previously associated with obesity in mice (*LRP2*) and humans (*UCP2*), among other intriguing mutations affecting ciliary function (*DNAAF1*). A gene ontology and pathway analysis of genes harbouring mutations resulted in the significant identification of overrepresented pathways related to ATP/ITP (adenosine/inosine triphosphate) metabolism and, in general, to the regulation of lipid metabolism. We discuss the clinical and physiological consequences of these mutations and the importance of these findings for either the clinical assessment or eventual treatment of morbid obesity.

Keywords: DNAAF1; genetics; leptin-melanocortin; LRP2; megalin; monogenic; obesity; pituitary; UCP2; whole-exome sequencing

---

1. Introduction

Obesity is a global epidemic: the World Health Organization (WHO) estimates that half a billion people over the age of twenty worldwide are obese [1]. Global projections estimate that worldwide, 1.2 billion individuals will be obese by 2030 [2].

Obesity and its related traits have high estimates of heritability (h^2 between 40% and 70%) [3]. The investigation of candidate genes and genome-wide association studies have identified more than 60 obesity susceptibility genes that predispose to increased body weight, waist circumference, waist-hip ratio, body mass index (BMI) and fat percentage or fat mass. However, mutations in these genes account for a very small fraction of the obesity phenotypic variance [4,5]. It has been estimated that at least 7% of children with severe early-onset obesity (defined by an onset before the age of 10 years and BMI over three standard deviations (SD) above normal) have a single locus sequence variant determining obesity [6]. Nine susceptibility genes, determinants of non-syndromic Mendelian forms of human obesity, are involved in the hypothalamic control of energy balance via the leptin-melanocortin pathway and/or in neural development [4]: brain-derived neurotrophic factor (BDNF), leptin (LEP), leptin receptor (LEPR), melanocortin-4 receptor (MC4R), neurotrophic tyrosine kinase receptor type 2 (NTRK2), prohormone convertase 1 (PCSK1), proopiomelanocortin (POMC), single-minded homolog 1 (SIM1) and, more recently, melanocortin 2 receptor accessory protein 2 (MRAP2) [7].

Exome sequencing is rapidly becoming the first-line approach for monogenic disorders [8]. The use of whole-exome capture and the complete sequencing of the coding genome of parent-child trios is a highly effective approach for identifying homozygous, compound heterozygous and *de novo* coding sequence variants, as multiple *de novo* sequence variants occurring within a specific gene (or within a gene family or pathway) are extremely implausible events [8]. Its rationale is based on the fact that gene variants located in exons are more likely to be pathogenic than those located in introns or between genes. The power of this strategy has increased with the access to large numbers of publicly available exome sequences that allow the controlled comparison of frequencies, as well as the identification of *de novo* variants and stratification by ethnicity. This strategy has been used to identify candidate genes for several Mendelian and complex traits [9–12]. In the assessment of obesity, whole-exome sequencing has identified sequence variants in the leptin receptor gene [13], in the *ADCY3* gene [14] and in the *BBIP1* gene in patients with Bardet–Biedl syndrome [15]. However, no novel pathogenic genes or pathways associated with obesity have been identified through this approach yet.

In this study, by employing whole-exome capture and sequencing in the assessment of two patients with severe early-onset obesity (and their parents), we identified *de novo* mutations and the compound heterozygous status of several damaging variants. Intriguingly, some of these variants were harboured in genes involved in the pathophysiology of obesity (such as *LRP2* and *UCP2*), providing the foundation for future research in this field. Thus far, we emphasize that the networking of clinical case-reports and genetic analyses would be crucial to finding the major loci underpinning complex disorders.

2. Experimental

Two family trios, the probands of which had severe early-onset obesity (onset before the age of 10 years and BMI over three SD above normal) were included in this study. All parents and capable patients provided written informed consent for the genetic research studies, which were performed in accordance with the study protocol approved by the Australian National University Human Research Ethics Committee (Protocol 2011/108, approved on the 6 May, 2011) and in concordance with the Helsinki Declaration of 1975, as revised in 2008. DNA was extracted from peripheral blood from the patients and parents for genetic analysis.

2.1. DNA Library Preparation, Exome Capture and Sequencing Protocol

Libraries were constructed from 1 μg of genomic DNA using an Illumina TruSeq genomic DNA library kit (Illumina Inc., San Diego, CA, USA). Libraries were multiplexed with 6 samples pooled together (500 ng of each library). Exons were enriched from the pooled 3 μg of library DNA using an Illumina TruSeq Exome enrichment kit (Illumina Inc.). Each exome-enriched pool was run on a 100-base-pair paired-end run on an Illumina HiSeq 2000 sequencer (Illumina Inc.). We surveyed 201,071 genomic regions in total using the exome capture platform. Ninety percent of the bases in approximately 197,000 of these targeted regions had at least one read coverage. All regions were sampled at approximately 50× coverage.

2.2. Sequence Read Processing, Alignment, Bioinformatics and Genetic Analyses

The sequencing image data were processed in real time using Illumina Real Time Analysis (RTA) software (Illumina Inc.) and converted to fastq files containing DNA base calls (A, C, G and T) and quality scores using the Illumina CASAVA pipeline (a software program that converts raw image data into sequences). The resulting fastq files were further processed for variant analysis.

The entire workflow of data curation and analysis for variant-calling was developed by the Genome Discovery Unit (GDU) at The Australian National University. Key components of the workflow include: (i) quality assessment; (ii) read alignment; (iii) local realignment around the known and novel indel regions to refine indel boundaries; (iv) recalibration of base qualities; (v) variant calling; and (vi) assigning quality scores to variants (detailed workflow information is in the Supplemental Material).

Subsequently, we included a filtering phase (using information from dbSNP and the 1K Exome Project), with the following sequential steps: (1) identification of rare or *de novo* variants (a lower minor allele frequency cut-off (MAF) in the window of 0.1%–1.0%); (2) filtering of variants to include those that are potentially pathogenic or are specific variants associated with disease susceptibility using several tools, namely, SIFT, PolyPhen2, Mutation Taster, Mutation Assessor and Functional Analysis through Hidden Markov Models (FATHMM), as implemented by the DNA-seq Analysis Package (SVS7.7.6, Golden Helix, Bozeman, MT, USA) (variants were not excluded if classified as potentially damaging by at least one of these filtering tools); (3) filtering of damaging variants based on genes known to be associated with human disease; and (4) independent confirmation of selected variants by Sanger sequencing (Supplemental Material). The definition of *de novo* sequence variants, compound heterozygous polymorphisms and rare recessive homozygous polymorphisms was performed with different modules of the DNA-seq Analysis Package (SVS7.7.6, Golden Helix, Bozeman, MT, USA).

To identify potential enriched endocrine-physiological pathways, a genetic ontology pathway analysis was performed. For constructing the pathways, variants with potential functional changes detected by the *de novo* and the compound heterozygous analysis were examined with the set of algorithms implemented in MetaCore (Thomson Reuters, New York, NY, USA) for the heuristic interpretation of maps, networks and rich ontologies for diseases.

3. Clinical Reports

Patient 1: A Brazilian male teenager with a history of excessive weight gain starting at age 3, decelerated growth since age 11 and delayed puberty was first evaluated at age 14 y, 3 m. His body weight was 105.0 kg (+5.32 SD score), his height 152.5 cm (−1.45 SD score) and his BMI 45.2 kg/m^2 (+11.03 SD score) (Figure 1). The patient had no complaints of hearing deficits or vision loss. Testicular volumes were <2 mL bilaterally, and he was at Tanner pubertal stage P2–P3. He had cubitus valgus and round facies. During physical examination, profuse sweating was noted. Other physical signs were unremarkable.

Figure 1. Height, weight and BMI of Patient 1, from birth to age 15 y, 10 m.

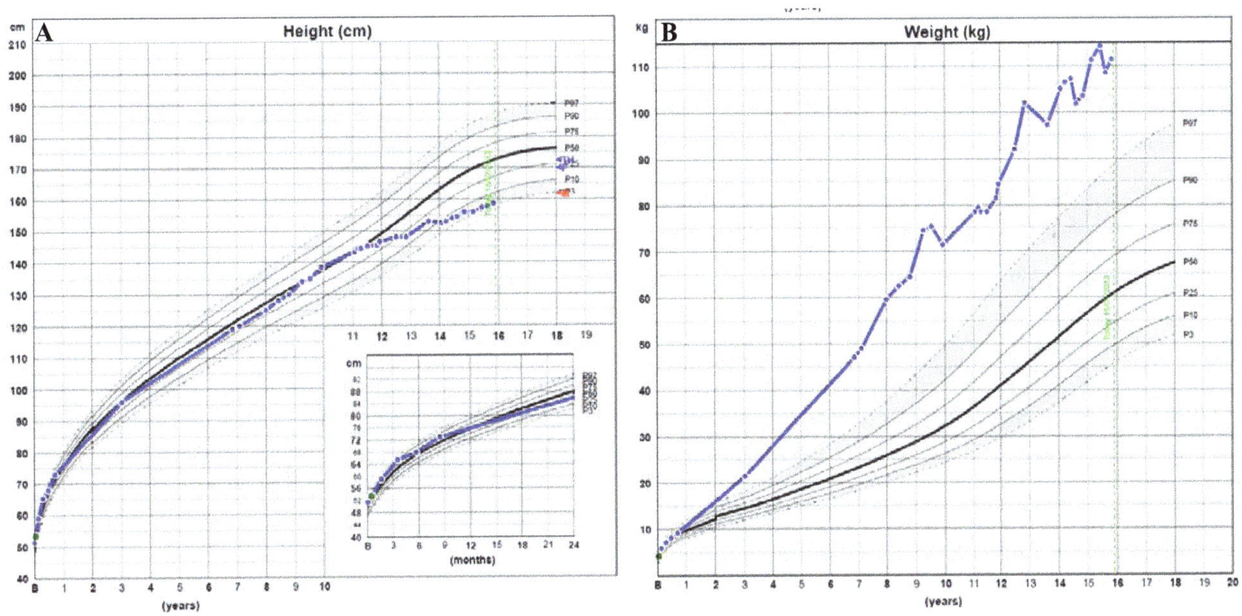

Figure 1. *Cont.*

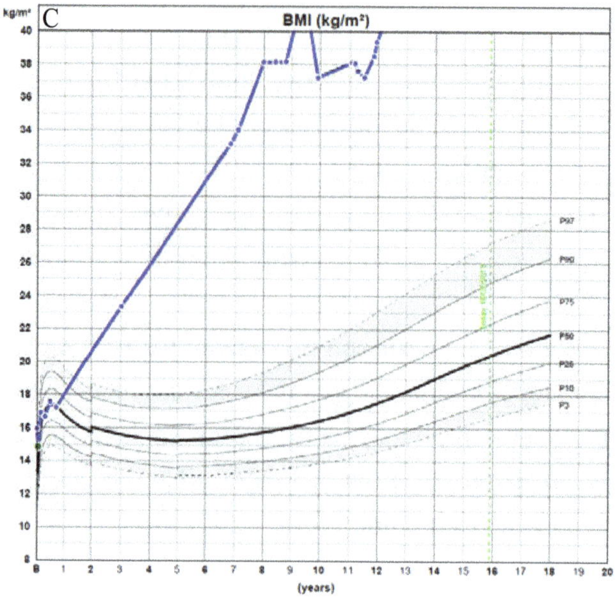

Growth charts illustrating changes in height (**A**), weight (**B**) and BMI (**C**). M, maternal height; P, paternal height; TH, target height. From WHO Child Growth Standard: Methods and Developments.

Obesity was a common finding in his family, but all individuals had normal height. His father is obese (BMI 36.7 kg/m^2), as are his paternal grandfather (BMI 44.6 kg/m^2) and paternal uncle (BMI 50.5 kg/m^2). His mother's BMI is 30.1 kg/m^2, and two maternal aunts are also obese (BMI 31.6 and 30.1 kg/m^2). His older brother is overweight (BMI 29 kg/m^2). None of his family members have a history of severe early-onset obesity. There was no history of consanguinity in the family. Pregnancy was uneventful, and size at birth was 4.2 kg and 51.5 cm.

He had been previously diagnosed with central hypothyroidism at age 9, with thyroid-stimulating hormone (TSH) of 7.4 mU/L and free T4 of 9.5 pmol/L, with undetectable titres of antithyroglobulin and antithyroperoxidase antibodies and a normal thyroid ultrasound. He also had elevated serum prolactin levels of 2,908 pmol/L, measured for the first time at age 9. Magnetic resonance enhanced by the contrast gadolinium showed a pituitary gland of normal volume, with no evidence of pituitary adenoma and without any structural abnormalities in the brain. The search for macroprolactin was negative (71% recovery after polyethylene glycol precipitation). He had been on treatment with levothyroxine 88 μg/day since age 9 y, 5 m, and cabergoline 0.5 mg every 10–15 days since age 13 y, 2 m, which had normalized his serum TSH and prolactin levels.

During the previous five years, he was treated with hypocaloric mixed diets, frequent physical activity, sibutramine 10 mg/day and orlistat 120 mg after meals. This approach resulted in only a 6-kg noncontinuous weight loss.

The patient's serum triglycerides and total cholesterol were elevated (3.11 and 4.84 mmol/L, respectively), with low HDL-cholesterol of 0.85 mmol/L and normal calculated LDL-cholesterol of 1.99 mmol/L. Fasting plasma glucose and insulin were 5.33 mmol/L and 13 μU/mL, respectively, with the homeostasis model assessment-estimated insulin resistance (HOMA-IR) index equal to 3.07. His serum insulin-like growth factor 1 (IGF-1) level was below the age reference range (71.25 nmol/L; reference value 130–563 nmol/L). Growth hormone (GH) secretion was evaluated during a standard insulin provocative test, with no development of hypoglycaemia during the test (lowest glucose level of 3.55 mmol/L and respective GH level of 0.06 µg/L). Morning cortisol and adrenocorticotropic hormone (ACTH) levels were normal (0.68 µmol/L and 1.60 pmol/L, respectively). His serum leptin levels were 8.1 and 18.0 µg/L at age 9 and appropriately elevated at age 13 (69.7 µg/L). At age 13, his total testosterone levels were pre-pubertal (0.90 nmol/L), and follicle-stimulating hormone (FSH) and luteinizing hormone (LH) were undetectable. At age 13 y, 9 m, his bone age was 15.6 years. Table 1 summarizes the laboratory test results and their respective reference range values.

Table 1. Laboratory test results for Patient 1.

Test	Value	Normal reference range
Thyroid-stimulating hormone (TSH) *	7.4 mU/L	0.3–5.0 mU/L
Free T4 *	9.5 pmol/L	10.3–25.7 pmol/L
Antithyroglobulin (ATG) and antithyroperoxidase (ATPO) antibodies *	Both negative	<9.0 IU/mL (ATG) <116 IU/mL (ATPO)
Prolactin *	2908 pmol/L	82–504 pmol/L
Macroprolactin *	Negative (71% recovery)	>50% recovery
Total cholesterol $	4.84 mmol/L	4.4 mmol/L
HDL cholesterol $	0.85 mmol/L	>1.16 mmol/L
LDL cholesterol $	1.99 mmol/L	<2.84 mmol/L
Triglycerides $	3.11 mmol/L	<1.02 mmol/L
Fasting plasma glucose $	5.33 mmol/L	3.89–5.5 mmol/L
Fasting insulin $	13 µU/mL	1.8–4.6 µU/mL
Insulin-like growth factor 1 (IGF-1) &	71.25 nmol/L	130–563 nmol/L
Growth hormone (GH)/glucose &#	0.06 µg/L/3.55 mmol/L	>5 µg/L/<1.94 mmol/L
Adrenocorticotropic hormone (ACTH) (morning)	1.60 pmol/L	2.2–13.2 pmol/L
Cortisol (morning)	0.68 µmol/L	0.14–0.70 µmol/L
Leptin	8.1 * and 69.7 µg/L &	Detectable
Total testosterone &	0.90 nmol/L	3.47–41.60 nmol/L
Follicle-stimulating hormone (FSH) &	Undetectable	0.5–10.5 IU/L
Luteinizing hormone (LH) &	Undetectable	0.5–7.9 IU/L
Total calcium ^	2.62 mmol/L	2.40–2.64 mmol/L
Inorganic phosphate ^	173.4 mmol/L	108.4–164.2 mmol/L
Magnesium ^	1.1 mmol/L	0.7–0.9 mmol/L
Alkaline phosphatise ^	114 U/L	66–571 U/L
25-hydroxy vitamin D ^	85 mmol/L	>75 mmol/L
Parathyroid hormone (PTH) ^	2.6 pmol/L	1.0–5.5 pmol/L
Selenium @	0.03 µmol/L	0.25–2.4 µmol/L
Total urinary protein @	0.08 g/24 hours	<0.15 g/24 hours

* Measured at age 9, not treated with levothyroxine and cabergoline; $ measured at age 12; # GH and lowest glucose level measured during a standard insulin provocative test; & measured at age 13; ^ measured at age 14; @ measured at age 16.

Radiographic studies showed lumbar scoliosis convex to the right, mild reduction of intervertebral spaces at L4-S1, mild shortening of L1, as well as a short fourth metacarpal. In order to exclude the diagnosis of Albright's hereditary osteodystrophy, serum electrolytes, alkaline phosphatase, vitamin D and parathyroid hormone (PTH) were measured, which were all unremarkable for Albright's hereditary osteodystrophy (total calcium 2.62 mmol/L; inorganic phosphate 173.4 mmol/L; magnesium 1.1 mmol/L, alkaline phosphatase 114 U/L, 25-hydroxy vitamin D 85 mmol/L and PTH 2.6 pmol/L). His serum levels of selenium were 0.03 μmol/L, 10-fold lower than the lowest limit of the normal range (reference values 0.25–2.4 μmol/L). Total urinary protein levels were 0.08 g/24 hours.

Intramuscular injections of testosterone esters (70 mg every four weeks) were initiated for puberty induction. Biosynthetic GH was initiated in a dose of 0.33 mg/day and increased to 0.66 mg/day according to IGF-1 levels. Four months after starting GH therapy, he developed episodes of fever of unexplained origin, which resolved spontaneously after six months. During that period, investigation for infection disease was negative, including a normal PET-scan, with leucocytosis as the only observed abnormality.

Patient 2: A two year-old Brazilian girl was evaluated for severe early-onset obesity. Her body weight was 23 kg (+4.79 SD score), her height was 93 cm (+2.32 SD score) and her BMI was 26.6 kg/m^2 (+4.49 SD score). She was born with 2.9 kg and 46 cm, from an uneventful pregnancy. Excessive weight gain was noted upon a few weeks after birth. Neurologic development was normal, with no evidence of Prader–Willi or Bardet–Biedl syndromes. She had normal serum leptin levels of 18 μg/L. A history of recurrent bacterial and viral respiratory tract infections was noted, which warranted the need for antibiotic therapy almost every month. There was no significant familial history of obesity or consanguinity. The physical examination was unremarkable.

4. Results

We called a total of 455,342 variants, 336,652 of them polymorphic, 21,613 matched at the dbNSFP, 12,286 with potential pathogenic effects and 2,291 with a minor allele frequency <1% when compared to the 1 kG phase 1. Our filtering approach by *de novo* functional mutation screening reported three *de novo* sequence variants with a potential damaging effect (Table 2). These sequence variants were found in Patient 1 (*UCP2*) and in Patient 2 (*AICDA* and *FAM71E2*).

The compound heterozygous analysis identified 20 variants with potential functional effects associated with eight genes, namely: *LRP2*, *AMPD3*, *OR8U8-OR9G1* and *SLC22A6* in Patient 1 and *TTN*, *APEH*, *DNAAF1* and *KIR3DL3* in Patient 2 (Table 3). From a literature search on the compound heterozygous variants that were classified as damaging, we identified two sequence variants in the *LRP2* gene that may potentially affect LRP2 protein function in Patient 1: a genomic variant G→T (NC_000002.11:g.170009391G>T), resulting in a nonsynonymous substitution on codon 4127 (NM_004525.2:c.12379C>A; NP_004516.2:p.Arg4127Ser), and a genomic variant C→T (NC_000002.11:g.170030506C>T), resulting in a nonsynonymous substitution on codon 3646 (NM_004525.2:c.10937G>A; (NP_004516.2:p.Arg3646His). The *LRP2* gene encodes a multi-ligand endocytic receptor (also known as megalin or glycoprotein 330) involved in the regulation of the leptin-melanocortin pathway.

Table 2. *De novo* damaging sequence variants.

Patient	Variant	Chr	Position	Ref All	Alt All	Identifier	Classification	Gene	Transcript	Exon	HGVS Coding	HGVS Protein
1	11:73689104-SNV	11	73,689,104	G	A	rs660339	Nonsyn SNV	UCP2	NM_003355	4	c.164C>T	p.Ala55Val
2	12:8757523-Ins	12	8,757,523	-	A	rs5796316	Splicing	AICDA	NM_020661	4	c.428-5_428-4insT	
2	19:55873642-SNV	19	55,873,642	C	T	rs4252574	Nonsyn SNV	FAM71E2	NM_001145402	3	c.535G>A	p.Glu179Lys

Chr: chromosome; Ref All: reference allele; Alt All: alternate allele; HGVS: Human Genome Variation Society nomenclature.

Table 3. Sequence rare variants (allele frequency in the general population <1%) with potential damaging effect by compound heterozygous analysis.

Patient	Single Nucleotide Variant	Chromosome	Position	Identifier	Gene	dbSNP MAF Frequency	Alleles	Reference Allele	Reference Aminoacid	Altered Aminoacid	HGVS Protein
1	2:170009391-SNV	2	170009391	rs148356370	LRP2	0.005	G/T	G	R	S	p.R4127S
1	2:170030506-SNV	2	170030506	rs142549310	LRP2	0.002	C/T	C	R	H	p.R3646H
1	11:10518373-SNV	11	10518373	rs144107914	AMPD3	0.001	C/T	C	S	L	p.S323L
1	11:10527316-SNV	11	10527316	N/A	AMPD3	***	A/G	G	R	Q	p.R571Q
1	11:56468198-SNV	11	56468198	rs4990194	OR8U8-OR9G1	0.069	A/G	A	Y	C	p.Y112C
1	11:56468212-SNV	11	56468212	rs591369	OR8U8-OR9G1	**	A/G	G	V	M	p.V117M
1	11:56468554-SNV	11	56468554	rs12420076	OR8U8-OR9G1	0.061	A/C	A	K	Q	p.K231Q
1	11:56468560-SNV	11	56468560	rs10896516	OR8U8-OR9G1	**	C/T	T	Y	H	p.Y233H
1	11:56468561-SNV	11	56468561	rs10896517	OR8U8-OR9G1	0.047	A/G	A	Y	C	p.Y233C
1	11:62748503-SNV	11	62748503	rs150409056	SLC22A6	*	G/T	G	R	S	p.R331S
1	11:62749384-SNV	11	62749384	rs200609617	SLC22A6	***	C/T	C	A	T	p.A243T
2	2:179507021-SNV	2	179507021	N/A	TTN	***	G/A	G	R	C	p.R4436C
2	2:179577628-SNV	2	179577628	N/A	TTN	***	C/T	C	V	I	p.V7798I
2	2:179634421-SNV	2	179634421	rs200875815	TTN	***	T/G	T	T	P	c.8749A>C
2	3:49716372-SNV	3	49716372	N/A	APEH	***	A/G	G	R	H	p.R383H
2	3:49720698-SNV	3	49720698	N/A	APEH	***	A/G	G	A	T	p.A708T
2	16:84203467-SNV	16	84203467	rs143322223	DNAAF1	***	C/G	C	E	Q	p.E345Q
2	16:84208329-SNV	16	84208329	rs139519641	DNAAF1	*	A/G	A	?	?	Splicing
2	19:55239223-SNV	19	55239223	rs117372288	KIR3DL3	***	A/G	A	V	I	p.V168I
2	19:55241240-SNV	19	55241240	rs111516669	KIR3DL3	**	A/G	A	V	I	p.V313I

Note: The first two rows show the sequence variants that affect function, harboured at the *LRP2* gene (NC_000002.11:g.170009391G>T and NC_000002.11:g.170030506C>T). N/A: not available. * Monomorphic in available data from dbSNP database; ** MAF only available for 2 chromosomes; *** There is no frequency data.

Sanger sequencing of the two amplicons containing the two *LRP2* sequence variants validated the findings obtained from the whole-exome capture and sequencing analysis (Figure 2). These results confirmed that each parent is heterozygous for one of the sequence variants (the mother is heterozygous for the mentioned sequence variant (g/t) (NC_000002.11:g.170009391G>T), and the father is heterozygous for the sequence variant (c/t) (NC_000002.11:g.170030506C>T), whereas the patient is heterozygous for both *LRP2* sequence variants. Therefore, the patient is compound heterozygous for the *LRP2* gene as initially described in the whole-exome capture and sequencing analysis (Figure 2).

Figure 2. Sanger sequencing results from Patient 1 and parents.

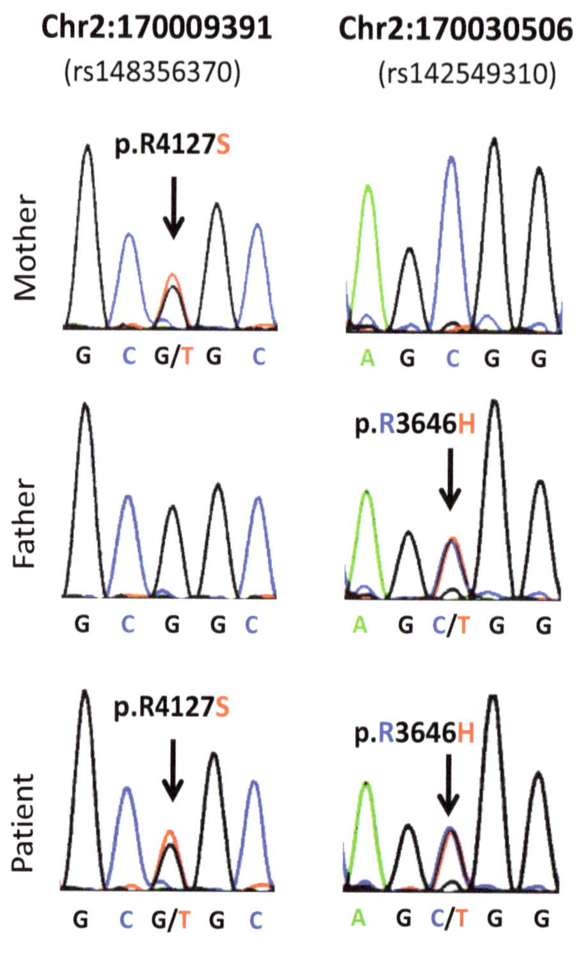

Sanger sequencing of the *LRP2* gene shows that both parents are heterozygous for one of the mutations. The mother is heterozygous for the Chr2:170009391 sequence variant (g/t) (NC_000002.11:g.170009391G>T), and the father is heterozygous for the Chr2: 170030506 variant (c/t) (NC_000002.11:g.170030506C>T). The patient is compound heterozygous and has both sequence variants.

MetaCore analysis, including those genes harbouring functional mutations, namely *UCP2*, *AICDA*, *FAM71E2*, *TTN*, *APEH*, *DNAAF1*, *KIR3DL3*, *LRP2*, *AMPD3*, *OR8U8-OR9G1* and *SLC22A6*, defined six pathways significantly overrepresented (after false discovery correction, FDR), e.g.,: (1) ATP/ITP (adenosine/inosine triphosphate) metabolism; (2) regulation of lipid metabolism/peroxisome proliferator-activated receptor (PPAR) regulation of lipid metabolism; (3) development of insulin,

IGF-1 and TNF-alpha in brown adipocyte differentiation; (4) mitochondrial dysfunction in neurodegenerative diseases; (5) oxidative stress role of Sirtuin1 and PGC1 alpha in the activation of the defence system; and (6) CTP/UTP (cytidine/uridine triphosphate) metabolism (Table 4).

Table 4. Significant pathways from MetaCore analysis of candidate morbid obesity genes.

Pathways of Candidate Morbid Obesity Genes	Genes from Input List in Pathway	*p*-Value	FDR
ATP, ITP metabolism	AMPD3	1.107e-3	6.640e-3
Regulation of lipid metabolism PPAR regulation of lipid metabolism	UCP2	1.851e-2	3.164e-2
Development of insulin, IGF-1 and TNF-alpha in brown adipocyte differentiation	UCP2	2.332e-2	3.164e-2
Mitochondrial dysfunction in neurodegenerative diseases	UCP2	2.594e-2	3.164e-2
Oxidative stress role of Sirtuin1 and PGC1 alpha in the activation of the defence system	UCP2	2.637e-2	3.164e-2
CTP UTP metabolism	AICDA	4.709e-2	4.709e-2

Note: *p*-values and false discovery (FDR) correction adjusting for multiple comparisons, representing the probability that these pathways generated from our candidate gene list would appear by coincidental chance from feeding a random set of genes.

5. Conclusions

In this study, by employing whole-exome capture and sequencing, we identified novel sequence variants in the *LRP2* gene that might be associated with the phenotype of severe early-onset obesity in Patient 1. Similar to the other genes associated with monogenic forms of obesity, *LRP2* is also involved in the regulation of the leptin-melanocortin pathway [16]. In addition, in Patient 1, we found a *de novo* sequence variant in the *UCP2* gene, a transporter protein present in the mitochondrial inner membrane that is a key regulator of energy balance, the variants of which have already been associated with obesity [17]. In Patient 2, we identified sequence variants in the *DNAAF1* gene, which might be related to the second patient's phenotype. The *DNAAF1* gene is required for the stability of the ciliary architecture, and it has been demonstrated that ciliary dysfunction is associated with the pathogenesis of obesity [18,19].

Recently, LRP2 has been nominated as a novel appetite regulator responsible for generating satiety signals in hypothalamic neurons [16]. It is a multiligand endocytic receptor, a member of the low density lipoprotein receptor gene family, which binds a large variety of ligands. Leptin is one of its ligands; LRP2 mediates its reuptake in renal tubules [20] and promotes leptin transport across the choroid plexus [21]. LRP2 also binds to the long-form leptin receptor (LepRb), forming a complex that is co-localized and subjected to endocytosis in the hypothalamic neurons. Subsequently, the endocytosis of this co-localized complex leads to the activation of signal transducer and activator of transcription 3 (STAT3) signalling in hypothalamic neurons, including proopiomelanocortin (POMC)- and LepRb-expressing neurons. As a consequence, food intake and body weight are decreased. In the absence of functional LRP2, STAT3 signalling is decreased. Therefore, hunger is stimulated and satiety decreased [16].

The binding of LRP2 and LepRb is enhanced by clusterin, a sulphated glycoprotein widely expressed in hypothalamic areas involved in the regulation of food intake and energy metabolism [22]. Chronic intracerebroventricular (icv) administration of clusterin causes the reduction of food intake, body weight and epididymal fat mass [16]. LRP2 is expressed in rodent hypothalamus, and its inhibition by small interfering RNA significantly blunts the effects of clusterin icv injections on food intake and Stat3 activation [16]. Therefore, LRP2 acts as a key mediator of the food intake-suppressing effects of clusterin, and its absence can cause obesity in rodents (as previously demonstrated) [16].

Sequence variants in the *LRP2* gene have been previously associated with Donnai–Barrow/facio-oculo-acoustico-renal (DB/FOAR) syndrome [23]. The phenotype of this syndrome includes agenesis of the corpus callosum, developmental delay, enlarged anterior fontanelle, high myopia, hypertelorism, proteinuria and sensorineural hearing loss, but not obesity. In a review by Pober *et al.*, sensorineural hearing loss, high myopia and proteinuria were present in 100% of DB/FOAR syndrome cases. None of those features were present in Patient 1; therefore, we ruled this diagnosis out. Serum selenium levels in patients with DB/FOAR syndrome have not been reported, but it has been shown that LRP2 mediates the reuptake of selenoproteins in the kidney and that LRP2-mutant mice have low selenium serum levels due to the increased urinary excretion of selenoproteins [24]. Since the patient's serum levels of selenium were 10-fold lower than those of the reference range, this finding supports the impairment of the biological function of LRP2.

Besides suffering from severe early-onset obesity, Patient 1 also had pituitary dysfunction characterized by GH deficiency, central hypothyroidism and hypogonadotropic hypogonadism, with concomitant hyperprolactinaemia. It is unclear whether sequence variants in the *LRP2* gene can directly affect pituitary development and function. LRP2 is essential for brain development [25], as knock-out mice exhibit holoprosencephaly [26] and *Lrp2*-mutant mice have abnormal cortical axon development [27]. Humans with DB/FOAR syndrome have structural brain abnormalities, mainly agenesis of the corpus callosum and, in one reported case, empty sella turcica [28]. These support the hypothesis that *LRP2* sequence variants might lead to abnormalities in pituitary development and hypopituitarism.

In addition, Patient 1 had a *de novo* variant in the *UCP2* gene. As a transporter protein that is expressed in the mitochondrial inner membrane, UCP2 decreases mitochondrial ATP production by mediating H^+ leak across the inner membrane [17]. Patient 1 has a G to A substitution at rs660339, resulting in an Ala55Val substitution, which has been associated with obesity in diverse settings [29,30]. Moreover, polymorphisms at rs660339 may also affect metabolic efficiency in terms of energy expenditure [31]. It is noteworthy to mention that this *de novo* variant has a reported minor allele frequency of about 0.5% and that it is a site that mutates recurrently.

The other *de novo* sequence variants that we found in Patient 2 (AICDA, Activation-Induced Cytidine Deaminase; and FAM71E2, Family with Sequence Similarity 71, Member E2) are protein coding genes for a RNA-editing deaminase and for a protein of unknown function, respectively. The variant rs5796316 in AICDA has been reported in one previous study. Whereas AICDA is possibly not implicated in the pathophysiology of obesity, the function of FAM71E2 is unknown to date. However, rs4252574 in FAM71E2 is quite common, and the variant A allele is the major allele reported at about 70%.

In Patient 2, we did not find gene variants that would strongly explain the phenotype, as we did for Patient 1. However, we identified sequence variants harboured in the *DNAAF1* that are possibly implicated in the patient's history of recurrent infections and severe obesity. That gene is responsible for encoding a protein that is cilium-specific and is required for the stability of the ciliary architecture. *DNAFF1* is one of the 21 genes in which mutations are associated with primary ciliary dyskinesia (PCD) [32]. It is unlikely that Patient 2 has PCD, given the absence of its clinical manifestations (bronchiectasis, defects in body situs and, later in life, infertility). However, it is possible that identified *DNAAF1* sequence variants are causing a milder form of PCD with recurrent airway infections and severe obesity. Ciliary dysfunction has been associated with severe early-onset obesity, and currently, it is known that two obesity syndromes are caused by mutations in genes regulating ciliary function: Bardet–Biedl syndrome and Alström syndrome [19]. It has been demonstrated that ciliary dysfunction leads to the development of obesity in animal models, due to diverse alterations in central and peripheral pathways regulating energy metabolism [18,33–38]. For Patient 2, electron microscopy would be useful in the assessment of the effect of the DNAAF1 gene variant on ciliary structure. Furthermore, as we have not confirmed the DNAAF1 variant by Sanger sequencing, we cannot confirm that it is in *trans* in the patient (*i.e.*, each parent contributing one of the variants).

It is unlikely that the other gene variants with potential damaging effect (Table 3) also play a role in the pathogenesis of obesity, since they are not related to the regulation of the leptin-melanocortin pathway. Particularly, *TTN* is a large gene, the variants of which are frequently unrelated to disease; and *OR8U8-OR9G1,* the variants of whichmost likely represent artefacts, possibly due to alignment problems.

By performing MetaCore pathway analysis of the genes harbouring functional mutations, we observed that six pathways are significantly overrepresented, all of them involving energy or ATP/ITP/CTP/UCP metabolism. The importance of UCP2 on energy metabolism was further strengthened by the observation that four of these pathways were centred on the *UCP2* gene.

In whole-genome association studies (GWAS), the *LRP2* and the *DNAAF1* genes were previously associated with increased BMI in a British population, without reaching a level of significance that is relevant for GWAS [39]. Curiously, a higher significance level for a single nucleotide polymorphism (SNP) within the *LRP2* gene ($p = 8.68 \times 10^{-6}$) was found in a GWAS of patients with anorexia nervosa [40]. Although variants at *UCP2* rs660339 have been associated with increased BMI in Europeans [41], this finding was not replicated in a recent meta-analysis in populations representing four ethnicities [42].

Whole-exome capture and sequencing can be used with family-based phenotype ascertainment strategies (nuclear and extended families) to exploit parent-child transmission and relative-relative sharing/not sharing patterns, as well as with arbitrary strategies of phenotype dichotomization to increase efficiency. In an extreme phenotype study design, individuals who are at both ends of a phenotype distribution are selected for sequencing. It is assumed that alleles contributing to the trait in individuals who are at both ends of the phenotype distribution are enriched, and sequencing even a modest sample size can potentially identify novel candidate alleles. The same consideration is also applicable to additional alternatives aimed to identify *de novo* variants that involve the sequencing of parent-offspring trios in which only the offspring is affected. The clinical use of whole-exome capture and sequencing is promising, as demonstrated in a recent study that evaluated 250 patients with undiagnosed diseases: the success rate in obtaining a genetic diagnosis was as high as 25% in that study [12].

Recently, guidelines for investigating and reporting the causality of sequence variants in human disease have been published, to avoid an acceleration of false-positive reports of causality [43]. In our study, we comply with those guidelines, but we acknowledge that our findings are not sufficient to implicate those gene variants as determinants of the obese phenotype. The significance of our results can be limited due to the fact that the number of probands is very small. However, a similar approach has already been validated and published by other studies, such as the Finding of Rare Disease Genes (FORGE) Canada Consortium [44]. In addition, WES has been applied for the diagnosis of several diseases, as observed in many case reports with very small sample sizes [45]. Furthermore, despite the fact that our results have not been replicated yet in other obese individuals, they are important to raise awareness of the *LRP2* gene as a possible candidate as a novel monogenic cause of obesity. In addition, our results lack confirmation through functional data, which should be pursued in future studies.

Whole-exome capture and sequencing analysis is a time- and resource-intense endeavour. Currently, we employ software that allows rapid selection of any genetic variant according to variant type, novelty (via screening public and private databases) and predicted protein effect. However, linking these results to phenotypic manifestations in a particular person is currently performed by a mixture of manual analysis using a number of additional databases (e.g., Human Genome Mutation Database, Online Mendelian Inheritance in Man (OMIM), PubMed and UCSC, among others). We built on existing analytic tools in order to rapidly detect and annotate genomic variants associated with human disease. We are aware that analytical criteria for filtering need to be flexible and up-to-date; therefore, we undertook a systematic upgrade and iterative processes of database evaluation by considering each filter.

In conclusion, by employing a novel and unique strategy for whole-exome capture and sequencing analysis of two trios comprised of patients with severe early-onset obesity, we have identified sequence variants in the *LRP2* and in the *UCP2* genes that might explain the phenotype of a patient with severe early-onset obesity, central hypothyroidism, hypogonadotropic hypogonadism, growth hormone deficiency and idiopathic hyperprolactinaemia. In addition, we identified a sequence variant in the *DNAAF1* gene that might be implicated in the development of severe obesity associated with ciliary dysfunction. Whereas *de novo* variants in the *UCP2* gene have already been associated with obesity, the role of *LRP2* and *DNAAF1* sequence variants in human obesity needs to be further investigated by functional studies, and the frequency and distribution of those sequence variants need to be evaluated in a larger number of obese individuals.

Acknowledgments

This work was funded by the Australian National University and the John James Foundation.

Author Contributions

Obtained clinical data and described the phenotypes: Margaret C.S. Boguszewski, Cesar L. Boguszewski, Gilberto Paz-Filho. Supervised the exome sequencing procedures and performed bioinformatics analyses: Oscar M. Arcos-Burgos, Gavin A. Huttley, Hardip R. Patel, Angad Johar, Aaron Chuah. Conceptualized the study and experimental design: Julio Licinio, Ma-Li Wong, Oscar M. Arcos-Burgos, Gilberto Paz-Filho. Performed Sanger sequencing and analysed data: Claudio A. Mastronardi. Wrote the manuscript: Margaret C.S. Boguszewski, Claudio A. Mastronardi, Gilberto Paz-Filho, Oscar M. Arcos-Burgos, Gavin A. Huttley, Julio Licinio, Ma-Li Wong.

References

1. WHO. Obesity and Overweight. Available online: http://www.who.int/mediacentre/factsheets/fs311/en/index.html (accessed on 23 July 2014).
2. Malik, V.S.; Willett, W.C.; Hu, F.B. Global obesity: Trends, risk factors and policy implications. *Nat. Rev. Endocrinol.* **2013**, *9*, 13–27.
3. Maes, H.H.; Neale, M.C.; Eaves, L.J. Genetic and environmental factors in relative body weight and human adiposity. *Behav. Genet.* **1997**, *27*, 325–351.
4. El-Sayed Moustafa, J.S.; Froguel, P. From obesity genetics to the future of personalized obesity therapy. *Nat. Rev. Endocrinol.* **2013**, *9*, 402–413.
5. Loos, R.J. Genetic determinants of common obesity and their value in prediction. *Best Pract. Res. Clin. Endocrinol. Metab.* **2012**, *26*, 211–226.
6. Farooqi, S.; O'Rahilly, S. Genetics of obesity in humans. *Endocr. Rev.* **2006**, *27*, 710–718.
7. Asai, M.; Ramachandrappa, S.; Joachim, M.; Shen, Y.; Zhang, R.; Nuthalapati, N.; Ramanathan, V.; Strochlic, D.E.; Ferket, P.; Linhart, K.; *et al.* Loss of function of the melanocortin 2 receptor accessory protein 2 is associated with mammalian obesity. *Science* **2013**, *341*, 275–278.
8. Bamshad, M.J.; Ng, S.B.; Bigham, A.W.; Tabor, H.K.; Emond, M.J.; Nickerson, D.A.; Shendure, J. Exome sequencing as a tool for mendelian disease gene discovery. *Nat. Rev. Genet.* **2011**, *12*, 745–755.
9. Liew, W.K.; Ben-Omran, T.; Darras, B.T.; Prabhu, S.P.; de Vivo, D.C.; Vatta, M.; Yang, Y.; Eng, C.M.; Chung, W.K. Clinical application of whole-exome sequencing: A novel autosomal recessive spastic ataxia of charlevoix-saguenay sequence variation in a child with ataxia. *JAMA Neurol.* **2013**, *70*, 788–791.
10. Need, A.C.; Shashi, V.; Hitomi, Y.; Schoch, K.; Shianna, K.V.; McDonald, M.T.; Meisler, M.H.; Goldstein, D.B. Clinical application of exome sequencing in undiagnosed genetic conditions. *J. Med. Genet.* **2012**, *49*, 353–361.
11. Veltman, J.A.; Brunner, H.G. *De novo* mutations in human genetic disease. *Nat. Rev. Genet.* **2012**, *13*, 565–575.
12. Yang, Y.; Muzny, D.M.; Reid, J.G.; Bainbridge, M.N.; Willis, A.; Ward, P.A.; Braxton, A.; Beuten, J.; Xia, F.; Niu, Z.; *et al.* Clinical whole-exome sequencing for the diagnosis of mendelian disorders. *N. Engl. J. Med.* **2013**, *369*, 1502–1511.
13. Gill, R.; Him Cheung, Y.; Shen, Y.; Lanzano, P.; Mirza, N.M.; Ten, S.; Maclaren, N.K.; Motaghedi, R.; Han, J.C.; Yanovski, J.A.; *et al.* Whole-exome sequencing identifies novel lepr mutations in individuals with severe early onset obesity. *Obesity* **2013**, *22*, 576–584.
14. Hendricks, A.E. *Whole Exome Sequencing Case-Control Using 1000 Severe Obesity Cases Identifies Putative New Loci and Replicates Previously Established Loci*; American Society of Human Genetics: Boston, MA, USA, 2013.

15. Scheidecker, S.; Etard, C.; Pierce, N.W.; Geoffroy, V.; Schaefer, E.; Muller, J.; Chennen, K.; Flori, E.; Pelletier, V.; Poch, O.; *et al.* Exome sequencing of bardet-biedl syndrome patient identifies a null mutation in the bbsome subunit bbip1 (bbs18). *J. Med. Genet.* **2014**, *51*, 132–136.
16. Gil, S.Y.; Youn, B.S.; Byun, K.; Huang, H.; Namkoong, C.; Jang, P.G.; Lee, J.Y.; Jo, Y.H.; Kang, G.M.; Kim, H.K.; *et al.* Clusterin and lrp2 are critical components of the hypothalamic feeding regulatory pathway. *Nat. Commun.* **2013**, *4*, 1862.
17. Fleury, C.; Neverova, M.; Collins, S.; Raimbault, S.; Champigny, O.; Levi-Meyrueis, C.; Bouillaud, F.; Seldin, M.F.; Surwit, R.S.; Ricquier, D.; *et al.* Uncoupling protein-2: A novel gene linked to obesity and hyperinsulinemia. *Nat. Genet.* **1997**, *15*, 269–272.
18. Berbari, N.F.; Pasek, R.C.; Malarkey, E.B.; Yazdi, S.M.; McNair, A.D.; Lewis, W.R.; Nagy, T.R.; Kesterson, R.A.; Yoder, B.K. Leptin resistance is a secondary consequence of the obesity in ciliopathy mutant mice. *Proc. Natl. Acad. Sci. USA* **2013**, *110*, 7796–7801.
19. Mok, C.A.; Heon, E.; Zhen, M. Ciliary dysfunction and obesity. *Clin. Genet.* **2010**, *77*, 18–27.
20. Hama, H.; Saito, A.; Takeda, T.; Tanuma, A.; Xie, Y.; Sato, K.; Kazama, J.J.; Gejyo, F. Evidence indicating that renal tubular metabolism of leptin is mediated by megalin but not by the leptin receptors. *Endocrinology* **2004**, *145*, 3935–3940.
21. Dietrich, M.O.; Spuch, C.; Antequera, D.; Rodal, I.; de Yebenes, J.G.; Molina, J.A.; Bermejo, F.; Carro, E. Megalin mediates the transport of leptin across the blood-csf barrier. *Neurobiol. Aging* **2008**, *29*, 902–912.
22. Aronow, B.J.; Lund, S.D.; Brown, T.L.; Harmony, J.A.; Witte, D.P. Apolipoprotein j expression at fluid-tissue interfaces: Potential role in barrier cytoprotection. *Proc. Natl. Acad. Sci. USA* **1993**, *90*, 725–729.
23. Pober, B.R.; Longoni, M.; Noonan, K.M. A review of donnai-barrow and facio-oculo-acoustico-renal (db/foar) syndrome: Clinical features and differential diagnosis. *Birth Defects Res. Part A Clin. Mol. Teratol.* **2009**, *85*, 76–81.
24. Chiu-Ugalde, J.; Theilig, F.; Behrends, T.; Drebes, J.; Sieland, C.; Subbarayal, P.; Kohrle, J.; Hammes, A.; Schomburg, L.; Schweizer, U. Mutation of megalin leads to urinary loss of selenoprotein p and selenium deficiency in serum, liver, kidneys and brain. *Biochem. J.* **2010**, *431*, 103–111.
25. May, P.; Woldt, E.; Matz, R.L.; Boucher, P. The ldl receptor-related protein (lrp) family: An old family of proteins with new physiological functions. *Ann. Med.* **2007**, *39*, 219–228.
26. Willnow, T.E.; Hilpert, J.; Armstrong, S.A.; Rohlmann, A.; Hammer, R.E.; Burns, D.K.; Herz, J. Defective forebrain development in mice lacking gp330/megalin. *Proc. Natl. Acad. Sci. USA* **1996**, *93*, 8460–8464.
27. Ha, S.; Stottmann, R.W.; Furley, A.J.; Beier, D.R. A forward genetic screen in mice identifies mutants with abnormal cortical patterning. *Cereb. Cortex* **2013**, doi:10.1093/cercor/bht209.
28. Kantarci, S.; Al-Gazali, L.; Hill, R.S.; Donnai, D.; Black, G.C.; Bieth, E.; Chassaing, N.; Lacombe, D.; Devriendt, K.; Teebi, A.; *et al.* Mutations in lrp2, which encodes the multiligand receptor megalin, cause donnai-barrow and facio-oculo-acoustico-renal syndromes. *Nat. Genet.* **2007**, *39*, 957–959.

29. Chen, H.H.; Lee, W.J.; Wang, W.; Huang, M.T.; Lee, Y.C.; Pan, W.H. Ala55val polymorphism on ucp2 gene predicts greater weight loss in morbidly obese patients undergoing gastric banding. *Obes. Surg.* **2007**, *17*, 926–933.
30. Oktavianthi, S.; Trimarsanto, H.; Febinia, C.A.; Suastika, K.; Saraswati, M.R.; Dwipayana, P.; Arindrarto, W.; Sudoyo, H.; Malik, S.G. Uncoupling protein 2 gene polymorphisms are associated with obesity. *Cardiovasc. Diabetol.* **2012**, *11*, 41.
31. Buemann, B.; Schierning, B.; Toubro, S.; Bibby, B.M.; Sorensen, T.; Dalgaard, L.; Pedersen, O.; Astrup, A. The association between the val/ala-55 polymorphism of the uncoupling protein 2 gene and exercise efficiency. *Int. J. Obes. Relat. Metab. Disord.* **2001**, *25*, 467–471.
32. Knowles, M.R.; Daniels, L.A.; Davis, S.D.; Zariwala, M.A.; Leigh, M.W. Primary ciliary dyskinesia. Recent advances in diagnostics, genetics, and characterization of clinical disease. *Am. J. Respir. Crit. Care Med.* **2013**, *188*, 913–922.
33. Bell, C.G.; Meyre, D.; Samson, C.; Boyle, C.; Lecoeur, C.; Tauber, M.; Jouret, B.; Jaquet, D.; Levy-Marchal, C.; Charles, M.A.; *et al.* Association of melanin-concentrating hormone receptor 1 5' polymorphism with early-onset extreme obesity. *Diabetes* **2005**, *54*, 3049–3055.
34. Cota, D.; Proulx, K.; Smith, K.A.; Kozma, S.C.; Thomas, G.; Woods, S.C.; Seeley, R.J. Hypothalamic mtor signaling regulates food intake. *Science* **2006**, *312*, 927–930.
35. Davenport, J.R.; Watts, A.J.; Roper, V.C.; Croyle, M.J.; van Groen, T.; Wyss, J.M.; Nagy, T.R.; Kesterson, R.A.; Yoder, B.K. Disruption of intraflagellar transport in adult mice leads to obesity and slow-onset cystic kidney disease. *Curr. Biol. CB* **2007**, *17*, 1586–1594.
36. Marion, V.; Stoetzel, C.; Schlicht, D.; Messaddeq, N.; Koch, M.; Flori, E.; Danse, J.M.; Mandel, J.L.; Dollfus, H. Transient ciliogenesis involving bardet-biedl syndrome proteins is a fundamental characteristic of adipogenic differentiation. *Proc. Natl. Acad. Sci. USA* **2009**, *106*, 1820–1825.
37. Sen Gupta, P.; Prodromou, N.V.; Chapple, J.P. Can faulty antennae increase adiposity? The link between cilia proteins and obesity. *J. Endocrinol.* **2009**, *203*, 327–336.
38. Szabo, N.E.; Zhao, T.; Cankaya, M.; Theil, T.; Zhou, X.; Alvarez-Bolado, G. Role of neuroepithelial sonic hedgehog in hypothalamic patterning. *J. Neurosci.* **2009**, *29*, 6989–7002.
39. British 1958 birth cohort DNA Collection. Available online: http://www.b58cgene.sgul.ac.uk/ (accessed on 23 July 2014).
40. Wang, K.; Zhang, H.; Bloss, C.S.; Duvvuri, V.; Kaye, W.; Schork, N.J.; Berrettini, W.; Hakonarson, H.; Price Foundation Collaborative Group. A genome-wide association study on common snps and rare cnvs in anorexia nervosa. *Mol. Psychiatry* **2011**, *16*, 949–959.
41. Brondani, L.A.; Assmann, T.S.; de Souza, B.M.; Boucas, A.P.; Canani, L.H.; Crispim, D. Meta-analysis reveals the association of common variants in the uncoupling protein (ucp) 1–3 genes with body mass index variability. *PLoS One* **2014**, *9*, e96411.
42. Tan, L.J.; Zhu, H.; He, H.; Wu, K.H.; Li, J.; Chen, X.D.; Zhang, J.G.; Shen, H.; Tian, Q.; Krousel-Wood, M.; *et al.* Replication of 6 obesity genes in a meta-analysis of genome-wide association studies from diverse ancestries. *PLoS One* **2014**, *9*, e96149.
43. MacArthur, D.G.; Manolio, T.A.; Dimmock, D.P.; Rehm, H.L.; Shendure, J.; Abecasis, G.R.; Adams, D.R.; Altman, R.B.; Antonarakis, S.E.; Ashley, E.A.; *et al.* Guidelines for investigating causality of sequence variants in human disease. *Nature* **2014** *508*, 469–476.

44. Beaulieu, C.L.; Majewski, J.; Schwartzentruber, J.; Samuels, M.E.; Fernandez, B.A.; Bernier, F.P.; Brudno, M.; Knoppers, B.; Marcadier, J.; Dyment, D.; *et al.* Forge canada consortium: Outcomes of a 2-year national rare-disease gene-discovery project. *Am. J. Hum. Genet.* **2014**, *94*, 809–817.
45. Rabbani, B.; Tekin, M.; Mahdieh, N. The promise of whole-exome sequencing in medical genetics. *J. Hum. Genet.* **2014**, *59*, 5–15.

4

Genes and Genetic Testing in Hereditary Ataxias

Erin Sandford [1] and Margit Burmeister [1,2,3,4,*]

1 Molecular & Behavioral Neuroscience Institute, University of Michigan, Ann Arbor, MI 48109, USA; E-Mail: esandfor@umich.edu
2 Departments of Human Genetics, University of Michigan, Ann Arbor, MI 48109, USA
3 Department of Computational Medicine & Bioinformatics, University of Michigan, Ann Arbor, MI 48109, USA
4 Department of Psychiatry, University of Michigan, Ann Arbor, MI 48109, USA

* Author to whom correspondence should be addressed; E-Mail: margit@umich.edu;

Abstract: Ataxia is a neurological cerebellar disorder characterized by loss of coordination during muscle movements affecting walking, vision, and speech. Genetic ataxias are very heterogeneous, with causative variants reported in over 50 genes, which can be inherited in classical dominant, recessive, X-linked, or mitochondrial fashion. A common mechanism of dominant ataxias is repeat expansions, where increasing lengths of repeated DNA sequences result in non-functional proteins that accumulate in the body causing disease. Greater understanding of all ataxia genes has helped identify several different pathways, such as DNA repair, ubiquitination, and ion transport, which can be used to help further identify new genes and potential treatments. Testing for the most common mutations in these genes is now clinically routine to help with prognosis and treatment decisions, but next generation sequencing will revolutionize how genetic testing will be done. Despite the large number of known ataxia causing genes, however, many individuals with ataxia are unable to obtain a genetic diagnosis, suggesting that more genes need to be discovered. Utilization of next generation sequencing technologies, expression studies, and increased knowledge of ataxia pathways will aid in the identification of new ataxia genes.

Keywords: ataxia; genetics; pathways; testing

---

1. Introduction

Ataxia is a neurological sign that involves a lack of coordinated muscle movement, which impacts walking, speech, and vision. Ataxia can present as an isolated symptom, or present as one of many symptoms of a more complex disease. Acquired ataxias may be temporary or permanent, and can be caused by environmental factors, such as alcohol, trauma, or exposure to toxins, or by other underlying medical conditions such as stroke, infection, tumors, or vitamin deficiencies. However, many ataxias have an underlying genetic cause. Hereditary ataxias are a group of highly heterogeneous diseases, but each usually follows a typical Mendelian dominant, recessive, or X-linked inheritance. The prevalence of hereditary ataxias varies by population and has been estimated at 1–9 per 100,000 people [1–4]. Many hereditary diseases also present with ataxia as one symptom of a more complex phenotype. This review will focus on disorders classified primarily as ataxia, along with those ataxias that result in other symptoms like intellectual disability, with known genetic association.

Early work on the genetic origins of ataxia began in 1993 with the discovery of a CAG repeat responsible for spinocerebellar ataxia (SCA) type 1 [5]. Continued screening for CAG repeat expansions identified several additional dominant SCAs that are caused by the same mechanism [6–10]. With the advancement of next generation sequencing technology, genome and exome sequencing have become an affordable option for screening for disease genes. Exome sequencing for Mendelian diseases first gained prominence in 2010 with the discovery of the disease gene for Miller syndrome and since then, mutations in several new ataxia genes have been identified utilizing exome sequencing, including *ATP2B3*, *KCND3*, *DNMT1*, *UCHL1*, and *TPP1*, illustrating the utility of the technology [11–17]. While mutations in many of these new genes were found in only one family ("private" mutations), thus far, mutations in *KCND3* were found in multiple different families on several continents [13,14]. Despite these advances, it is estimated that up to 40% of those with ataxia do not know the genetic cause, illustrating the need to continue research into the identification of ataxia genes in order to provide a diagnosis and potentially a treatment [18].

2. Phenotypes of Hereditary Ataxias

Hereditary ataxias exhibit a wide range of phenotypes, in both clinical features and age of onset. Some ataxias are described as "pure cerebellar", where symptoms are all related to cerebellar control of muscle movement. This can include ataxic gait and movement of body and limbs, along with nystagmus, dysarthia, and hypotonia. Many of these features are easily observed by external examination. Magnetic resonance imaging often provides the clearest explanation for the ataxia through the identification of cerebellar atrophy, but may appear normal in some cases [19,20]. Other ataxias can present with more extensive additional neurological symptoms, such as Parkinsonism, epilepsy, dementia, and neuropathy. Multisystem involvement can include symptoms such as deafness and intellectual disability. These symptoms may be progressive, gradually becoming more severe over time, or non-progressive, where the symptoms are stable.

The age of symptom onset in affected individuals can vary dramatically, both within and across different ataxias, with symptoms present from birth through onset in the 7th and 8th decades of life. Late onset ataxias are more commonly progressive and can result in patients becoming wheelchair bound or even experience a reduced lifespan. Congenital ataxias display symptoms within the first year of life and are often non-progressive, however many congenital ataxias more often present as multisystem diseases. These children may display muscular hypotonia prior to onset of ataxia symptoms, resulting in “floppy baby syndrome”.

A common phenomenon in the dominantly inherited ataxias is anticipation, where the younger generation exhibits symptoms at an earlier age. The rate of anticipation can vary, depending on genetic and environmental factors, but differences in age of onset, up to 20 years, have been reported. Much, but not all, of anticipation can be explained by increasing repeat length of the CAG expansions (see Section 3.1.1). Anticipation can be difficult for clinicians to correctly diagnose, as younger individuals with a family history of ataxia may describe more psychosomatic symptoms in the expectation of developing symptoms later in life.

3. Ataxia Genetics

Hereditary ataxias are genetically and phenotypically heterogeneous. Similar phenotypes may be caused by mutations in many different genes, and several genes cause different types of ataxia depending upon the mutation. While many ataxias appear worldwide, such as Friedreich’s ataxia or SCA3, others are more common in one population. Dentatorubral-pallidoluysian atrophy (DRPLA) is most common in Japan and SCA2 is prevalent in Cuba. Other ataxias may be completely restricted to certain populations, such as Cayman ataxia on the Cayman Islands. Knowledge of a patient’s ethnic origin can, therefore, be helpful, along with phenotype and family history. In most newly diagnosed cases with ataxia, screening panels for many ataxia genes is recommended. As ataxia can be misdiagnosed as multiple sclerosis or Parkinson’s, finding a genetic cause often solidifies a diagnosis, not only for an individual, but for the whole family.

3.1. Autosomal Dominant

Many dominant ataxias have been classified as SCAs or episodic ataxias (EA). At least 34 different SCAs and seven EAs have been described clinically, with 28 having known associated genetic mutations. Dominant ataxias tend to have an onset later in life and be slowly progressive. SCAs, particularly those caused by repeat expansions, can exhibit a larger range of symptom onset and a faster rate of progression. A detailed review of the clinical characteristics of SCAs was published in 2009 [21]. Individuals with EA experience episodes of ataxia that can range from minutes to hours in duration and are triggered by environmental stimuli such as stress, alcohol, or exercise [22–24].

In some cases, the causative gene was identical in several previously reported SCAs, so SCA15, SCA16, and SCA29 all are caused by mutations in *ITPR1* and SCA19 and SCA22 are caused by mutations in *KCND3* [13,14,25–27]. Repeat expansion in *CACNA1A* results in SCA6 while single nucleotide variants (SNV), insertions, and deletions result in EA2. Known autosomal dominant (AD) ataxia genes are reported in Table S1.

3.1.1. Repeat Expansions

The most common forms of dominant ataxias are caused by repeat expansion. Short repeats, typically three to six bases long, appear at variable repeat number within many genes. Occasionally these repeat regions become unstable during replication, leading to either deletions of repeats, which rarely causes problems, or to expansion of the number of repeats. Typically within a repeat region, there are instances of non-repeated bases, such as a CAA in a string of CAG. Mutations that convert these imperfections in the repeat region to match the surrounding repeats result in an unstable sequence and increased likelihood of expansion. In ataxias, the number of repeats may increase anywhere from less than 2 to over 100 fold, depending on the gene. The most common repeat expansions are CAG expansions. As CAG encodes glutamine, these are also referred to as a polyglutamine or polyQ repeats, as these repeats form strings of glutamines (Q) in the coding region. There are currently seven known AD ataxias caused by CAG polyglutamine expansions: SCA1, SCA2, SCA3 (also known as Machado Joseph disease or MJD), SCA6, SCA7, SCA17, and DRPLA. In addition, repeat expansions outside the coding region, in introns or the untranslated regions of the gene, also can cause ataxia without causing polyglutamine disease, but rather by interfering with the regulation of the gene: SCA8 (CTG), SCA10 (ATTCT), SCA12 (CAG), SCA31 (TGGAA), and SCA36 (GGCCTG).

The most common SCAs reported are SCA1, SCA2, SCA3, SCA6, and SCA7. Rates for each vary by population; the National Ataxia Foundation reports that SCA6 is responsible for up the 30% of dominant ataxia cases in Japan, but only 15% in the U.S. and 2% in Italy. Together these five SCA make up about 60% of the reported dominant cases of ataxia [21]. With the high frequency of these SCAs, it is not surprising that they were the first genetic mutations responsible for SCA that were identified.

Age of onset is highly variable with repeat expansion disorders, ranging from early childhood to the later decades of adulthood. There is an inverse correlation between repeat length and age of onset, with longer repeats resulting in symptoms at a younger age. Expansions often increase in length in each subsequent generation, leading to a phenomenon called anticipation, where the next generation starts exhibiting symptoms at an earlier age than the previous. Reduction of repeat length has been reported but this occurs more rarely. Many individuals have repeats at an intermediate length, resulting in incomplete penetrance of the disease, but these are more likely to expand in future generations. Expansion of repeat regions can therefore appear as sporadic cases when the repeat is newly expanded, as this individual may have no other affected family members.

Repeat expansions cause disease through toxic gain of function. This gain of function can allow expanded proteins to avoid degradation, exhibit changes in expression, and influence function of other interacting genes [28–30]. Recently, it has been demonstrated in SCA8 and FXTAS, an X-linked ataxia, that RNA can be translated independent of a traditional ATG start site, in a process referred to as repeat-associated non-ATG translation, contributing to the harmful effects of aberrant proteins [31–33].

3.1.2. Other Mutations in AD Ataxias

Several of the other more recently discovered dominant ataxias are caused by conventional mutations: SNVs, insertions, and deletions. Conventional mutations are much less common in AD ataxias than repeat expansions. Several mutations have only been reported in select populations or families, while others appear to exist worldwide. SCA28, for example, causes 1.5% of AD ataxia in Europeans, but has not been detected in other large populations such as Chinese [34,35]. Dominant mutations can result in disease through haploinsufficiency due to gene deletion or disruption of functionally important residues, or by dominant negative mechanisms. Although more rare than in repeat expansions, anticipation has been documented in cases of indel or SNV mutations. The mechanism behind anticipation in ataxia due to indel or SNV mutations is unknown.

The variety of mutation types present in dominant ataxias illustrates the need for careful attention to molecular assays used to screen for new mutations. *ITPR1* was initially discarded as a candidate gene but later reassessment of the same samples detected the disease-causing deletion [25,36]. The confirmation of *ITPR1* as an ataxia causing gene in humans led to the careful screening and discovery of mutations in SCA16 and SCA29 patients [26,27].

3.2. Autosomal Recessive

Autosomal recessive (AR) ataxias occur more frequently than AD ataxias. Known AR ataxia genes are reported in Table S2. Despite the greater frequency of AR ataxias, many of these cases go genetically undiagnosed. Often, only one individual in a family presents with recessive ataxia. These cases may appear sporadic or idiopathic, making it difficult to distinguish AR from a *de novo* AD mutation or a new expansion event. In addition, the number of genes causing AR ataxia is large, and often mutations are family-specific or private variants, which appear most frequently under conditions of a suspected founder effect or consanguineous union. Recessive ataxias, more often than dominant, have symptom onset from birth or in early childhood, but this may be due to ascertainment, and later onset recessive ataxias certainly also exist. Unlike AD, early onset AR are typically non-progressive in their symptoms, with more multisystem involvement leading to other symptoms such as intellectual disability [37,38].

The most common autosomal recessive ataxia, and the most common early onset ataxia, is Friedreich's ataxia (FRDA). FRDA is estimated to have a prevalence of 1 in 20–50,000. In certain regions of the world, carrier rates have been estimated to be as high as 1 in 11 [39]. It is most commonly seen in individuals of European ancestry but is present worldwide. FRDA is primarily caused by a GAA intronic repeat expansion of the frataxin gene, with rare conventional mutations also reported [40,41]. The intronic expansion interferes with transcription and results in suppression of gene expression [42,43]. FRDA is a prime example where understanding the cellular pathology has guided research towards treatment, with several groups exploring methods to therapeutically increase the expression of frataxin [44], some of which are in or nearing clinical trials

Ataxia telangiectasia (A-T) is an early onset ataxia affecting 1 in 40–100,000. As mutations disrupt DNA repair, individuals with A-T are susceptible to radiation and oxidative stress. Heterozygous carriers for mutated *ATM* gene have a greater susceptibility to developing cancer. Mutations in *ATM* are highly variable, with over 600 unique variants reported.

There are several other ataxias that exhibit clinical features and molecular pathology similar to A-T. A-T like disorder is caused by mutations in *MRE11A*, another DNA repair gene. Individuals with A-T like disorder share the same neurological defects, along with oculomotor apraxia, but lack the telangiectasia and other features. Four other genes have been identified to cause ataxia with oculomotor apraxia (AOA). AOA2 is caused by mutations in *SETX* and is predicted to be responsible for 8% of non-Friedreich recessive ataxias [45]. It is prevalent among French-Canadians, but also present in other populations [45]. AOA1 is common among Japanese and Portuguese, where it was additionally characterized with features of low serum albumin and high cholesterol levels [46]. Mutations in *GRID2* and *PIK3R5*, which cause AOA and AOA3, are much less common.

3.3. X-Linked

In contrast to AD and AR ataxias, there are comparatively few known X-linked ataxias. The most common X-linked ataxia is fragile X-associated tremor ataxia syndrome (FXTAS). FXTAS is caused by a CGG repeat expansion in the 5' untranslated region of the *FMR1* gene [47]. This ataxia-associated expansion is often referred to as a fragile X "premutation". The normal length of the *FMR1* repeat is less than 39 repeats, whereas 55 to 200 repeats are considered to be a premutation. Males with greater than 200 repeats have the full expansion mutation, which causes fragile X syndrome, a severe disease caused by expansion of the same repeat [48]. In the U.S., carrier rates for the *FMR1* premutation are estimated at 1 in 209 for females and 1 in 430 for males [49]. A study in a population from Quebec estimated premutation rates at 1 in 259 for females and 1 in 813 for males [50,51].

FXTAS is characterized by tremor and ataxia with late onset, usually past the fifth decade. As the gene is X-linked, males are far more commonly affected than females [52]. In female carriers, an estimated 20% experience symptoms of premature ovarian insufficiency, with onset of menopause before 40 and/or fertility issues [53]. Males with fragile X display a very different phenotype from FXTAS, with prominent intellectual disability and abnormal facial features. *FMR1* is a prime example of how subtle differences in mutations within the same gene can greatly impact the phenotype.

Other X-linked ataxias are rare, often restricted to a single family. A mutation in *ATP2B3*, also known as *PMCA3*, was recently associated with spinocerebellar ataxia in an Italian family. Researchers found a single point mutation disrupted calcium transport in the cell, resulting in a "pure cerebellar" phenotype, with congenital onset ataxia, cerebellar atrophy, hypotonia, and slow eye movements [12,54]. Although the phenotype reported is similar to that seen in other families, this is the only reported *ATP2B3* ataxia mutation to date. Sideroblastic anemia with ataxia (ASAT) is caused by mutations in *ABCB7*.

3.4. Mitochondrial

Mitochondrial DNA is maternally transmitted through mitochondria in the oocyte. Mutations in mitochondrial DNA genes tend to result in more multisystem diseases that can contain ataxia as a symptom. Neuropathy, ataxia, and retinitis pigmentosa (NARP) is caused by mutations in the mitochondrial DNA gene *MTATP6* [55].

Mutations in nuclear genes that function primarily in the mitochondria can also cause ataxia. Despite their association to the mitochondria, these mutations are inherited in an AR pattern. Mutations in *POLG*, which is a subunit of mitochondrial DNA polymerase, are responsible for ataxia and other multisystem features [56]. *C10orf2*, or twinkle, is necessary for proper mtDNA replication and is responsible for a variety of neurological phenotypes including infantile onset ataxia [57–60].

3.5. Multiple Systems Atrophy and other Multisystem Diseases that Include Ataxia

Multisystem diseases can be more difficult to diagnose due to the variability in presentation. More diverse neurological phenotypes, such as seizures and myopathy, and non-neurological symptoms such as hearing loss, cardiac problems, and diabetes can complicate these disorders. Multiple system atrophy (MSA) is a progressive neurodegenerative disorder. Individuals may initially present with Parkinsonism or ataxia, and progress to more severe cerebellar atrophy and nervous system dysfunctions. Mutations in *COQ2* shown to be responsible for MSA have been shown to be more common in the Japanese population [61]. Refsum disease can also cause cerebellar ataxia but ataxia is not present in all Refsum patients. Several members of the peroxisome biogenesis factor family are responsible for several peroxisome biogenesis disorders that can appear similar to Refsum. These diseases range in severity from resulting in early death to survival and functional ability in adulthood. The broad phenotypes displayed in these diseases can make them difficult to diagnose and classify.

4. Mutations in Conserved Pathways Cause Ataxia

Despite the great advances made in sequencing technology and the discovery of new genes, there is still a gap in the full understanding of the function of these gene products. Many of the functions of ataxia genes have yet to be discovered, despite overwhelming evidence that they are responsible for causing disease. Discovery of genetic pathways involved in ataxia genes is important to our understanding of disease pathogenesis, and may also impact some treatments. Expression studies and protein interaction assays focused on known ataxia genes have helped identify pathways and protein interactions [62]. Researching expression and pathways can be difficult, primarily due to the low availability of relevant tissue, as brain donation and biopsy are delicate topics for those with ataxia and their families. Knowledge of these pathways will not only be important for efforts for treatment development but aid in the discovery of new ataxia genes through the identification of common pathways and interactions. A success story for this approach is the identification of a new EA candidate gene, *UBR4*, which was selected as a candidate gene due to its role in ubiquitination and localization with another ataxia gene, *ITPR1* [63].

4.1. DNA Repair

The ability of a cell to repair damage to DNA is important in order to maintain proper function and avoid deleterious mutations. DNA damage can result in cell death by apoptosis or the formation of cancerous cells. Several ataxia genes have roles in DNA repair, with many involved in ataxia with oculomotor apraxia. *MRE11* acts in a complex to locate damaged DNA, where it recruits *ATM* to phosphorylate p53 and induce DNA repair [64]. In individuals with ataxia-causing *MRE11* mutations, *MRE11* fails to effectively form a complex and recruit *ATM* [65]. Mutations in *SETX*, responsible for ataxia with oculomotor apraxia, greatly decrease the ability of cells to repair double strand breaks caused by oxidative stress [66]. Single strand repair mechanisms are impaired by mutations in *TDP1* and *APTX* [67–69].

4.2. Channelopathies

Mutations in genes responsible for the transport of ions in and out of the cell result in channelopathies. Channelopathies have received the most attention as a common pathway in neurological disease, with several reviews focused on the role of channel genes in disease and neurological disorders. Defects in ion channel genes usually result in dominant negative mechanisms, as they can alter the current and exchange of ions across cell membranes, affecting cell signaling or causing intracellular accumulation. Ion voltage channels help to regulate the action potential of neurons and release neurotransmitters. EA1, EA2, and EA5 are all caused by mutations in channel genes, a potassium voltage gated channel and two calcium voltage dependent channels [24,70,71]. Two other potassium voltage gated channel genes, *KCNC3* and *KCND3*, are responsible for SCA13 and SCA19/22 [13,14,72]. Inositol 1,4,5 triphosphate binding to *ITPR1* mediates the release of calcium from intracellular stores in the endoplasmic reticulum [73,74]. Deletions in *ITPR1* are hypothesized to cause adult onset ataxia through haploinsufficiency, and mutations in conserved domains affect channel function resulting in congenital ataxia [25–27].

4.3. Ubiquitination

Ubiquitination serves multiple roles within the cell, including targeting proteins for degradation. Many ataxias result from a mutant protein escaping this degradation system. Disruption of ubiquitination systems can cause failures in many cellular processes, such as protein degradation pathways, membrane tracking, apoptosis, and immune system processes.

Several ataxia genes are in or interact with ubiquitination system proteins. Mutations in *RNF170*, an E3 ubiquitin ligase, are responsible for AD sensory ataxia [75,76]. *RNF170* was shown to associate with inositol 1,4,5-triphosphate receptors (*IP3*), while mutations in its receptor, *IP3R1* (*ITPR1*) also cause ataxia, providing a link between ubiquitination and ion channel signaling [76]. *ATXN3* is a de-ubiquitinating enzyme that interacts with parkin, an E3 ubiquitin ligase, resulting in more de-ubiquitinated parkin in the presence of *ATXN3* repeat expansion mutants [77]. Recently mutations in two different E3 ligases have been associated with ataxia and hypogonadism: *RNF216* and *STUB1* [78,79].

4.4. Transcription/Translation

The ability to control gene expression and protein abundance is important for proper function in the cell and organism. Failure in proper transcriptional mechanism and regulation can result in a variety of diseases including cancer, autoimmune, and neurological [80] disorders. *ATXN1* forms a complex with the transcriptional repressor capicua and may interact with the transcription factor *RORα* [29,81]. Nemo-like kinase (*NLK*) has been shown to interact with *ATXN1* transcriptional complex, and decreased expression of *NLK* positively modulates the phenotype in SCA1 models, providing another biological target for future treatments [82]. The transcription factor *RORα* exhibits decreased levels in SCA1 and SCA3, with null and mutant mice for *Rora* showing cerebellar defects and ataxia [81,83,84]. Along with DNA repair, mutations in *SETX* also interfere with transcription, highlighting interactions between senataxin and proteins involved in transcription and RNA processing [85].

5. Genetic Testing of Ataxias and Personalized Medicine

5.1. Is Genetic Testing of Ataxia Useful?

Rare forms of ataxia respond to Vitamin E or Coenzyme Q10 (AVED and SCAR9/ARCA2), but for most ataxias, only symptomatic treatment is available. Genetic testing for diseases with no treatment is controversial and in the U.S., is usually considered a personal choice, whereas developing countries do not routinely offer testing. Thus, what are the reasons to test a patient with ataxia for known genes? Indeed, some of those with ataxia have commented upon finding out that they had e.g., SCA2, so now what? There is little difference in treatment or prognosis, so why all the expense of testing?

One reason given for genetic testing is family planning. This most often arises in dominant ataxias where the disease is seen in prior generations, but can apply to recessive ataxias, especially in isolated populations where disease alleles may be at a higher frequency. Individuals who are carriers of a disease mutation may make different reproductive choices to avoid passing the disease on to the next generation. They may choose to avoid passing on genetic material by not having children, choosing adoption, or use of egg/sperm donors. Those wanting biological children may utilize pre-implantation screening with *in vitro* fertilization or termination of pregnancy after prenatal diagnosis [86]. Pre-implantation testing and testing of children have resulted in a new ethical conundrum of "genetic ignorance", where parents may decide to remain ignorant about their own results, but wish to test offspring or embryos, possibly unnecessarily [87].

A more complex situation is that of FXTAS. This diagnosis in an older male with ataxia implies a very high risk of the more severe fragile X syndrome in any grandsons or nephews through his female relatives. Since the *FMR1* expansion is on the X chromosome, females can be asymptomatic carriers. For example, a female with a male relative with FXTAS may be a carrier, and hence be at risk of having a son with fragile X syndrome. A survey on genetic screening for *FMR1* mutations showed that while individuals are concerned about finding out they are carriers, and the emotional stress that may accompany that, many note the value in being able to make informed reproductive choices and possible benefits to other family members [88].

Genetic testing also will allow pre-symptomatic testing. One man, after finding out the genetic cause of his mother's ataxia, decided to get tested himself before purchasing a house—he reasoned that if he had the mutation, the house should accommodate his future potential disability needs such as walker and wheel chair accessibility. Pre-symptomatic testing can result in unintended consequences for those tested, which may explain why for some neurodegenerative diseases, like Huntington's, a minority of those at risk are tested [89,90]. There are reports of greater instances of depression in individuals with positive test results, possible stigmatization by peers or family members, or having difficulty obtaining life insurance.

Genetic testing offers a definitive diagnostic confirmation for patients. Some individuals with ataxia are first diagnosed as having amyotrophic lateral sclerosis, multiple sclerosis, MSA, or Parkinson's. Genetic testing will become an increasingly important part of differential diagnosis in these individuals. Desire for knowledge and closure about what is causing their symptoms can be comforting to affected individuals and family members. Genetic testing also has clear clinical ramifications for prognosis. As there are clinical differences in progression rates between different forms of ataxia [21], genetic testing can help patients and their physicians understand their own prognosis. For example, SCA7 leads to severe vision problems or blindness, and SCA6 also leads to some vision problems, whereas SCA2 symptoms also include neuropathy, tremors and cramps. Rarely, Vitamin-E responsive ataxia may be confused clinically with Friedreich's ataxia, and hence this diagnosis will open up a new treatment with large doses of vitamin E. In turn, sometimes spastic paraplegias are misdiagnosed as ataxias, and treating the spasticity aspect may bring relief [91].

5.2. Genetic Testing Now and in the Future Era of Cheap Sequencing

Currently, genetic testing is performed by a number of academic and commercial laboratories. For recessive ataxias, usually Friedreich's ataxia is tested first, although depending on the symptoms, other ataxias are included. For dominant ataxias, the five most common SCAs are often tested first [92]. If these are negative, comprehensive panels for most known dominant or recessive ataxia genes, or all, are available from commercial sources [93]. Comprehensive panels are helpful when the mode of inheritance is not clear—e.g., if a parent died before symptom onset, or has some neurological symptoms that are not identical, or had been diagnosed with a different disorder.

Next generation exome sequencing has shown some success in clinical environments, demonstrating that it may be more efficient than testing for mutations in ataxia genes individually [94–98]. Several academic clinical laboratories offer targeted sequencing of hundreds of genes, whereas others, such as Baylor and University of Chicago, offer exome sequencing, but may specifically evaluate ataxia genes [99,100]. At what point it is best to move from sequencing genes one at a time to large panels, and when from large panels to whole exome, is currently not clear. In addition, whole exome or genome sequencing can identify mutations in genes unrelated to ataxia, such as genes associated with early onset breast cancer. This is not unique to ataxia, and how to deal with such secondary or "incidental" findings is currently actively debated by ethicists and clinicians.

In addition, there are current limitations to next generation sequencing technology. Currently large repeat expansions cannot be accurately sequenced or mapped to identify the common repeat expansion mutations. This is a limitation of the short reads captured by the sequencing technology; reads comprised of entirely repeats cannot be aligned to accurately determine placement or length. New computational methods are being developed in an attempt to tackle this problem. It is important for clinicians and genetic counselors to consider that next generation sequencing does not guarantee a diagnosis and should address this point with patients desiring sequencing.

Hence, a fully comprehensive genome analysis that covers all ataxia gene mutations is not currently available. Given the heterogeneity of ataxias, and the large number of genes still being detected each year, a comprehensive genetic test will be a challenge for researchers and clinicians.

6. Conclusions

The variability in phenotypic symptoms and genetic causes provide a challenge for clinicians and geneticists in studying ataxia. Advancements in sequencing technology have greatly increased our rate of discovery of new ataxia genes and ability to screen for known genes. With the price of whole exome sequencing and soon whole genome sequencing falling below $1000 a sample, it has become the cost effective approach to screen multiple genes at the same time. Continuing expression studies and investigation into the role of genes will help identify shared pathways and functions. A challenge in this movement towards next generation sequencing technology is the discovery of new repeat expansions, which are difficult to detect using this new technology. The number of known genes mutations responsible for ataxia keeps growing every year; however we still do not have well defined functions or pathways for many of these genes. With greater understanding of the pathways these genes are involved in and how each mutation causes disease, we may be able to generate more targeted and effective treatments in the future.

Acknowledgments

We would like to acknowledge the National Ataxia Foundation (NAF) for their assistance in recruitment and opportunity to meet and discuss with those affected by ataxia. We would like to thank our many clinical colleagues for recruitment and discussions, and all of the families that have taken time to donate samples and personal stories. Our ataxia research is funded by the National Institutes of Health (NIH) grant R01-NS078560.

Author Contributions

All authors contributed to and wrote this review.

References

1. Durr, A. Autosomal dominant cerebellar ataxias: Polyglutamine expansions and beyond. *Lancet Neurol.* **2010**, *9*, 885–894.
2. Joo, B.-E.; Lee, C.-N.; Park, K.-W. Prevalence rate and functional status of cerebellar ataxia in Korea. *Cerebellum Lond. Engl.* **2012**, *11*, 733–738.
3. Coutinho, P.; Ruano, L.; Loureiro, J.L.; Cruz, V.T.; Barros, J.; Tuna, A.; Barbot, C.; Guimarães, J.; Alonso, I.; Silveira, I.; *et al.* Hereditary ataxia and spastic paraplegia in Portugal: A population-based prevalence study. *JAMA Neurol.* **2013**, *70*, 746–755.
4. Koht, J.; Tallaksen, C.M.E. Cerebellar ataxia in the eastern and southern parts of Norway. *Acta Neurol. Scand. Suppl.* **2007**, *187*, 76–79.
5. Orr, H.T.; Chung, M.Y.; Banfi, S.; Kwiatkowski, T.J., Jr.; Servadio, A.; Beaudet, A.L.; McCall, A.E.; Duvick, L.A.; Ranum, L.P.; Zoghbi, H.Y. Expansion of an unstable trinucleotide CAG repeat in spinocerebellar ataxia type 1. *Nat. Genet.* **1993**, *4*, 221–226.
6. Sanpei, K.; Takano, H.; Igarashi, S.; Sato, T.; Oyake, M.; Sasaki, H.; Wakisaka, A.; Tashiro, K.; Ishida, Y.; Ikeuchi, T.; *et al.* Identification of the spinocerebellar ataxia type 2 gene using a direct identification of repeat expansion and cloning technique, DIRECT. *Nat. Genet.* **1996**, *14*, 277–284.

7. Pulst, S.M.; Nechiporuk, A.; Nechiporuk, T.; Gispert, S.; Chen, X.N.; Lopes-Cendes, I.; Pearlman, S.; Starkman, S.; Orozco-Diaz, G.; Lunkes, A.; *et al.* Moderate expansion of a normally biallelic trinucleotide repeat in spinocerebellar ataxia type 2. *Nat. Genet.* **1996**, *14*, 269–276.
8. Kawaguchi, Y.; Okamoto, T.; Taniwaki, M.; Aizawa, M.; Inoue, M.; Katayama, S.; Kawakami, H.; Nakamura, S.; Nishimura, M.; Akiguchi, I. CAG expansions in a novel gene for Machado-Joseph disease at chromosome 14q32.1. *Nat. Genet.* **1994**, *8*, 221–228.
9. Haberhausen, G.; Damian, M.S.; Leweke, F.; Müller, U. Spinocerebellar ataxia, type 3 (SCA3) is genetically identical to Machado-Joseph disease (MJD). *J. Neurol. Sci.* **1995**, *132*, 71–75.
10. Zhuchenko, O.; Bailey, J.; Bonnen, P.; Ashizawa, T.; Stockton, D.W.; Amos, C.; Dobyns, W.B.; Subramony, S.H.; Zoghbi, H.Y.; Lee, C.C. Autosomal dominant cerebellar ataxia (SCA6) associated with small polyglutamine expansions in the alpha 1A-voltage-dependent calcium channel. *Nat. Genet.* **1997**, *15*, 62–69.
11. Ng, S.B.; Buckingham, K.J.; Lee, C.; Bigham, A.W.; Tabor, H.K.; Dent, K.M.; Huff, C.D.; Shannon, P.T.; Jabs, E.W.; Nickerson, D.A.; *et al.* Exome sequencing identifies the cause of a mendelian disorder. *Nat. Genet.* **2010**, *42*, 30–35.
12. Zanni, G.; Calì, T.; Kalscheuer, V.M.; Ottolini, D.; Barresi, S.; Lebrun, N.; Montecchi-Palazzi, L.; Hu, H.; Chelly, J.; Bertini, E.; *et al.* Mutation of plasma membrane Ca2+ ATPase isoform 3 in a family with X-linked congenital cerebellar ataxia impairs Ca2+ homeostasis. *Proc. Natl. Acad. Sci. USA* **2012**, *109*, 14514–14519.
13. Lee, Y.-C.; Durr, A.; Majczenko, K.; Huang, Y.-H.; Liu, Y.-C.; Lien, C.-C.; Tsai, P.-C.; Ichikawa, Y.; Goto, J.; Monin, M.-L.; *et al.* Mutations in KCND3 cause spinocerebellar ataxia type 22. *Ann. Neurol.* **2012**, *72*, 859–869.
14. Duarri, A.; Jezierska, J.; Fokkens, M.; Meijer, M.; Schelhaas, H.J.; den Dunnen, W.F.A.; van Dijk, F.; Verschuuren-Bemelmans, C.; Hageman, G.; van de Vlies, P.; *et al.* Mutations in potassium channel kcnd3 cause spinocerebellar ataxia type 19. *Ann. Neurol.* **2012**, *72*, 870–880.
15. Winkelmann, J.; Lin, L.; Schormair, B.; Kornum, B.R.; Faraco, J.; Plazzi, G.; Melberg, A.; Cornelio, F.; Urban, A.E.; Pizza, F.; *et al.* Mutations in DNMT1 cause autosomal dominant cerebellar ataxia, deafness and narcolepsy. *Hum. Mol. Genet.* **2012**, *21*, 2205–2210.
16. Bilguvar, K.; Tyagi, N.K.; Ozkara, C.; Tuysuz, B.; Bakircioglu, M.; Choi, M.; Delil, S.; Caglayan, A.O.; Baranoski, J.F.; Erturk, O.; *et al.* Recessive loss of function of the neuronal ubiquitin hydrolase UCHL1 leads to early-onset progressive neurodegeneration. *Proc. Natl. Acad. Sci. USA* **2013**, *110*, 3489–3494.
17. Sun, Y.; Almomani, R.; Breedveld, G.J.; Santen, G.W.E.; Aten, E.; Lefeber, D.J.; Hoff, J.I.; Brusse, E.; Verheijen, F.W.; Verdijk, R.M.; *et al.* Autosomal recessive spinocerebellar ataxia 7 (SCAR7) is caused by variants in TPP1, the gene involved in classic late-infantile neuronal ceroid lipofuscinosis 2 disease (CLN2 disease). *Hum. Mutat.* **2013**, *34*, 706–713.
18. Sailer, A.; Houlden, H. Recent advances in the genetics of cerebellar ataxias. *Curr. Neurol. Neurosci. Rep.* **2012**, *12*, 227–236.
19. Lhatoo, S.D.; Rao, D.G.; Kane, N.M.; Ormerod, I.E. Very late onset Friedreich's presenting as

spastic tetraparesis without ataxia or neuropathy. *Neurology* **2001**, *56*, 1776–1777.

20. Castelnovo, G.; Biolsi, B.; Barbaud, A.; Labauge, P.; Schmitt, M. Isolated spastic paraparesis leading to diagnosis of Friedreich's ataxia. *J. Neurol. Neurosurg. Psychiatry* **2000**, *69*, 693.
21. Paulson, H.L. The spinocerebellar ataxias. *J. Neuroophthalmol.* **2009**, *29*, 227–237.
22. Baloh, R.W. Episodic ataxias 1 and 2. *Handb. Clin. Neurol.* **2012**, *103*, 595–602.
23. Jen, J.C.; Wan, J.; Palos, T.P.; Howard, B.D.; Baloh, R.W. Mutation in the glutamate transporter EAAT1 causes episodic ataxia, hemiplegia, and seizures. *Neurology* **2005**, *65*, 529–534.
24. Escayg, A.; de Waard, M.; Lee, D.D.; Bichet, D.; Wolf, P.; Mayer, T.; Johnston, J.; Baloh, R.; Sander, T.; Meisler, M.H. Coding and noncoding variation of the human calcium-channel beta4-subunit gene CACNB4 in patients with idiopathic generalized epilepsy and episodic ataxia. *Am. J. Hum. Genet.* **2000**, *66*, 1531–1539.
25. Van de Leemput, J.; Chandran, J.; Knight, M.A.; Holtzclaw, L.A.; Scholz, S.; Cookson, M.R.; Houlden, H.; Gwinn-Hardy, K.; Fung, H.-C.; Lin, X.; *et al.* Deletion at ITPR1 underlies ataxia in mice and spinocerebellar ataxia 15 in humans. *PLoS Genet.* **2007**, *3*, e108.
26. Iwaki, A.; Kawano, Y.; Miura, S.; Shibata, H.; Matsuse, D.; Li, W.; Furuya, H.; Ohyagi, Y.; Taniwaki, T.; Kira, J.; *et al.* Heterozygous deletion of ITPR1, but not SUMF1, in spinocerebellar ataxia type 16. *J. Med. Genet.* **2008**, *45*, 32–35.
27. Huang, L.; Chardon, J.W.; Carter, M.T.; Friend, K.L.; Dudding, T.E.; Schwartzentruber, J.; Zou, R.; Schofield, P.W.; Douglas, S.; Bulman, D.E.; *et al.* Missense mutations in ITPR1 cause autosomal dominant congenital nonprogressive spinocerebellar ataxia. *Orphanet J. Rare Dis.* **2012**, *7*, 67.
28. Cummings, C.J.; Reinstein, E.; Sun, Y.; Antalffy, B.; Jiang, Y.; Ciechanover, A.; Orr, H.T.; Beaudet, A.L.; Zoghbi, H.Y. Mutation of the E6-AP ubiquitin ligase reduces nuclear inclusion frequency while accelerating polyglutamine-induced pathology in SCA1 mice. *Neuron* **1999**, *24*, 879–892.
29. Lam, Y.C.; Bowman, A.B.; Jafar-Nejad, P.; Lim, J.; Richman, R.; Fryer, J.D.; Hyun, E.D.; Duvick, L.A.; Orr, H.T.; Botas, J.; *et al.* ATAXIN-1 interacts with the repressor Capicua in its native complex to cause SCA1 neuropathology. *Cell* **2006**, *127*, 1335–1347.
30. Suzuki, K.; Zhou, J.; Sato, T.; Takao, K.; Miyagawa, T.; Oyake, M.; Yamada, M.; Takahashi, H.; Takahashi, Y.; Goto, J.; *et al.* DRPLA transgenic mouse substrains carrying single copy of full-length mutant human DRPLA gene with variable sizes of expanded CAG repeats exhibit CAG repeat length- and age-dependent changes in behavioral abnormalities and gene expression profiles. *Neurobiol. Dis.* **2012**, *46*, 336–350.
31. Zu, T.; Gibbens, B.; Doty, N.S.; Gomes-Pereira, M.; Huguet, A.; Stone, M.D.; Margolis, J.; Peterson, M.; Markowski, T.W.; Ingram, M.A.C.; *et al.* Non-ATG-initiated translation directed by microsatellite expansions. *Proc. Natl. Acad. Sci. USA* **2011**, *108*, 260–265.
32. Cleary, J.D.; Ranum, L.P.W. Repeat-associated non-ATG (RAN) translation in neurological disease. *Hum. Mol. Genet.* **2013**, *22*, R45–R51.
33. Todd, P.K.; Oh, S.Y.; Krans, A.; He, F.; Sellier, C.; Frazer, M.; Renoux, A.J.; Chen, K.; Scaglione, K.M.; Basrur, V.; *et al.* CGG repeat-associated translation mediates neurodegeneration in fragile X tremor ataxia syndrome. *Neuron* **2013**, *78*, 440–455.

34. Cagnoli, C.; Stevanin, G.; Brussino, A.; Barberis, M.; Mancini, C.; Margolis, R.L.; Holmes, S.E.; Nobili, M.; Forlani, S.; Padovan, S.; *et al.* Missense mutations in the AFG3L2 proteolytic domain account for ~1.5% of European autosomal dominant cerebellar ataxias. *Hum. Mutat.* **2010**, *31*, 1117–1124.
35. Jia, D.; Tang, B.; Chen, Z.; Shi, Y.; Sun, Z.; Zhang, L.; Wang, J.; Xia, K.; Jiang, H. Spinocerebellar ataxia type 28 (SCA28) is an uncommon cause of dominant ataxia among Chinese kindreds. *Int. J. Neurosci.* **2012**, *122*, 560–562.
36. Knight, M.A.; Kennerson, M.L.; Anney, R.J.; Matsuura, T.; Nicholson, G.A.; Salimi-Tari, P.; Gardner, R.J.M.; Storey, E.; Forrest, S.M. Spinocerebellar ataxia type 15 (sca15) maps to 3p24.2-3pter: Exclusion of the ITPR1 gene, the human orthologue of an ataxic mouse mutant. *Neurobiol. Dis.* **2003**, *13*, 147–157.
37. Embiruçu, E.K.; Martyn, M.L.; Schlesinger, D.; Kok, F. Autosomal recessive ataxias: 20 types, and counting. *Arq. Neuropsiquiatr.* **2009**, *67*, 1143–1156.
38. Fogel, B.L.; Perlman, S. Clinical features and molecular genetics of autosomal recessive cerebellar ataxias. *Lancet Neurol.* **2007**, *6*, 245–257.
39. Zamba-Papanicolaou, E.; Koutsou, P.; Daiou, C.; Gaglia, E.; Georghiou, A.; Christodoulou, K. High frequency of Friedreich's ataxia carriers in the Paphos district of Cyprus. *Acta Myol.* **2009**, *28*, 24–26.
40. Campuzano, V.; Montermini, L.; Moltò, M.D.; Pianese, L.; Cossée, M.; Cavalcanti, F.; Monros, E.; Rodius, F.; Duclos, F.; Monticelli, A.; *et al.* Friedreich's ataxia: Autosomal recessive disease caused by an intronic GAA triplet repeat expansion. *Science* **1996**, *271*, 1423–1427.
41. Bidichandani, S.I.; Ashizawa, T.; Patel, P.I. Atypical Friedreich ataxia caused by compound heterozygosity for a novel missense mutation and the GAA triplet-repeat expansion. *Am. J. Hum. Genet.* **1997**, *60*, 1251–1256.
42. Bidichandani, S.I.; Ashizawa, T.; Patel, P.I. The GAA triplet-repeat expansion in Friedreich ataxia interferes with transcription and may be associated with an unusual DNA structure. *Am. J. Hum. Genet.* **1998**, *62*, 111–121.
43. Campuzano, V.; Montermini, L.; Lutz, Y.; Cova, L.; Hindelang, C.; Jiralerspong, S.; Trottier, Y.; Kish, S.J.; Faucheux, B.; Trouillas, P.; *et al.* Frataxin is reduced in Friedreich ataxia patients and is associated with mitochondrial membranes. *Hum. Mol. Genet.* **1997**, *6*, 1771–1780.
44. Chapdelaine, P.; Coulombe, Z.; Chikh, A.; Gerard, C.; Tremblay, J.P. A Potential New Therapeutic Approach for Friedreich Ataxia: Induction of Frataxin Expression with TALE Proteins. *Mol. Ther. Nucleic Acids* **2013**, *2*, e119.
45. Le Ber, I.; Bouslam, N.; Rivaud-Péchoux, S.; Guimarães, J.; Benomar, A.; Chamayou, C.; Goizet, C.; Moreira, M.-C.; Klur, S.; Yahyaoui, M.; *et al.* Frequency and phenotypic spectrum of ataxia with oculomotor apraxia 2: A clinical and genetic study in 18 patients. *Brain J. Neurol.* **2004**, *127*, 759–767.
46. Moreira, M.C.; Barbot, C.; Tachi, N.; Kozuka, N.; Uchida, E.; Gibson, T.; Mendonça, P.; Costa, M.; Barros, J.; Yanagisawa, T.; *et al.* The gene mutated in ataxia-ocular apraxia 1 encodes the new HIT/Zn-finger protein aprataxin. *Nat. Genet.* **2001**, *29*, 189–193.

47. Hagerman, R.J.; Leehey, M.; Heinrichs, W.; Tassone, F.; Wilson, R.; Hills, J.; Grigsby, J.; Gage, B.; Hagerman, P.J. Intention tremor, parkinsonism, and generalized brain atrophy in male carriers of fragile X. *Neurology* **2001**, *57*, 127–130.
48. Fu, Y.H.; Kuhl, D.P.; Pizzuti, A.; Pieretti, M.; Sutcliffe, J.S.; Richards, S.; Verkerk, A.J.; Holden, J.J.; Fenwick, R.G., Jr.; Warren, S.T. Variation of the CGG repeat at the fragile X site results in genetic instability: resolution of the Sherman paradox. *Cell* **1991**, *67*, 1047–1058.
49. Tassone, F.; Iong, K.P.; Tong, T.-H.; Lo, J.; Gane, L.W.; Berry-Kravis, E.; Nguyen, D.; Mu, L.Y.; Laffin, J.; Bailey, D.B.; *et al.* FMR1 CGG allele size and prevalence ascertained through newborn screening in the United States. *Genome Med.* **2012**, *4*, 100.
50. Rousseau, F.; Rouillard, P.; Morel, M.L.; Khandjian, E.W.; Morgan, K. Prevalence of carriers of premutation-size alleles of the FMRI gene—And implications for the population genetics of the fragile X syndrome. *Am. J. Hum. Genet.* **1995**, *57*, 1006–1018.
51. Dombrowski, C.; Lévesque, S.; Morel, M.L.; Rouillard, P.; Morgan, K.; Rousseau, F. Premutation and intermediate-size FMR1 alleles in 10572 males from the general population: Loss of an AGG interruption is a late event in the generation of fragile X syndrome alleles. *Hum. Mol. Genet.* **2002**, *11*, 371–378.
52. Rodriguez-Revenga, L.; Madrigal, I.; Pagonabarraga, J.; Xunclà, M.; Badenas, C.; Kulisevsky, J.; Gomez, B.; Milà, M. Penetrance of FMR1 premutation associated pathologies in fragile X syndrome families. *Eur. J. Hum. Genet. EJHG* **2009**, *17*, 1359–1362.
53. Schwartz, C.E.; Dean, J.; Howard-Peebles, P.N.; Bugge, M.; Mikkelsen, M.; Tommerup, N.; Hull, C.; Hagerman, R.; Holden, J.J.; Stevenson, R.E. Obstetrical and gynecological complications in fragile X carriers: a multicenter study. *Am. J. Med. Genet.* **1994**, *51*, 400–402.
54. Bertini, E.; des Portes, V.; Zanni, G.; Santorelli, F.; Dionisi-Vici, C.; Vicari, S.; Fariello, G.; Chelly, J. X-linked congenital ataxia: A clinical and genetic study. *Am. J. Med. Genet.* **2000**, *92*, 53–56.
55. Holt, I.J.; Harding, A.E.; Petty, R.K.; Morgan-Hughes, J.A. A new mitochondrial disease associated with mitochondrial DNA heteroplasmy. *Am. J. Hum. Genet.* **1990**, *46*, 428–433.
56. Synofzik, M.; Srulijes, K.; Godau, J.; Berg, D.; Schöls, L. Characterizing POLG ataxia: Clinics, electrophysiology and imaging. *Cerebellum Lond. Engl.* **2012**, *11*, 1002–1011.
57. Faruq, M.; Narang, A.; Kumari, R.; Pandey, R.; Garg, A.; Behari, M.; Dash, D.; Srivastava, A.; Mukerji, M. Novel mutations in typical and atypical genetic loci through exome sequencing in autosomal recessive cerebellar ataxia families. *Clin. Genet.* **2013**, doi:10.1111/cge.12279.
58. Nikali, K.; Suomalainen, A.; Saharinen, J.; Kuokkanen, M.; Spelbrink, J.N.; Lönnqvist, T.; Peltonen, L. Infantile onset spinocerebellar ataxia is caused by recessive mutations in mitochondrial proteins Twinkle and Twinky. *Hum. Mol. Genet.* **2005**, *14*, 2981–2990.
59. Hartley, J.N.; Booth, F.A.; del Bigio, M.R.; Mhanni, A.A. Novel Autosomal Recessive c10orf2 Mutations Causing Infantile-Onset Spinocerebellar Ataxia. *Case Rep. Pediatr.* **2012**, *2012*, 303096.
60. Wanrooij, S.; Falkenberg, M. The human mitochondrial replication fork in health and disease. *Biochim. Biophys. Acta* **2010**, *1797*, 1378–1388.
61. Multiple-System Atrophy Research Collaboration Mutations in COQ2 in familial and sporadic multiple-system atrophy. *N. Engl. J. Med.* **2013**, *369*, 233–244.

62. Lim, J.; Hao, T.; Shaw, C.; Patel, A.J.; Szabó, G.; Rual, J.-F.; Fisk, C.J.; Li, N.; Smolyar, A.; Hill, D.E.; *et al.* A protein-protein interaction network for human inherited ataxias and disorders of Purkinje cell degeneration. *Cell* **2006**, *125*, 801–814.
63. Conroy, J.; McGettigan, P.; Murphy, R.; Webb, D.; Murphy, S.M.; McCoy, B.; Albertyn, C.; McCreary, D.; McDonagh, C.; Walsh, O.; *et al.* A novel locus for episodic ataxia:UBR4 the likely candidate. *Eur. J. Hum. Genet.* **2014**, *22*, 505–510.
64. Lee, J.-H.; Paull, T.T. ATM activation by DNA double-strand breaks through the Mre11-Rad50-Nbs1 complex. *Science* **2005**, *308*, 551–554.
65. Regal, J.A.; Festerling, T.A.; Buis, J.M.; Ferguson, D.O. Disease-associated MRE11 mutants impact ATM/ATR DNA damage signaling by distinct mechanisms. *Hum. Mol. Genet.* **2013**, *22*, 5146–5159.
66. Suraweera, A.; Becherel, O.J.; Chen, P.; Rundle, N.; Woods, R.; Nakamura, J.; Gatei, M.; Criscuolo, C.; Filla, A.; Chessa, L.; *et al.* Senataxin, defective in ataxia oculomotor apraxia type 2, is involved in the defense against oxidative DNA damage. *J. Cell Biol.* **2007**, *177*, 969–979.
67. Takahashi, T.; Tada, M.; Igarashi, S.; Koyama, A.; Date, H.; Yokoseki, A.; Shiga, A.; Yoshida, Y.; Tsuji, S.; Nishizawa, M.; *et al.* Aprataxin, causative gene product for EAOH/AOA1, repairs DNA single-strand breaks with damaged 3'-phosphate and 3'-phosphoglycolate ends. *Nucleic Acids Res.* **2007**, *35*, 3797–3809.
68. Zhou, T.; Lee, J.W.; Tatavarthi, H.; Lupski, J.R.; Valerie, K.; Povirk, L.F. Deficiency in 3'-phosphoglycolate processing in human cells with a hereditary mutation in tyrosyl-DNA phosphodiesterase (TDP1). *Nucleic Acids Res.* **2005**, *33*, 289–297.
69. El-Khamisy, S.F.; Saifi, G.M.; Weinfeld, M.; Johansson, F.; Helleday, T.; Lupski, J.R.; Caldecott, K.W. Defective DNA single-strand break repair in spinocerebellar ataxia with axonal neuropathy-1. *Nature* **2005**, *434*, 108–113.
70. Browne, D.L.; Gancher, S.T.; Nutt, J.G.; Brunt, E.R.; Smith, E.A.; Kramer, P.; Litt, M. Episodic ataxia/myokymia syndrome is associated with point mutations in the human potassium channel gene, KCNA1. *Nat. Genet.* **1994**, *8*, 136–140.
71. Ophoff, R.A.; Terwindt, G.M.; Vergouwe, M.N.; van Eijk, R.; Oefner, P.J.; Hoffman, S.M.; Lamerdin, J.E.; Mohrenweiser, H.W.; Bulman, D.E.; Ferrari, M.; *et al.* Familial hemiplegic migraine and episodic ataxia type-2 are caused by mutations in the Ca^{2+} channel gene CACNL1A4. *Cell* **1996**, *87*, 543–552.
72. Waters, M.F.; Minassian, N.A.; Stevanin, G.; Figueroa, K.P.; Bannister, J.P.A.; Nolte, D.; Mock, A.F.; Evidente, V.G.H.; Fee, D.B.; Müller, U.; *et al.* Mutations in voltage-gated potassium channel KCNC3 cause degenerative and developmental central nervous system phenotypes. *Nat. Genet.* **2006**, *38*, 447–451.
73. Streb, H.; Irvine, R.F.; Berridge, M.J.; Schulz, I. Release of Ca^{2+} from a nonmitochondrial intracellular store in pancreatic acinar cells by inositol-1,4,5-trisphosphate. *Nature* **1983**, *306*, 67–69.
74. Berridge, M.J. Inositol trisphosphate and calcium signalling mechanisms. *Biochim. Biophys. Acta* **2009**, *1793*, 933–940.
75. Valdmanis, P.N.; Dupré, N.; Lachance, M.; Stochmanski, S.J.; Belzil, V.V.; Dion, P.A.; Thiffault, I.; Brais, B.; Weston, L.; Saint-Amant, L.; *et al.* A mutation in the RNF170 gene causes autosomal dominant sensory ataxia. *Brain J. Neurol.* **2011**, *134*, 602–607.

76. Lu, J.P.; Wang, Y.; Sliter, D.A.; Pearce, M.M.P.; Wojcikiewicz, R.J.H. RNF170 protein, an endoplasmic reticulum membrane ubiquitin ligase, mediates inositol 1,4,5-trisphosphate receptor ubiquitination and degradation. *J. Biol. Chem.* **2011**, *286*, 24426–24433.
77. Durcan, T.M.; Kontogiannea, M.; Thorarinsdottir, T.; Fallon, L.; Williams, A.J.; Djarmati, A.; Fantaneanu, T.; Paulson, H.L.; Fon, E.A. The Machado-Joseph disease-associated mutant form of ataxin-3 regulates parkin ubiquitination and stability. *Hum. Mol. Genet.* **2011**, *20*, 141–154.
78. Margolin, D.H.; Kousi, M.; Chan, Y.-M.; Lim, E.T.; Schmahmann, J.D.; Hadjivassiliou, M.; Hall, J.E.; Adam, I.; Dwyer, A.; Plummer, L.; *et al.* Ataxia, dementia, and hypogonadotropism caused by disordered ubiquitination. *N. Engl. J. Med.* **2013**, *368*, 1992–2003.
79. Shi, C.-H.; Schisler, J.C.; Rubel, C.E.; Tan, S.; Song, B.; McDonough, H.; Xu, L.; Portbury, A.L.; Mao, C.-Y.; True, C.; *et al.* Ataxia and hypogonadism caused by the loss of ubiquitin ligase activity of the U box protein CHIP. *Hum. Mol. Genet.* **2014**, *23*, 1013–1024.
80. Lee, T.I.; Young, R.A. Transcriptional regulation and its misregulation in disease. *Cell* **2013**, *152*, 1237–1251.
81. Serra, H.G.; Duvick, L.; Zu, T.; Carlson, K.; Stevens, S.; Jorgensen, N.; Lysholm, A.; Burright, E.; Zoghbi, H.Y.; Clark, H.B.; *et al.* RORalpha-mediated Purkinje cell development determines disease severity in adult SCA1 mice. *Cell* **2006**, *127*, 697–708.
82. Ju, H.; Kokubu, H.; Todd, T.W.; Kahle, J.J.; Kim, S.; Richman, R.; Chirala, K.; Orr, H.T.; Zoghbi, H.Y.; Lim, J. Polyglutamine disease toxicity is regulated by Nemo-like kinase in spinocerebellar ataxia type 1. *J. Neurosci.* **2013**, *33*, 9328–9336.
83. Konno, A.; Shuvaev, A.N.; Miyake, N.; Miyake, K.; Iizuka, A.; Matsuura, S.; Huda, F.; Nakamura, K.; Yanagi, S.; Shimada, T.; *et al.* Mutant Ataxin-3 with an Abnormally Expanded Polyglutamine Chain Disrupts Dendritic Development and Metabotropic Glutamate Receptor Signaling in Mouse Cerebellar Purkinje Cells. *Cerebellum* **2013**, *13*, 29–41.
84. Dussault, I.; Fawcett, D.; Matthyssen, A.; Bader, J.A.; Giguère, V. Orphan nuclear receptor ROR alpha-deficient mice display the cerebellar defects of staggerer. *Mech. Dev.* **1998**, *70*, 147–153.
85. Suraweera, A.; Lim, Y.; Woods, R.; Birrell, G.W.; Nasim, T.; Becherel, O.J.; Lavin, M.F. Functional role for senataxin, defective in ataxia oculomotor apraxia type 2, in transcriptional regulation. *Hum. Mol. Genet.* **2009**, *18*, 3384–3396.
86. Cruz-Mariño, T.; Velázquez-Pérez, L.; González-Zaldivar, Y.; Aguilera-Rodríguez, R.; Velázquez-Santos, M.; Vázquez-Mojena, Y.; Estupiñán-Rodríguez, A.; Reynaldo-Armiñán, R.; Almaguer-Mederos, L.E.; Laffita-Mesa, J.M.; *et al.* Couples at risk for spinocerebellar ataxia type 2: the Cuban prenatal diagnosis experience. *J. Community Genet.* **2013**, *4*, 451–460.
87. Erez, A.; Plunkett, K.; Sutton, V.R.; McGuire, A.L. The right to ignore genetic status of late onset genetic disease in the genomic era; Prenatal testing for Huntington disease as a paradigm. *Am. J. Med. Genet. A* **2010**, *152A*, 1774–1780.
88. Archibald, A.D.; Hickerton, C.L.; Jaques, A.M.; Wake, S.; Cohen, J.; Metcalfe, S.A. "It's about having the choice": Stakeholder perceptions of population-based genetic carrier screening for fragile X syndrome. *Am. J. Med. Genet. A* **2013**, *161A*, 48–58.
89. Andersson, P.L.; Juth, N.; Petersén, Å.; Graff, C.; Edberg, A.-K. Ethical aspects of undergoing a predictive genetic testing for Huntington's disease. *Nurs. Ethics* **2013**, *20*, 189–199.

90. Tanaka, K.; Sekijima, Y.; Yoshida, K.; Tamai, M.; Kosho, T.; Sakurai, A.; Wakui, K.; Ikeda, S.; Fukushima, Y. Follow-up nationwide survey on predictive genetic testing for late-onset hereditary neurological diseases in Japan. *J. Hum. Genet.* **2013**, *58*, 560–563.
91. Doi, H.; Ohba, C.; Tsurusaki, Y.; Miyatake, S.; Miyake, N.; Saitsu, H.; Kawamoto, Y.; Yoshida, T.; Koyano, S.; Suzuki, Y.; *et al.* Identification of a novel homozygous SPG7 mutation in a Japanese patient with spastic ataxia: making an efficient diagnosis using exome sequencing for autosomal recessive cerebellar ataxia and spastic paraplegia. *Intern. Med. Tokyo Jpn.* **2013**, *52*, 1629–1633.
92. UW Laboratory Medicine Clinical Test Information. Available online: http://web.labmed.washington.edu/tests/genetics/SCAPN/ (accessed on 18 April 2014).
93. Athena Diagnostics: Test Catalog. Available online: http://www.athenadiagnostics.com/content/test-catalog/ (accessed on 18 April 2014).
94. Yang, Y.; Muzny, D.M.; Reid, J.G.; Bainbridge, M.N.; Willis, A.; Ward, P.A.; Braxton, A.; Beuten, J.; Xia, F.; Niu, Z.; *et al.* Clinical whole-exome sequencing for the diagnosis of mendelian disorders. *N. Engl. J. Med.* **2013**, *369*, 1502–1511.
95. Hammer, M.B.; Eleuch-Fayache, G.; Gibbs, J.R.; Arepalli, S.K.; Chong, S.B.; Sassi, C.; Bouhlal, Y.; Hentati, F.; Amouri, R.; Singleton, A.B. Exome sequencing: an efficient diagnostic tool for complex neurodegenerative disorders. *Eur. J. Neurol. Off. J. Eur. Fed. Neurol. Soc.* **2013**, *20*, 486–492.
96. Sawyer, S.L.; Schwartzentruber, J.; Beaulieu, C.L.; Dyment, D.; Smith, A.; Warman Chardon, J.; Yoon, G.; Rouleau, G.A.; Suchowersky, O.; Siu, V.; *et al.* Exome sequencing as a diagnostic tool for pediatric-onset ataxia. *Hum. Mutat.* **2014**, *35*, 45–49.
97. Németh, A.H.; Kwasniewska, A.C.; Lise, S.; Parolin Schnekenberg, R.; Becker, E.B.E.; Bera, K.D.; Shanks, M.E.; Gregory, L.; Buck, D.; Zameel Cader, M.; *et al.* Next generation sequencing for molecular diagnosis of neurological disorders using ataxias as a model. *Brain J. Neurol.* **2013**, *136*, 3106–3118.
98. Ohba, C.; Osaka, H.; Iai, M.; Yamashita, S.; Suzuki, Y.; Aida, N.; Shimozawa, N.; Takamura, A.; Doi, H.; Tomita-Katsumoto, A.; *et al.* Diagnostic utility of whole exome sequencing in patients showing cerebellar and/or vermis atrophy in childhood. *Neurogenetics* **2013**, *14*, 225–232.
99. Baylor College of Medicine: Medical Genetics Laboratories. Available online: https://www.bcm.edu/research/medical-genetics-labs/test_detail.cfm?testcode=1500/ (accessed on 18 April 2014).
100. The University of Chicago: Ataxia Exome Panel. Available online: http://dnatesting.uchicago.edu/tests/676/ (accessed on 18 April 2014).

5

Delivery of a Clinical Genomics Service

William G. Newman [1,2,*] and Graeme C. Black[1,2,*]

1 Manchester Centre for Genomic Medicine, University of Manchester, Manchester, M13 9WL, UK
2 Manchester Centre for Genomic Medicine, Central Manchester University Hospitals, NHS Foundation Trust, Manchester, M13 9WL, UK

* Authors to whom correspondence should be addressed;
E-Mails: William.newman@manchester.ac.uk (W.G.N.); Graeme.black@manchester.ac.uk (G.C.B.);

Abstract: Over the past five years, next generation sequencing has revolutionised the discovery of genes responsible for rare inherited diseases previously resistant to traditional discovery techniques. This review considers how this new technology is being introduced into clinical practice to aid diagnosis and improve the clinical management of individuals and families affected by rare diseases where access to genetic testing was previously limited. We compare and contrast the different approaches that have been adopted including panel based tests, exome and genome sequencing. We provide insights from our own clinical practice demonstrating the challenges and benefits of this new technology.

Keywords: next generation sequencing; rare disease; exome; panel testing

---

1. Clinical Genetic Testing for Inherited Disorders

Rare diseases are individually rare but affect a large number of individuals—for example, it is estimated that collectively they affect around 1 in 17 individuals in Western populations [1]. The identification of a specific genetic variant, in a patient DNA sample, that is responsible for a rare inherited disease can establish or confirm a clinical diagnosis, inform screening programmes and the implementation of personalised approaches to medical management. The information also facilitates risk assessment for affected families and enables reproductive decision-making.

Molecular genetic testing for rare diseases has been managed by a small number of expert clinicians, Clinical geneticists, over the past three to four decades, but is now becoming relevant to more patients seen across all clinical specialties. However, for clinicians within such so-called "mainstream" specialties (*i.e.*, outside of clinical genetics) it is often difficult to know how to access genetic testing for their patients. Over the past thirty years, Medical Genetics laboratories have been providing mutation testing for a relatively small number of inherited disorders due to variants in single genes. Of the approximately 7000 rare inherited disorders that have been defined, 3500 have so far been characterized at a molecular level [2]. The Genetic Testing Registry has collated the details on 16,000 tests for 4200 conditions analysing 2800 different genes [3]. The majority of these tests are still undertaken on a research basis. Clinically accredited testing provided by diagnostic laboratories is often limited.

The traditional testing model has been driven by clinical hypotheses (Figure 1, using congenital cataract, a genetically heterogeneous condition, as an exemplar). A clinician usually defines, through detailed clinical investigation, a specific phenotype and subsequently develops a testable clinical hypothesis. The resulting clinical question leads to the request of a specific (usually single gene) test or at most the testing of a very small number of potentially relevant genes. This aims to confirm or refute the clinician's suspicions and historically has been limited in great part by the technological limitations of nucleic acid sequencing. The pick up rate of such a testing approach varies considerably from approximately 0.6% for Fragile X syndrome [4] to over 40% for CHARGE syndrome [5]. In general, this has been a highly targeted approach, that is expensive, iterative and inefficient because of the limited number of target genes that can be tested and by the tendency to institute a large number of simultaneous investigations. By its very nature, it has also been limited to patients, and their relatives, with clinical features indicative of a specific genetic disease. Even where genetic testing is well established in familial breast cancer, genetic testing for *BRCA1* and *BRCA2* mutations has been limited to those with a very strong family history of the condition. Genetic molecular analysis has been especially challenging for genetically heterogeneous conditions, that is those conditions of identical phenotype cause by mutations in a wide range of genes, including intellectual and developmental delay, deafness, retinal dystrophies, congenital cataract, neuropathies and cerebellar ataxias.

2. Next Generation Sequencing as a Diagnostic Tool

In 2009, the first proof of principle studies were published exploring the application of massively paralleled or so called "next generation" sequencing (NGS) to identify the novel causes of rare inherited diseases [6,7]. These conditions had previously not been amenable to standard gene discovery approaches, e.g., *de novo* autosomal dominant disorders could not be refined by linkage analysis and/or candidate gene approaches had proved unsuccessful. The technology and bioinformatic approach demonstrated an extremely powerful ability to identify disease-causing genes from large genomic regions using small patient cohorts. This technology has led to the molecular characterization of numerous rare disorders and has been hailed as a revolution in medical research and practice [8]. NGS has already been applied in many disciplines across medicine, including in microbiology, virology, transplantation medicine and in the identification of acquired (somatic) mutations in tumours. However, this paper considers the use of clinical application of NGS in the molecular diagnosis of rare diseases.

Figure 1. Classical clinical hypothesis-driven diagnostic approach. Traditional investigation of genetically and clinically heterogeneous conditions, such as congenital cataract, require an inefficient and iterative process based upon the development and testing of multiple clinical hypotheses, which leads to testing of many genes in series.

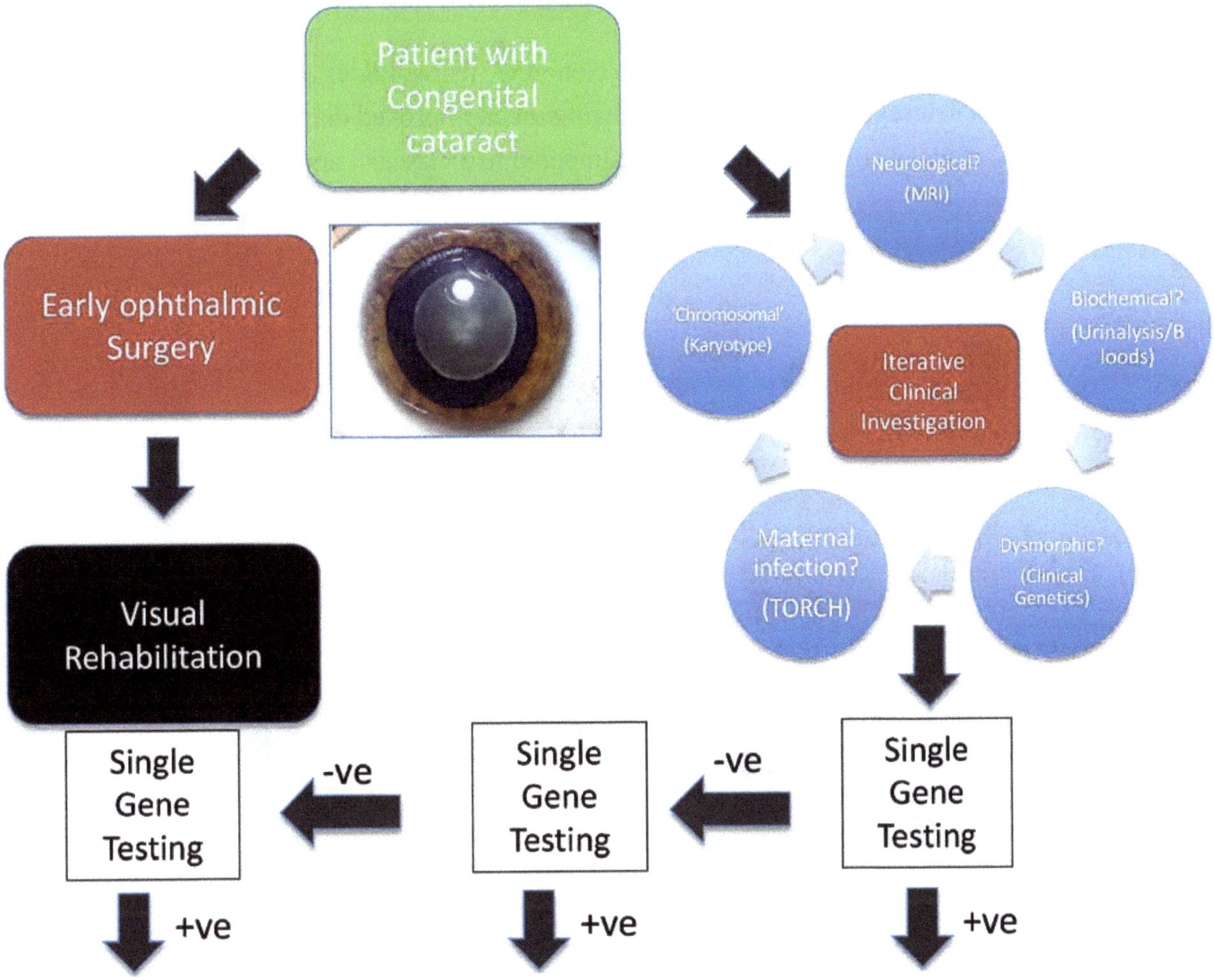

NGS, when first applied to Mendelian disorders focussed on gene discovery where the majority of studies used either approaches focussing of targeted sequencing of genomic regions or most commonly on whole exome sequencing (WES). The WES approach is focused on approximately 1% of the genome, which includes coding and non-coding exons, some intronic and untranslated regions and promoters. It is important to note that the terms whole exome and whole genome sequencing are misnomers as the entire sequence of the exome or genome is not covered using the currently available techniques [9]. Focussing on the protein-coding DNA sequence such an approach generates manageable datasets; although large when compared to conventional sequencing, these are comparatively small when compared to the data from complete genomes. These present challenging, but surmountable, computing challenges [8].

In the clinical setting, the commonest initial approach—that has been introduced by many clinical laboratories—is the targeted sequencing of a panel of genes relevant to a specific disease indication (Figure 2). Here, NGS has already had a major impact. Our own experience with testing of a panel of 105 inherited retinal dystrophy (IRD) genes has seen an increase in detection of the causal variant from

14% to 60% over the past two years of providing this service, allowing earlier implementation of genetic diagnosis and a reduction in the use of other diagnostic options [10]. More recently an "exome" approach to clinical diagnostic NGS sequencing has been adopted due to considerable practical advantages from the ability to develop a single diagnostic pathway for a huge range of clinical indications [11]. Please provide the original file (in ppt or other format) or a copy in tiff format of Figure 2 with high resolution.

Figure 2. Genomic diagnostic approach Genomic technologies allow early genetic investigation of heterogeneous disorders, allowing much improved diagnostic pick-up, early diagnosis and reduced cost of investigation compared to a classical approach.

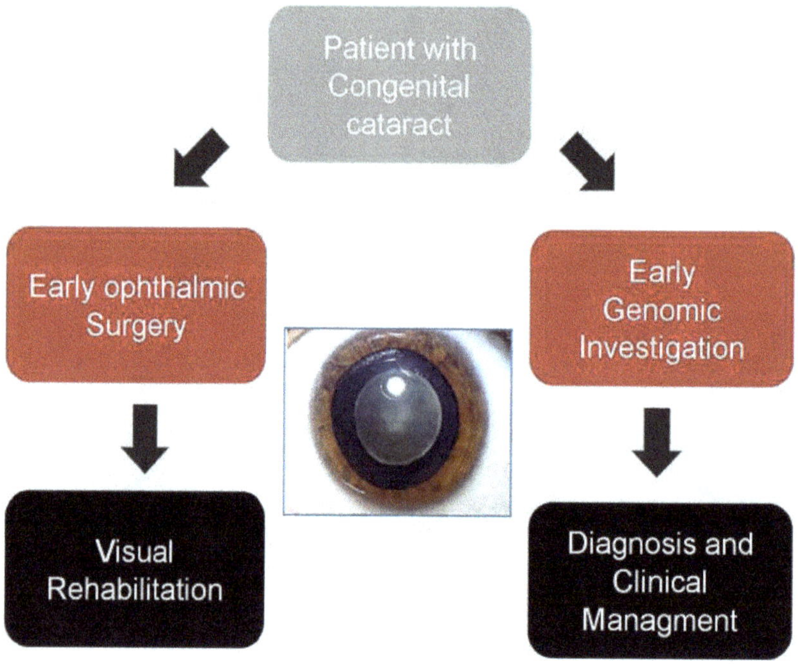

2.1. Targeted Next Generation Sequencing for Diagnostic Molecular Testing

To sequence relatively small numbers of genes many approaches have been introduced to harness the power of NGS to improve throughput, reduce costs and improve turnaround times. For example, long range PCR has been used in our laboratory for *BRCA1* and *BRCA2* mutation analysis to generate large overlapping amplicons, which can then be sequenced. For panels of genes many technologies have been used to target the specific sequences, e.g., amplicon generation, Haloplex and hybridisation capture. Each of these methods has advantages and disadvantages in terms of labor intensity, cost and the specificity of the sequence generated. Further, each method has some limitations when identifying small insertion/deletion mutations has meant that, when using it *as a replacement for Sanger-based diagnosis,* care needs to be taken in using it as an equally effective diagnostic mechanism of excluding mutations in given genes [12].

Testing of panels of genes sequenced by NGS has been introduced with considerable success. The capture of selected sequences has been employed on both a research and diagnostic basis to study groups of genes—focused around biochemical pathways or those known to cause specific phenotypes, usually to analyse 20–200 genes in genetically heterogeneous disorders, such as IRD [10].

Such panels have many advantages over exome-based approaches: since they sequence fewer targets than genome-wide approaches they currently remain cheaper in absolute cost terms, although not when cost per base is used as the basis for evaluation. Panel-based approaches can now achieve even and very high levels of coverage of the targets selected, ≥99% coverage, allowing considerable clarity in diagnostic reporting. Importantly, many of the methodologies for capture-based exome sequencing, in particular at lower levels of overall coverage, have resulted in patchy coverage with a significant dropout of many, in particular GC-rich exons [13]. From a diagnostic viewpoint this dropout has been seen as challenging as it hinders the ability of the clinical scientist to deliver a report that provides a confident definition that a comprehensive screening of the selected genes has been undertaken and that, for exonic and flanking sequences, it is unlikely that a given individual carries a pathogenic variant. Such clarity is important as there are increasing numbers of clinicians, who are unfamiliar with complex genetic terminology and mechanisms, requesting genetic testing to inform their clinical practice.

Gene panel testing has other attractions when applied in the diagnostic sphere. The ability to define the genes that are screened lowers the likelihood that unexpected and potentially actionable findings may be encountered. However, it should be recognised that even amongst panel testing unexpected findings will nonetheless be found: for example, for two panels designed for ophthalmic disorders by our group [10,14], a wide range of conditions are covered by approximately 100–200 genes, many of broad pleiotropic effect [15]. Taking the example of retinal disease, it should be remembered that the ability to diagnose, in those with apparently isolated retinal dystrophy, syndromic conditions such as Senior-Loken and Bardet-Biedl syndromes can be—*for the patient*—unexpected and can result in altered management. When compared to single gene testing this then requires a more detailed approach to consent and counselling when implementing NGS testing.

When compared to exome (WES) or whole genome (WGS) sequencing, gene panel testing thus offers apparent simplicity and has consequently been employed to improve diagnosis of genetically heterogeneous monogenic diseases (e.g., retinitis pigmentosa, congenital deafness, cardiomyopathy). The relatively small numbers of potentially novel or pathogenic variants identified enable a detailed and focussed approach to variant interpretation that is more manageable for clinical scientists analysing and interpreting variants discovered through WES. However, even at this level of complexity, there are challenges presented in patient reporting. For example, in a series of 700 patients with IRD that has been evaluated for variation in 105 genes, in 40 (approximately 12%) of those for whom a molecular cause for their condition was found were shown to be heterozygous carriers of a pathogenic variant in another gene known to cause autosomal recessive IRD (Black; personal communication [16]). Here, the issue of disclosure is not straightforward and requires clear policy decisions by the diagnostic team delivering NGS [17]. Furthermore, it is essential that these policies are complementary to, and understood by, the clinicians consenting to testing. Since families with higher levels of consanguinity may not be identified to clinical scientists, it may be necessary to report all incidental carriers in such circumstances. Such an approach may differ from WES, where the numbers of heterozygous recessive variants present in each individual is high and the identification of carrier status relating to conditions not similar to the primary indication for testing is potentially more complex.

As gene panels are adopted for clinically and phenotypically heterogeneous disorders, it becomes possible for gene testing to be employed earlier in the diagnostic pathway (Figure 2). The breadth of variants that are identified—even in small gene panels—means that interpretation is highly context dependent and in our experience this has led to the development of dialogue between the clinical reporting scientist and the diagnostic clinician. The development of multidisciplinary reporting processes allows sharing of complex phenotypic, family, clinical and genomic data. For example, amongst the multi-systemic diseases that cause congenital cataract, such as cerebroteninous xanthomatosis, Stickler or Cockayne syndromes, genomic discoveries may uncover unexpected or overlooked clinical features that require re-evaluation in the clinic. In addition, genomic discoveries in conditions such as inborn errors of metabolism may define a range of secondary clinical investigations that support genomic findings and facilitate precise diagnosis [18]. In delivering NGS multi-gene panels, the identification of variants of uncertain significance is common. While, from a clinician's viewpoint these cannot necessarily be acted upon, they represent a considerable workload for the team reporting genomic sequencing. At the current time the laboratory methodologies for NGS and the informatics tools to process the data have been honed substantially and have reached a point where this can be relatively easily automated. However, variant interpretation is both gene and phenotype specific. While there may well be certain guiding principles that are generally applicable, nonetheless this remains a labour intensive and complex aspect of NGS panel testing that must be factored into the costs of delivering testing in a healthcare setting. The diagnostic power of gene panel testing via NGS is remarkable and, alongside research developments, has allowed NGS very rapidly to contribute to clinical care. However, in planning the adoption of such processes, the hidden costs of testing are easily overlooked including the need for segregation studies, increased uptake of cascade testing and the need to evaluate the increasing demand for testing. These may be offset, potentially, by reduced adoption of clinical investigations that are superseded by NGS testing, but in many circumstances the costs of such tests are held in separate budgets to other aspects of clinical care. Finally, when considering multigene panels it is important to realise that, while compared to genome-wide approaches there is less data generated and analysed, there is nonetheless a considerable need for IT support, including sufficient computing hardware for data analysis and storage. Ensuring that data governance—in the diagnostic setting—fulfils those required in a healthcare setting immediately places a significant extra financial burden.

2.2. The Use of Genome-Wide NGS Approaches as a Diagnostic Tool

The speed of technological advance in NGS is remarkable, and has led to the technology being described as "disruptive" [19]. Recapitalisation and standardisation of approaches that are key to secure delivery of accredited diagnostics remain challenging in an environment that is yet to fully mature. The panel-based approaches, discussed above, are inherently prone to redundancy as new genes relevant to a particular condition are discovered. As wet lab sequencing and bioinformatic processing and analysis become standardized and provided by increasing numbers of diagnostic laboratories, a single test and pipeline that leads to rapid diagnosis is appealing, with economies of scale and resultant rapid turnaround. Consequently genome-wide approaches, which facilitate sequencing of all known genes, are increasingly seen to be an important step in the delivery of genomic medicine—and we will now consider both exome-based and genome based approaches (Figure 3).

Figure 3. Diagnostic approaches using next generation sequencing.

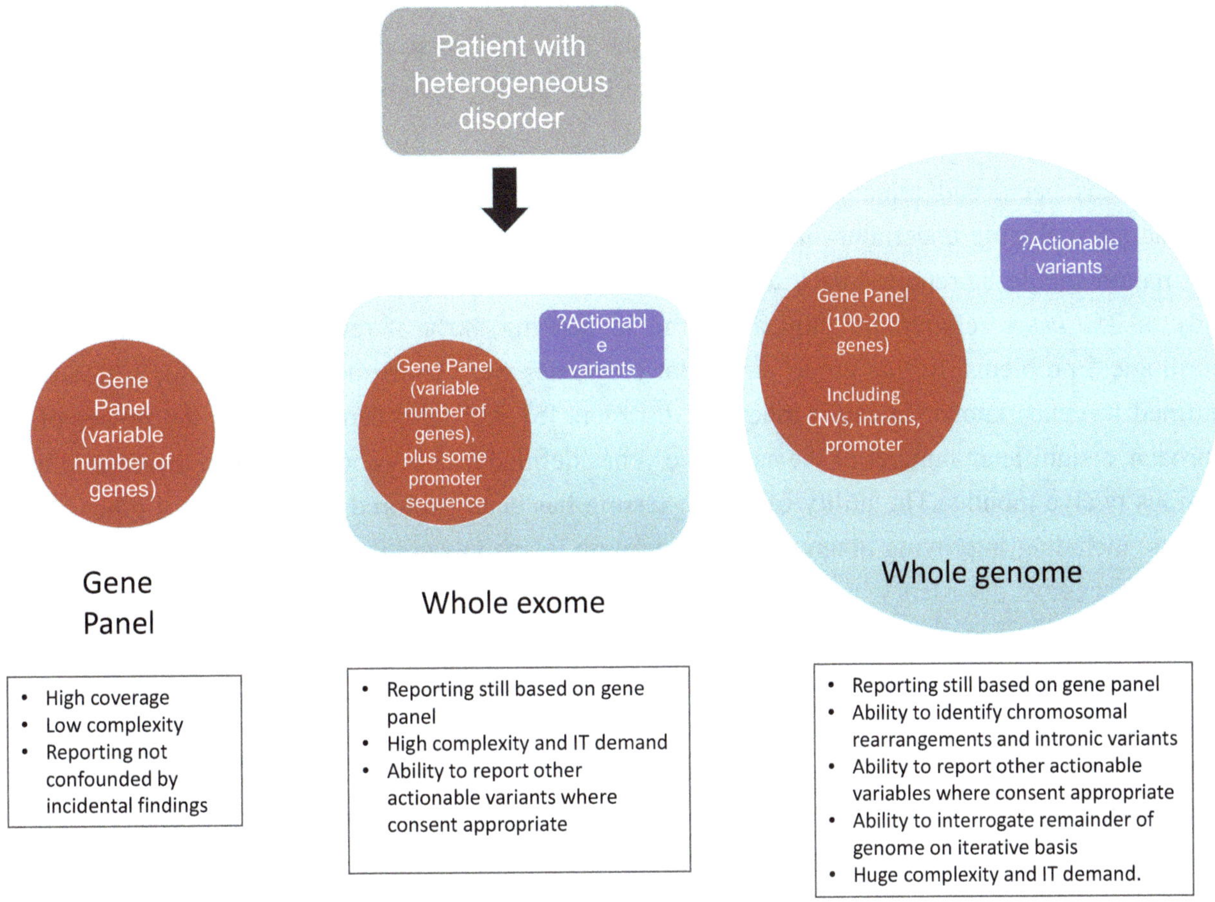

2.2.1. Clinical Exome Sequencing

The ability of NGS to sequencing the entire exome—that is all of the coding exons of the expressed component of the genome, has fuelled gene discovery and accelerated the understanding of the pathogenesis of many monogenic diseases. As a result, *clinical exome sequencing* has been launched at a number of Centres in the United States [5], Australia and Europe and is being actively developed by clinical laboratories across the world [20]. Interestingly, in order to develop workable pipelines and a cost effective manner, at present in most clinical centres clinical reports are generated providing genetic sequencing data that is directly related to the specific phenotype of the tested individual—that is such an approach is based upon an *in silico* panel of genes that are analysed bioinformatically and reported (Figure 3). Such a targeted approach to analysis reduces substantially the cost of analysis, validation and variant interpretation. However, as discussed above, it is important in the consent process that patients and their families understand such focused analytical approaches.

In addition to a focussed approach, extended clinical reports may also be delivered that can provide information about:

(i) Carrier status for a range of recessive disorders to inform future reproductive risks.
(ii) Inherited disorders that are not predicted on the basis of family history or clinical presentation and for which treatment or preventive screening may be appropriate—so-called actionable variants. An example would be the detection of a variant in the low density lipoprotein receptor

(*LDLR*) that would consistent with a diagnosis of familial hypercholesterolemia for which dietary intervention and statin treatment can reduce the risk of cardiovascular disease.

(iii) Pharmacogenetic data that may reduce the risk of adverse drug reactions, e.g., detection of variants in thiopurine methyl transferase that predict adverse response to thiopurines, e.g., azathiopurine.

There has been, and remains, extensive debate about the optimal approach to clinical exome sequencing, including uncertainty in defining the optimal population who should be tested and what information should be reported back to health care professionals and tested individuals. In one recent study of 250 cases referred for clinical exome sequencing, 80% of referrals were of children with neurological problems. In this group molecular diagnoses were confirmed in 62 (25%) with analysis confined to genes known to cause inherited disorders [5]. Of note, demonstrating the power of this approach, a significant number of the causative genes defined in this cohort had been discovered in the previous twelve months. The utility of exome testing has been explored in a number of other clinical settings, including improving diagnosis of children on intensive care units [21] or in children affected by likely recessive disorders when born to consanguineous parents [22].

Overall, exome sequencing lends itself to a high diagnostic yield in a range of clinical scenarios, including the molecular diagnosis of heterogeneous disorders, including primary immunodeficiencies and metabolic disorders. This precise diagnosis will result in reduced expenditure on alternative diagnostic tests and importantly provide patients and parents of affected children with diagnostic certainty. In addition to providing diagnostic information, reports are emerging of exome sequencing that has led to successful changes in clinical management—for example in the diagnosis and treatment of early onset inflammatory bowel disease [23] and in sepiapterin reductase deficiency in twins leading to supplementation of L-dopa therapy with 5-hydroxytryptophan [24].

2.2.2. Whole Genome Sequencing

Whole genome sequencing (WGS) is considered to be the most comprehensive form of genetic test currently available [25]. In contrast to exome sequencing relatively few studies have used (WGS) in rare disease gene discovery. Initially successes have mainly been confined to use of WGS in combination with other sequencing approaches [26] or to identify non-coding mutations that have an effect on genes known previously to cause the specific phenotype [27]. Combination approaches allow refinement of the data analysis from tens of gigabytes to megabyte levels. The control datasets for non-coding variants are less mature and the functional assays to determine the potential phenotypic effects of non-coding variants are challenging to undertake and interpret, such that confident identification of pathogeneic mutations in the non-coding genome for rare diseases remains a formidable challenge. However, WGS presents considerable technological advantages over exome sequencing in that, because it is not based around biased capture-based enrichment approaches, it generates data on an entire genome, often with a consistent average coverage. Consequently, coverage of GC rich regions is improved and there is a considerably improved ability to determine rearrangements and copy number variants. Most recently this has been applied to a cohort of 50 individuals with a diagnosis of severe learning disability (LD) [28], a series of conditions that are associated with extraordinary genetic heterogeneity that are frequently undiagnosed. The conditions can be associated with macroscopic and/or submicroscopic chromosomal rearrangements as well as *de novo* copy number variations (CNVs) and single-nucleotide variations (SNVs). These are currently diagnosed using combined microarray/NGS (targeted panel or exome) sequencing approaches and has been demonstrated that WGS represents a single genetic test that can characterize the full range of genetic variants and enable a clinician to reach a genetic diagnosis in the majority of patients with severe LD.

However, WGS is yet to be introduced widely into routine clinical practice due in large part to the technological and practical hurdles presented by the technology. The generation of terabytes of sequence data that require massive computing capacity to analyse means that WGS is mainly confined to large-scale research or commercial laboratories where it has been applied in disease gene discovery studies. Advances in computing will ensure that WGS will be introduced rapidly over the coming years to supercede both gene panel and WES.

2.2.3. Methodological Considerations of Different NGS Approaches

In adopting genomic technologies—from panel-based testing to WGA—the standardisation and full *clinical* validation of downstream processing will be essential. Here, a challenge is in ensuring that clinicians and clinical scientists are fully aware of the capabilities, limitations and overall design of analysis pipelines. For example, for WES we currently use a library preparation that results in, an average read depth of 140× across the exome which results in 94% coverage of the reference exome at 30× depth and generates approximately 13 Gb of data. Consequently, there is somewhat uneven coverage across many genes, a limitation that is important to stress to clinicians who may need to understand why a negative test may be received. For many capture technologies—that are used for both panel-based NGS approaches and for WES—the ability to assess dosage is limited and means that CNV analysis remains very challenging, potentially requiring reflex dosage testing. This is likely to be a limitation that WGS overcomes.

Analyses of raw data include data generation, collection and processing, followed by application-specific clustering, parsing and visualisation. Here there has been a pragmatic need to adopt research-designed bioinformatic analyses, which are often performed "in-house" using custom freeware designed pipelines for variant calling. Standardisation and full clinical stress testing will be key to ensuring that testing is of high quality, is reliably adopted and also to enabling effective data sharing across different diagnostic centres and platforms.

Variant interpretation remains extremely time-consuming and highly specialised. The process currently relies—in a diagnostic setting—on trained clinical scientists and has, to date, been far less automated than other areas of the NGS pipeline. *In silico* analyses determine whether sequence alterations are predicted to cause disruption of conserved residues. In diagnostic laboratories potential causal variants are often confirmed (currently, at least) by Sanger sequencing and segregation analyses, where possible, are undertaken to provide further evidence of pathogenicity. The definition of novel and pathogenic variants use sequence comparisons of sequences with (i) published data (themselves of highly variable reliability) (ii) databases of known mutations such as the Human Gene Mutation Database or publically available exome data resources such as Exome Variant Server [29] and (iii) the use of in-house databases of exome data. Such a labour intensive process remains important since a trained understanding of the technology and a high index of clinical suspicion can lead to re-evaluation of sequence data to define a causative mutation. For example, in a young child with severe triglyceridemia in whom a heterozygous, previously reported, mutation in *LPL* was identified. There was no sequence variant evident on the second allele to support a diagnosis of the autosomal recessive condition, lipoprotein lipase deficiency. However, the number of sequence reads was diminished across exons 4 and 5 of *LPL* (Figure 4) in comparison with an exome on the same sequence run. Subsequent, cDNA sequencing confirmed a heterozygous deletion of exons 4 and 5, confirming the diagnosis of lipoprotein lipase deficiency.

Figure 4. Copy number variation detected by exome sequencing. Decreased numbers of sequence reads are present in exons 4 (e.g., see arrow) and 5 of *LPL* in the individual with lipoprotein lipase deficiency (top panel) compare to exons 3 and 6 (similar number of reads in patient DNA sequence and that of a control individual below). This indicates a heterozygous deletion of exons 4 and 5 of *LPL*, which is confirmed in the bottom panel by sequencing of cDNA generated from RNA extracted from lymphocytes from the affected individual.

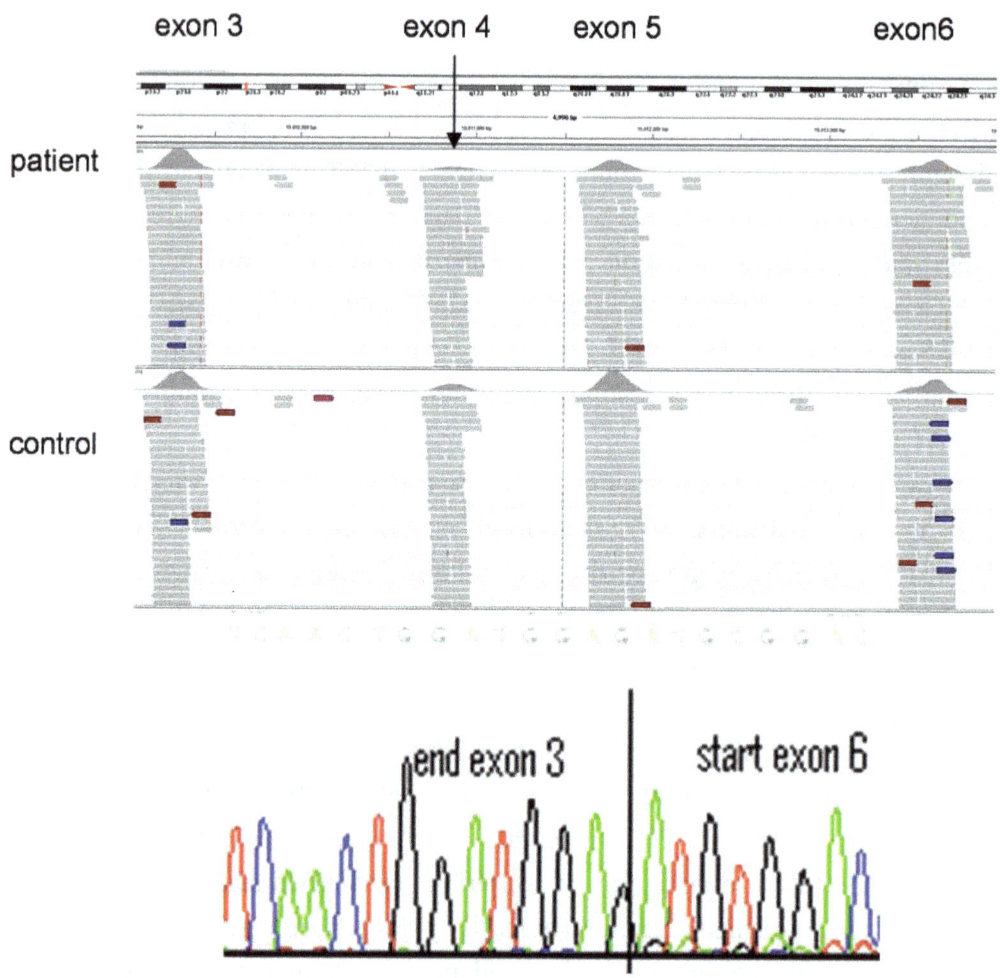

The limitations of current databasing are well understood amongst bioinformaticians and many clinical geneticists but will need to be more widely understood to enable secure variant interpretation across the clinical spectrum [30]. Of course, such resources are becoming more mature and informative as additional data is deposited and the ability to interpret exome-derived data is improving rapidly. By contrast, WGS will generate significant numbers of novel variants which will be difficult to interpret for pathogenicity and indeed it is likely that most early clinical analyses of WGS data will be focused on the *in silico* exome. For a thorough understanding of variant pathogenicity, high throughput functional studies including reporter assays, expression analyses, biochemical tests or *in vivo* assays will need to be developed to complement the emergent sequence data to allow full interpretation. Finally, the strategy around testing only the affected individual or in some scenarios WES or WGS of parents or other relatives (affected/unaffected) may be informative to refine the bioinformatics analysis and reduce the number of potential candidate causative gene variants. A successful example has been the application of a trio sequencing approach of affected child and unaffected parents to identify *de novo* pathogenic mutations, especially for severe congenital/developmental disorders [31].

3. Adoption of Clinical Genomics into Routine Clinical Practice

Next generation sequencing presents an exciting opportunity to revolutionise the diagnosis of rare disease and improve the effectiveness of healthcare delivery across all specialties. A number of specific areas will require focus if this is to be realised in a safe and effective manner:

3.1. Training

NGS is applicable across the healthcare spectrum—that is, it has been shown to be disease-agnostic. It is already proven to be a fundamental tool both clinical and research spheres and also—as recent studies have shown—relevant to both rare and common diseases. For example the next generation sequencing era introduces exciting new possibilities for singling out genetic variants of large effect that contribute to common disease in individuals as demonstrated for age-related macular degeneration [32]. A consequence of this broad relevance will be the opportunity to introduce NGS testing into the mainstream medical disciplines, including cardiology, neurology, and gastroenterology where to date genetic testing has been used less extensively and where the experience in delivering it remains more limited.

A recent survey of over 130 physicians at our Hospital across a number of specialties, including medicine (21%), surgery (13%), paediatrics (18%), anaesthetics (16%), and ophthalmology (7.5%) indicated enthusiasm for exome testing as a diagnostic aid. Over 11% of respondents had already requested an exome and over 53% envisaged requesting a test within the next five years. Limitations of current testing were availability (23%), difficulty with interpretation (47%) and concerns regarding identification of unexpected complex predictive data on cancer or neurodegenerative disease (23%). Such concerns emphasize the importance of clear guidance being established by national professional organisations in concert with patient support groups and other relevant stakeholders. However, experience from the practice of genetic medicine suggests that there is a need for an understanding of genetics, such as mutational mechanism and of genomic architecture and that this is aligned to experience in working closely with families and in delivering the counselling required to ensure effective and safe adoption of testing. Overall, therefore there is an urgent need for training to facilitate the adoption of the types of genomic technologies discussed above. This will need to be applied across all aspects of healthcare, including subspecialty clinicians and counsellors—potentially including those in primary care—who will need to be comfortable in understanding the nature and capability of the tests they order. Furthermore, this creates pressure to increase the numbers of scientists and bioinformatics experts who will be required to process the increasing number of tests.

The comparative youth of NGS is itself an inhibitor to widespread adoption in the clinical arena. In such a rapidly changing environment, the choices of technology and approach are fluid; exome capture technologies continue to improve, WGA costs are reducing and platforms rapidly maturing/becoming obsolete. Many healthcare-facing laboratories have until now been exercised with the decision to invest in the development of panel-based NGS tests or genome based (exome) approaches which are already considered by some to be out-dated. It is likely that the high cost of computing and of capitalisation/ recapitalisation will either favour the larger healthcare organisations, or even lead to widespread outsourcing of sequencing. This is exemplified by the move by the 100000 Genome Project to a centralised and homogenised sequencing approach [33]. Both approaches will have a significant impact on how the technologies are introduced.

3.2. Standardised Phenotyping

The power of new genetic testing technologies to define the causes of rare inherited disorders has been remarkable. However, a limitation to further discovery has been the ability to share data generated on independent families with variants in the same gene with similar or different clinical phenotypes. Such data sharing will facilitate the definition of the ultra-rare conditions which, to date have remained undiagnosed [30]. Many research groups have identified potential causative genetic variants in single families where the burden of proof has not been satisfied to confirm causation as a mutation(s) in a second unrelated family has not been demonstrated. Many international efforts have been initiated to address this issue, including The Human Phenotype Ontology project [34] and databases that allow sharing of clinical and sequence data between clinical research groups, e.g., PhenomeCentral.

3.3. Ethical Issues

A range of complex ethical issues will influence a generalised introduction of genome-wide NGS.

At present, clinical reports from such genomic testing are generated to provide feedback relevant to the presentation of the tested individual. Thus, despite the breadth of genetic information available many centres, including our own, have decided initially to apply a bioinformatic filter based on the phenotypic features of the patient that predefines the panel of genes that will be analysed [21]. Such an approach significantly reduces, but does not abolish, the likelihood of identifying co-incidental genetic variants and speeds up the data analysis.

However, the potential to generate data that identify predisposition to conditions that are not predicted from family history or current health is significant. The American College of Medical Genetics [35] and European Society of Human Genetics [36] have considered how extra information potentially generated from genome analysis should be fed back to individuals. Information about increased risks of cardiac disease, cancer and rare inherited disorders (such as Marfan syndrome) potentially lend themselves to targeted interventions with improved outcomes. However, concerns have been raised about individual autonomy, inappropriate use of this information to discriminate in terms of employment and insurance and the burden placed upon health professionals to feedback accurate information that can have a measurable benefit [35–37].

A key area of future debate will be whether only those genes that are relevant to a specific patient phenotype are assessed and information relating to these fed back to the patient from their clinical exome—and if not, then precisely which so-called "actionable variants" are reported. The use of WES and WGS is a rapidly evolving area of medicine with different views emerging as to how this should be delivered. Our local patient advocate group has indicated that patients are keen for supplemental information that is derived from such testing to be used for patient advantage. However, the anecdotal feedback from patients interviewed in a clinic setting where exome testing has been offered, has suggested more reluctance in this regard.

Lastly, it is important to note the cautionary tales from newborn screening programmes. Tandem mass spectrometry has revolutionized the number of inborn errors of metabolism that can potentially be identified in the newborn period from blood spot analysis. However, results should only be fed back to parents where there is clear evidence of benefit for the newborn child through treatment or altered clinical management, or information that may influence future parental reproductive choices. The natural history of the metabolic disorder should be known, reference should be made to histidinemia and the inappropriate adoption of newborn screening when some children were exposed the risk of liver biopsy despite the condition having a benign natural history [38]. The results of any genome test should be societally and individually acceptable and understandable.

3.4. Economic and Societal Issues

The adoption of NGS—and ultimately WGS—will happen only if the diagnostic yield is sufficient to offset the costs of adoption. The 100000 Genome Project in the UK and similar initiatives across the world will start to address the technical and interpretative challenges posed by WGS and allow comparison with WES. However, it is challenging to measure the benefit of NGS as introduced across a population. Many groups, including our own, have numerous case reports of benefit through the identification of a previously unknown diagnosis. Clinical testing has already been introduced and so undertaking studies to establish improvements in outcome is difficult in this context. Randomized control trials will potentially provide the most compelling evidence of benefit and may be possible for defined groups of conditions, but it will be very challenging to interpret the benefits across heterogeneous groups of rare disorders. Such studies will be increasingly difficult to conduct if genome testing becomes the standard of care. Furthermore there are no universally agreed outcome measures in Genetic Medicine. Standard outcome measures such as the EQ-5D are not likely to capture the potential benefit of genetic testing, as they do not often result in an alteration in any of the measured parameters, e.g., mobility [39]. An alternative to randomized trials will be to make comparisons against historical data to determine potential benefit, but such analyses are beset by potential bias.

The point at which a genome test should be used in the diagnostic pathway is yet to be defined. Should a standard suite of diagnostic tests be used initially and sequencing applied as a second line or for certain clinical indications? Should the NGS test be the first line investigation? Studies to define these pathways are urgently required to ensure appropriate use of resources and to maximise patient benefit. At present genomic tests are used with rather limited scope within medical practice. This may reflect limited education of health care professionals about their utility, a lack of a robust evidence base for their routine adoption into clinical practice; and limited evidence that some genetic tests alter the clinical management.

4. Conclusions

NGS has already transformed the landscape for individuals and families with rare inherited disorders. Conditions previously resistant to research or accurate diagnosis are now the focus of study and amenable to routine diagnosis through panel based approaches or clinical exomes. The advances in genomic sequencing technology and computing will mean that such sophisticated tests will become the standard of care for individuals with rare inherited disorders. The obligation for geneticists and healthcare professionals to harness this genomic revolution for maximum patient benefit is a real one. The ethical, legal and social implications are complex and require an open vibrant dialogue and engagement from all members of society.

Acknowledgements

The authors thank the Manchester Biomedical Research Centre for support.

Author Contributions

The authors contributed equally to the writing of this manuscript.

References

1. Chief Medical Officer. Rare is common. In *Chief Medical Officer. Annual Report of the Chief Medical Officer 2009*; Department of Health: London, UK, 2009; pp. 38–45.
2. National Institutes of Health. Office of Rare Disease Research (ORDR). Available online: http://rarediseases.info.nih.gov/AboutUs.aspx (accessed on 10 August 2014).
3. Genetic Testing Registry (GTR). Available online: http://www.ncbi.nlm.nih.gov/gtr/ (accessed on 10 August 2014).
4. Macpherson, J.; Sawyer, H. Best practice guidelines for molecular diagnosis of Fragile X Syndrome. Available online: http://www.acgs.uk.com/quality-committee/best-practice-guidelines/ (accessed on 10 August 2014).
5. Yang, Y.; Muzny, D.M.; Reid, J.G.; Bainbridge, M.N.; Willis, A.; Ward, P.A.; Braxton, A.; Beuten, J.; Xia, F.; Niu, Z.; *et al.* Clinical whole-exome sequencing for the diagnosis of mendelian disorders. *N. Engl. J. Med.* **2013**, *369*, 1502–1511.
6. Ng, S.B.; Turner, E.H.; Robertson, P.D.; Flygare, S.D.; Bigham, A.W.; Lee, C.; Shaffer, T.; Wong, M.; Bhattacharjee, A.; Eichler, E.E.; *et al.* Targeted capture and massively parallel sequencing of 12 human exomes. *Nature* **2009**, *461*, 272–276.
7. Ng, S.B.; Buckingham, K.J.; Lee, C.; Bigham, A.W.; Tabor, H.K.; Dent, K.M.; Huff, C.D.; Shannon, P.T.; Jabs, E.W.; Nickerson, D.A.; *et al.* Exome sequencing identifies the cause of a mendelian disorder. *Nat. Genet.* **2010**, *42*, 30–35.
8. Biesecker, L.G. Exome sequencing makes medical genomics a reality. *Nat. Genet.* **2010**, *42*, 13–14.
9. Rehm, H.L. Disease-targeted sequencing: A cornerstone in the clinic. *Nat. Rev. Genet.* **2013**, *14*, 295–300.
10. O'Sullivan, J.; Mullaney, B.G.; Bhaskar, S.S.; Dickerson, J.E.; Hall, G.; O'Grady, A.; Webster, A.;

Ramsden, S.C.; Black, G.C. A paradigm shift in the delivery of services for diagnosis of inherited retinal disease. *J. Med. Genet.* **2012**, *49*, 322–326.

11. Gilissen, C.; Hoischen, A.; Brunner, H.G.; Veltman, J.A. Disease gene identification strategies for exome sequencing. *Eur. J. Hum. Genet.* **2012**, *20*, 490–497.
12. Braun, T.A.; Mullins, R.F.; Wagner, A.H.; Andorf, J.L.; Johnston, R.M.; Bakall, B.B.; Deluca, A.P.; Fishman, G.A.; Lam, B.L.; Weleber, R.G.; *et al.* Non-exomic and synonymous variants in ABCA4 are an important cause of Stargardt disease. *Hum. Mol. Genet.* **2013**, *22*, 5136–5145.
13. Harismendy, O.; Ng, P.C.; Strausberg, R.L.; Wang, X.; Stockwell, T.B.; Beeson, K.Y.; Schork, N.J.; Murray, S.S.; Topol, E.J.; Levy, S.; *et al.* Evaluation of next generation sequencing platforms for population targeted sequencing studies. *Genome Biol.* **2009**, *10*, doi:10.1186/gb-2009-10-3-r32.
14. Gillespie, R.L.; O'Sullivan, J.; Ashworth, J.; Bhaskar, S.; Williams, S.; Biswas, S.; Kehdi, E.; Ramsden, S.C.; Clayton-Smith, J.; Black, G.C.; *et al.* Personalized diagnosis and management of congenital cataract by next-generation sequencing. *Ophthalmology* **2014**, doi:10.1016/j.ophtha.2014.06.006.
15. Kocarnik, J.M.; Fullerton, S.M. Returning pleiotropic results from genetic testing to patients and research participants. *JAMA* **2014**, *311*, 795–796.
16. Black, G.C. University of Manchester, Manchester, UK. Personal Communication, 2014.
17. Clarke, A.J. Managing the ethical challenges of next-generation sequencing in genomic medicine. *Br. Med. Bull.* **2014**, *111*, 17–30.
18. Jones, M.A.; Rhodenizer, D.; da Silva, C.; Huff, I.J.; Keong, L.; Bean, L.J.; Coffee, B.; Collins, C.; Tanner, A.K.; He, M.; *et al.* Molecular diagnostic testing for congenital disorders of glycosylation (CDG): Detection rate for single gene testing and next generation sequencing panel testing. *Mol. Genet. Metab.* **2013**, *110*, 78–85.
19. Manyika, J.; Chui, M.; Bughin, J.; Dobbs, R.; Bisson, P.; Marrs, A. Disruptive technologies: Advances that will transform life, business, and the global economy. McKinsey Global Institute, 2013. Available online: http://www.mckinsey.com/insights/business_technology/disruptive_technologies (accessed on 10 August 2014).
20. Jacob, H.J. Next-generation sequencing for clinical diagnostics. *N. Engl. J. Med.* **2013**, *369*, 1557–1558.
21. Saunders, C.J.; Miller, N.A.; Soden, S.E.; Dinwiddie, D.L.; Noll, A.; Alnadi, N.A.; Andraws, N.; Patterson, M.L.; Krivohlavek, L.A.; Fellis, J.; *et al.* Rapid whole-genome sequencing for genetic disease diagnosis in neonatal intensive care units. *Sci. Transl. Med.* **2012**, *4*, doi:10.1126/scitranslmed.3004041.
22. Dixon-Salazar, T.J.; Silhavy, J.L.; Udpa, N.; Schroth, J.; Bielas, S.; Schaffer, A.E.; Olvera, J.; Bafna, V.; Zaki, M.S.; Abdel-Salam, G.H.; *et al.* Exome sequencing can improve diagnosis and alter patient management. *Sci. Transl. Med.* **2012**, *4*, doi:10.1126/scitranslmed.3003544.
23. Worthey, E.A.; Mayer, A.N.; Syverson, G.D.; Helbling, D.; Bonacci, B.B.; Decker, B.; Serpe, J.M.; Dasu, T.; Tschannen, M.R.; Veith, R.L.; *et al.* Making a definitive diagnosis: Successful clinical application of whole exome sequencing in a child with intractable inflammatory bowel disease. *Genet. Med.* **2011**, *13*, 255–262.
24. Bainbridge, M.N.; Wiszniewski, W.; Murdock, D.R.; Friedman, J.; Gonzaga-Jauregui, C.; Newsham, I.; Reid, J.G.; Fink, J.K.; Morgan, M.B.; Gingras, M.C.; *et al.* Whole-genome sequencing for optimized patient management. *Sci. Transl. Med.* **2011**, *3*, doi:10.1126/scitranslmed.3002243.

25. Lupski, J.R.; Reid, J.G.; Gonzaga-Jauregui, C.; Rio Deiros, D.; Chen, D.C.; Nazareth, L.; Bainbridge, M.; Dinh, H.; Jing, C.; Wheeler, D.A.; *et al.* Whole-genome sequencing in a patient with Charcot-Marie-Tooth neuropathy. *N. Engl. J. Med.* **2010**, *362*, 1181–1191.
26. Onoufriadis, A.; Shoemark, A.; Munye, M.M.; James, C.T.; Schmidts, M.; Patel, M.; Rosser, E.M.; Bacchelli, C.; Beales, P.L.; Scambler, P.J.; *et al.* Combined exome and whole-genome sequencing identifies mutations in ARMC4 as a cause of primary ciliary dyskinesia with defects in the outer dynein arm. *J. Med. Genet.* **2014**, *51*, 61–67.
27. Weedon, M.N.; Cebola, I.; Patch, A.M.; Flanagan, S.E.; de Franco, E.; Caswell, R.; Rodríguez-Seguí, S.A.; Shaw-Smith, C.; Cho, C.H.; Lango A.H.; *et al.* Recessive mutations in a distal PTF1A enhancer cause isolated pancreatic agenesis. *Nat. Genet.* **2014**, *46*, 61–64.
28. Gilissen, C.; Hehir-Kwa, J.Y.; Thung, D.T.; van de Vorst, M.; van Bon, B.W.; Willemsen, M.H.; Kwint, M.; Janssen, I.M.; Hoischen, A.; Schenck, A.; *et al.* Genome sequencing identifies major causes of severe intellectual disability. *Nature* **2014**, *511*, 344–347.
29. Exome Variant Server, NHLBI GO Exome Sequencing Project (ESP). Available online: http://evs.gs.washington.edu/EVS/ (accessed on 10 August 2014).
30. Gottlieb, B.; Beitel, L.K.; Trifiro, M. Changing genetic paradigms: Creating next-generation genetic databases as tools to understand the emerging complexities of genotype/phenotype relationships. *Hum. Genomics* **2014**, *8*, doi:10.1186/1479-7364-8-9.
31. Veltman, J.A.; Brunner, H.G. *De novo* mutations in human genetic disease. *Nat. Rev. Genet.* **2012**, *13*, 565–575.
32. Raychaudhuri, S.; Iartchouk, O.; Chin, K.; Tan, P.L.; Tai, A.K.; Ripke, S.; Gowrisankar, S.; Vemuri, S.; Montgomery, K.; Yu, Y.; *et al.* A rare penetrant mutation in CFH confers high risk of age-related macular degeneration. *Nat. Genet.* **2011**, *43*, 1232–1236.
33. Torjesen, I. Genomes of 100,000 people will be sequenced to create an open access research resource. *Br. Med. J.* **2013**, *347*, doi:10.1136/bmj.f6690.
34. Köhler, S.; Doelken, S.C.; Mungall, C.J.; Bauer, S.; Firth, H.V.; Bailleul-Forestier, I.; Black, G.C.; Brown, D.L.; Brudno, M.; Campbell, J.; *et al.* The Human Phenotype Ontology project: Linking molecular biology and disease through phenotype data. *Nucleic Acids Res.* **2014**, *42*, D966–D974.
35. Green, R.C.; Berg, J.S.; Grody, W.W.; Kalia, S.S.; Korf, B.R.; Martin, C.L.; McGuire, A.L.; Nussbaum, R.L.; O'Daniel, J.M.; Ormond, K.E.; *et al.* ACMG recommendations for reporting of incidental findings in clinical exome and genome sequencing. *Genet. Med.* **2013**, *15*, 565–574.
36. Van El, C.G.; Cornel, M.C.; Borry, P.; Hastings, R.J.; Fellmann, F.; Hodgson, S.V.; Howard, H.C.; Cambon-Thomsen, A.; Knoppers, B.M.; Meijers-Heijboer, H.; *et al.* Whole-genome sequencing in health care: Recommendations of the European Society of Human Genetics. *Eur. J. Hum. Genet.* **2013**, *21*, 580–584.
37. Green, R.C.; Lupski, J.R.; Biesecker, L.G. Reporting genomic sequencing results to ordering clinicians: Incidental, but not exceptional. *JAMA* **2013**, *310*, 365–366.
38. Levy, H.L.; Shih, V.E.; Madigan, P.M. Routine newborn screening for histidinemia. Clinical and biochemical results. *N. Engl. J. Med.* **1974**, *291*, 1214–1219.
39. Payne, K.; McAllister, M.; Davies, L.M. Valuing the economic benefits of complex interventions: When maximising health is not sufficient. *Health Econ.* **2013**, *22*, 258–271.

6

Discovery in Genetic Skin Disease: The Impact of High Throughput Genetic Technologies

Thiviyani Maruthappu †, **Claire A. Scott** † **and David P. Kelsell** *

Centre for Cutaneous Research, Blizard Institute, Barts and The London School of Medicine and Dentistry, Queen Mary University of London, 4 Newark Street, London E1 2AT, UK; E-Mails: t.maruthappu@qmul.ac.uk (T.M.); c.a.scott@qmul.ac.uk (C.A.S.)

† These authors contributed equally to this work and should be considered joint first authors.

* Author to whom correspondence should be addressed; E-Mail: d.p.kelsell@qmul.ac.uk;

Abstract: The last decade has seen considerable advances in our understanding of the genetic basis of skin disease, as a consequence of high throughput sequencing technologies including next generation sequencing and whole exome sequencing. We have now determined the genes underlying several monogenic diseases, such as harlequin ichthyosis, Olmsted syndrome, and exfoliative ichthyosis, which have provided unique insights into the structure and function of the skin. In addition, through genome wide association studies we now have an understanding of how low penetrance variants contribute to inflammatory skin diseases such as psoriasis vulgaris and atopic dermatitis, and how they contribute to underlying pathophysiological disease processes. In this review we discuss strategies used to unravel the genes underlying both monogenic and complex trait skin diseases in the last 10 years and the implications on mechanistic studies, diagnostics, and therapeutics.

Keywords: genodermatoses; exome; ichthyosis; atopic eczema; psoriasis

---

1. Introduction

The advent of high throughput single nucleotide polymorphism (SNP) genotyping and latterly, next generation sequencing (NGS) technology including whole exome sequencing (WES) have revolutionised our approach to genetic diagnostics and novel gene discovery in the genodermatoses—a group of inherited skin disorders.

Prior to this, technologies including linkage analysis using genome wide microsatellite panels in combination with candidate gene screening by PCR and Sanger sequencing have been the primary method for discerning new skin disease-associated loci. Successes with this approach include Hailey-Hailey Disease (OMIM #169600) [1], Netherton Syndrome (OMIM #256500) [2], Darier-Disease (OMIM #124200) [3], and Dyschromatosis symmetrica hereditaria (OMIM #127400) [4]. Candidate gene screening approaches have also yielded success, particularly in deciphering the keratin disorders [5]. However, clinical and likely genetic heterogeneity of skin diseases and the availability of DNA from probands only, or from small families, have hindered disease gene discovery for many disorders [6]. This can now be surmounted with high-density SNP homozygosity mapping for consanguineous recessive disorders, and in particular NGS and WES for dominant and recessive disorders, which has facilitated our understanding of some of the genetic make up of common diseases.

Skin diseases are ideal for determining genotype-phenotype correlations because of the relative ease with which clinical and histological examination can be made. In addition, inflammatory pathways involved in the pathogenesis of skin diseases such as psoriasis vulgaris (PV) are relevant to a number of other immune-mediated diseases including inflammatory bowel disease and rheumatoid arthritis [7]. The genetic bases of many monogenic skin diseases have been unravelled and in this review we focus on examples of discoveries in cutaneous genetics, applying different strategies such as SNP microarray, microsatellite linkage analysis, targeted NGS and WES. Equally, it has also been informative in understanding the significance of *de novo* mutations including the unusual phenomenon of revertant mosaicism in the skin, where spontaneous correction of a disease-causing mutation in a somatic cell occurs [8]. We have also gained insights into complex trait diseases and will explore what contributions these have made to mechanistic insights, diagnosis and treatment of common skin diseases including psoriasis, atopic dermatitis (AD) (eczema), and skin cancer.

2. Harlequin Ichthyosis

The discovery that *ABCA12* gene mutations are associated with the skin disease harlequin ichthyosis (HI) is an example of where SNP microarray technology was used successfully to elucidate the genetic locus associated with this disease [9].

The inherited ichthyoses are a heterogeneous group of disorders characterised by skin scaling, often of the whole surface, and hyperkeratosis [10]. Syndromic (affecting multiple tissues) as well as nonsyndromic forms of ichthyosis exist and mutations in multiple genes are associated with disease including *TGM1* (OMIM *190195), *NIPAL4* (OMIM *609383), *STS* (OMIM *300747), *ALOX12B* (OMIM *603741), *ALOXE3* (OMIM *607206), *CYP4F22* (OMIM *611495), and *FLG* (OMIM *135940) amongst others (reviewed in [10]). Autosomal recessive congenital ichthyosis (ARCI) is comprised of three main groups: congenital ichthyosiform erythroderma (CIE), lamellar ichthyosis (LI) and HI [10]. HI (OMIM #242500) is the most severe form of ichthyosis and has a high perinatal mortality, with babies presenting at birth with hard scale plates with deep fissures, eclabium, and bilateral ectropion (reviewed in [9, 11]).

The discovery of the genetic cause of HI was hampered by availability of DNA from only affected family members or from small families due to the severity of the condition, thus genetic linkage studies were unfeasible [9]. To investigate the genetic basis of HI, Kelsell *et al.* (2005) [9] used a SNP microarray to map a block of homozygosity on chromosome 2q35 and to identify a minimal region between HI patients from consanguineous parents, which contained the *ABCA12* gene. ABCA12 belongs to the ATP-binding cassette (ABC) A family of transporters, some members of which have been implicated in lipid transport (reviewed in [12]).

ABCA12 was a promising gene candidate for HI because patient skin displayed aberrant lipid distribution [9] and missense mutations in *ABCA12* were already known to be associated with another form of ARCI, LI [13]. PCR and Sanger sequencing of the *ABCA12* gene in HI patients confirmed that recessive mutations were associated with HI [9,14]. Mutations in *ABCA12* are now known to be associated with all three forms of ARCI (reviewed in [10]). However, unlike for LI and CIE, in which largely missense *ABCA12* mutations are associated with disease [13,15,16], HI is usually associated with loss of function gene mutations including nonsense, frameshift, and splice site mutations, which severely disrupt the cellular functions of ABCA12 [9,17–19]. However, there are reports of patients who have ABCA12 missense mutations [9,11,18,20–22]. HI patients with homozygous loss of function mutations have an increased risk of mortality, indicating a survival advantage for patients with compound heterozygous mutations [11].

ABCA12 is thought to transport lipids via lamellar granules where they are processed and released to form lipid lamellae constituting the stratum corneum in the epidermis [14,23]. A reduction in the number, and structural abnormalities, of lamellar granules has been observed in HI patient skin [14,24,25]. In addition, characterisation of HI patient skin has shown a loss of nonpolar lipids [26] and abnormal glucosylceramide localization [14], and experiments with patient-derived keratinocytes showed aberrant glucosylceramide accumulation in lamellar granules [27], which is indicative of a lipid transport defect as a result of loss of ABCA12 function [14,26,27].

Similarly, *Abca12* knockout mice models [28–30] and an *abca12* knockout zebrafish model [31] showed features of aberrant lipid transport compared to controls (reviewed in [32]). HI skin also shows features of premature terminal differentiation and a decreased expression of certain proteases, which suggests that loss of ABCA12 disrupts keratinocyte differentiation and epidermal desquamation, resulting in the formation of an aberrant epidermal barrier [26].

Prior to the discovery of the genetic cause of HI, prenatal diagnostic investigations depended on obtaining a foetal biopsy for analysis by electron microscopy, and on sonography [33,34] (reviewed in [35]). The discovery of the genetic cause of different ichthyoses, including HI, represents a major milestone in the ability to perform genetic diagnosis, carrier screening, genetic counselling, and prenatal diagnosis.

Current approaches to genetic screening for HI can involve screening specific exons, as there are some recurrent ethnic group mutations in *ABCA12* [18,19] and using WES, circumventing the need for performing PCR and Sanger sequencing of all 53 coding exons of the *ABCA12* gene.

3. Exfoliative Ichthyosis

The discovery of cystatin A (*CSTA*) gene mutations in association with exfoliative ichthyosis [36] is an example of the successful implementation of combining SNP microarray analysis with targeted NGS to determine the genetic cause of disease.

Autosomal recessive exfoliative ichthyosis (OMIM #607936) is characterised by palmoplantar skin peeling and dry scaly skin, with trauma and moisture aggravating the condition [36]. Microsatellite linkage analysis of two related Bedouin families initially suggested linkage of the disease to chromosome 12q13, which contains the type II keratin cluster [37].

Blaydon *et al.* (2011) [36] revisited this family and applied whole genome homozygosity mapping which revealed a common block of homozygosity between affected Bedouin patients on chromosome 3q21 as the likely disease gene location. Sequence capture and NGS of this region was then performed and revealed a splice site mutation in *CSTA*, which was found to segregate with exfoliative ichthyosis in the Bedouin family. This locus was missed in the microsatellite genome scans performed by Hatsell *et al.* (2003) [37] due to markers for this region being uninformative. Sanger sequencing of *CSTA* in a different family with exfoliative ichthyosis revealed a homozygous nonsense mutation which also segregated with disease [36]. In a subsequent study, WES revealed a novel homozygous nonsense mutation in *CSTA* in a large family with acral peeling skin syndrome [38] with similar clinical features to the patients reported in Blaydon *et al.* (2011) [36].

Cystatins are cysteine protease inhibitors which are thought to have a protective function against endogenous and external proteases, and to potentially modulate the degradation of intra- and extracellular proteins (reviewed in [39]). CSTA has been identified as a constituent of the cornified envelope [40] and is expressed in the suprabasal layers of the epidermis, the highest expression of which is in the granular layer [36,41]. CSTA is secreted by keratinocytes *in vitro* and has also been found in sweat, and is believed to have a protective role by inhibiting the proteolytic activity of dust mite allergens Der p 1 and Der f 1 [42]. CSTA levels have also been implicated as prognostic markers in different cancers [43–45].

Characterisation of skin from exfoliative ichthyosis patients with *CSTA* mutations revealed widened intercellular gaps in the lower epidermis, whereas the upper epidermal layers appeared normal with no evident barrier defect [36]. Experiments using an *in vitro* keratinocyte cell knockdown model showed an adhesion defect in response to mechanical stress, and an organotypic *CSTA* knockdown model showed similar abnormalities to the patient skin [36]. This finding is indicative of CSTA having a key role in keratinocyte adhesion in the basal epidermal layers and that loss of CSTA causes a predisposition to epidermal splitting. There were no obvious abnormalities in a murine model with a chromosomal deletion, which included the *Csta* gene [46], although investigation of a skin phenotype was not described.

4. Olmsted Syndrome

The genetic basis of various skin diseases (Table 1) has been determined using exome sequencing technology. One example where WES enabled the identification of the underlying causative genes is Olmsted syndrome (OS) [47,48]. OS (OMIM #614594) is a rare disorder characterised by mutilating palmoplantar keratoderma and periorificial keratosis. Additional clinical features include constriction of the digits, dystrophy of the nails, diffuse alopecia and a predisposition to infection and development of squamous cell carcinoma on keratotic lesions [47]. Different modes of inheritance have been hypothesised [47–50].

Table 1. Examples of genes associated with skin disease discovered using exome sequencing technology.

Gene	Disease	Mode of Inheritance	Reference
AAGAB	Punctate palmoplantar keratoderma Type I	AD	[51,52]
ADAM10	Reticulate acropigmentation of Kitamura	AD	[53]
AQP5	Nonepidermolytic palmoplantar keratoderma	AD	[54]
ENPP1	Cole disease	AD	[55]
EXPH5	Inherited skin fragility	AR	[56]
HOXC13	Pure hair and nail ectodermal dysplasia	AR	[57]
KANK2	Palmoplantar keratoderma and woolly hair	AR	[58]
MBTPS2	Olmsted syndrome	XLR	[48]
POFUT1	Dowling-Degos disease	AD	[59]
POGLUT1	Dowling-Degos disease	AD	[60]
SERPINB7	Nagashima-type palmoplantar keratosis	AR	[61]
TRPV3	Olmsted syndrome	AD/AR	[47]/[62]

AD: autosomal dominant; AR: autosomal recessive; XLR: X-linked recessive.

WES was used successfully to identify mutations in the Transient Receptor Potential Cation Channel, Subfamily V, Member 3 (*TRPV3*) gene [47], and the Membrane-Bound Transcription Factor Protease, Site 2 (*MBTPS2*) gene [48] to be associated with OS.

Lin *et al.* (2012) performed WES of an OS patient and her unaffected parents and identified a novel *de novo* heterozygous mutation p.G573S in *TRPV3* [47]. Screening for *TRPV3* mutations in five other OS patients revealed that three were heterozygous for p.G573S, one heterozygous for p.G573C and one heterozygous for p.W692G [47].

TRPV3 is a member of the TRPV cation channel family, and is known to be expressed in various tissue types including skin and hair follicles [63–65]. The murine *TRPV3* mutants p.G573S and p.G573C were discovered in spontaneous hairless rodent strains that develop dermatitis, a trait inherited in an autosomal dominant manner [66]. *Trpv3* knockout mice display wavy hair, curly whiskers and a defective skin barrier, and it is believed that TRPV3 associates with TGF-α/EGFR in a signalling pathway to modulate keratinocyte differentiation and hair morphogenesis [67].

In vitro functional studies with the three OS-associated TRPV3 mutants indicated that they are gain of function mutants, creating constitutively open channels and causing increased cell death of cells expressing the mutants [47]. Similar results were obtained in *in vitro* expression studies with the murine TRPV3 mutants p.G573S and p.G573C [68]. It has been hypothesised that *in vivo* the mutants may cause apoptosis and subsequent keratoderma in patients, and could contribute to their pruritis [47].

A subsequent study using WES revealed the recurrent *TRPV3* mutation p.G573S in sporadic OS [69]. Screening by Sanger sequencing has also revealed a homozygous mutation in an OS patient, indicating recessive inheritance [62]. Both recessive [70] and sporadic [71] *TRPV3* mutations have been associated with atypical OS with erythromelalgia.

Exome sequencing of two affected males reported previously in a consanguineous pedigree [72] in which OS followed a suggested X-linked recessive inheritance pattern, revealed a novel *MBTPS2* gene mutation which segregated with disease in the family [48]. This discovery expands the number of disorders attributed to *MBTPS2* gene mutations, as other mutations in this gene are associated with ichthyosis follicularis with atrichia and photophobia (IFAP) syndrome [73–75], BRESEK/BRESHECK syndrome [76], and keratosis follicularis spinulosa decalvans (KFSD) [77].

MBTPS1 and MBTPS2 are involved in activating signalling proteins such as the transcription factors SREBPS, enabling cells to respond to sterols [78,79] and in the processing of ATF6, which is a component of the unfolded protein response (UPR) [80]. *In vitro* functional studies with IFAP and KFSD MBTPS2 mutants revealed decreased sterol responsiveness compared to wild-type [73,77], and mutants which caused the greatest impairment of enzyme activity seemed to be associated with increased disease severity in patients [73].

5. Complex Traits of the Skin

In the last 10 years there have been landmark discoveries in our understanding of the genetic basis and pathophysiology of inflammatory skin diseases, most notably PV and AD. Both are common, complex diseases, in which a host of environmental factors can trigger disease in genetically susceptible individuals [81,82]. Inflammatory dermatoses are associated with both a significant burden on healthcare resources and patients' quality of life [83,84].

Identification of susceptibility loci for PV and AD have resulted from developments in genome wide association studies (GWAS), which have been applied to all common disorders. Information has been generated by the HapMap and 1000 Genomes projects, in parallel with the technology to genotype multiple individual DNA samples at one million or more loci, allowing SNPs to be reviewed and enabling comparisons of allele frequency between large numbers of cases and controls to identify those which confer risk of disease [85]. The development of DNA microarray based genotyping allows up to a million SNPs to be tested simultaneously.

6. Psoriasis

PV is a common and chronic inflammatory disease, which can affect the skin, nails and joints. It is characterised by immune-mediated epidermal hyperproliferation [86]. It is a highly heritable disease, with increased concordance in monozygotic *versus* dizygotic twins (65%–72% *versus* 15%–30% respectively) [87]. During the last 10 years, almost 40 GWAS-identified novel psoriasis-susceptibility loci have been identified and more recently, the genes within these loci and their significance to the pathophysiology of PV are becoming clearer [88]. Interestingly, several show clustering to a distinct segment of the inflammatory cascade [89]. Psoriasis susceptibility locus 1 (*PSORS1*), located on the MHC region on chromosome 6p21, has been most consistently identified in GWAS with a significant odds ratio of 3.0 [90]. Genes implicated within this 250 kb interval include *HLA-C* (human leukocyte antigen C), *CCHCR1* (coiled-coil α-helical rod protein 1), and *CDSN* (corneodesmosin). These were considered as potential disease-associated genes due to their function and the presence of disease-associated SNPs within their coding sequence [91]. Identification of the causal disease susceptibility allele was extremely challenging, ultimately Nair *et al.* (2006) sequenced the entire *PSORS1* region in individuals bearing different *HLA-C* alleles to identify SNPs unique to the *PSORS1* haplotype. They indicated that *HLA-Cw6* was the major *PSORS1* disease allele [92], reflecting the importance of antigen presentation in the pathophysiology of PV.

Identification of susceptibility loci has contributed to our understanding of PV pathogenesis, which appears to involve the innate and adaptive immune responses. Pathways that have been identified in various studies include IL12/IL17 axis activation (*IL23R*, *IL12B*, *IL23A*, and *TRAF31P2*), type 1 interferon induction (*IFIH1*, *RNF114*, and *TYK2*) and NF-κB signaling (*CARD14*, *REL*, *NFKBIA*, *TNFAIP3*, and *TNF1P*) [89,90,92–97]. Of particular interest is the Th1-Th17 axis, involving the recently described subset of IL17 expressing T cells (Th17) [98] which is thought to play a major role in the development and maintenance of psoriatic plaques [97].

IL12 and IL23 are cytokines that induce naïve $CD4^+$ lymphocytes to differentiate into type 1 helper cells and type 17 helper cells, both of which are key mediators of PV [97]. IL12 and IL23 share a common p40 subunit encoded by the *IL12B* gene. In mice, injection of IL23 results in epidermal hyperplasia, which is mediated by IL22 produced by Th17 cells. This shows similarities to phenomena observed in humans [99]. GWAS have identified three SNPs with strong evidence of association with PV mapping near *IL12B*, *IL23A* (encoding the p19 subunit of IL23) and *IL23R* (encoding a subunit of the IL23 receptor) [94] raising the possibility that dysregulated IL23 signaling could lead to chronic immune responses within epithelial cells. Ustekinumab (Stelara®) is a human IgG1κ monoclonal antibody against the p40 subunit of the IL12 and IL23 cytokines that has demonstrated significant improvement in outcome measures for the treatment of PV in Phase III clinical trials [100]. A significant proportion of patients had at least 90% improvement in their psoriasis area-and-severity index (PASI) score, with a proportion experiencing complete clearance by 12 weeks [100]. These findings also establish a central role for the IL12/IL23 p40 cytokines in the pathophysiology of PV.

Another approach to utilise the discoveries gained from GWAS studies is personalised medicine. For example, patients with PV who carry risk variants in *IL12B* may benefit preferably from a monoclonal antibody targeting its p40 subunit, e.g., Ustekinumab. Studies using molecular profiling of PV and clinical phenotyping to predict treatment response have shown promise [101] and larger studies are underway. This is one example of how PV has been used as a paradigm for autoimmune disease and for proof-of-principle studies of targeted biologic therapies, because of the ease of accessing the skin and objectively measuring disease severity and responses to treatment.

Rare variants with large effect have been observed in families where PV segregates as an apparent Mendelian trait. The psoriasis susceptibility locus 2 (*PSORS2*) was first mapped in 1994 to human chromosomal region 17q25-qter in a large family of European ancestry [102]. More recently, it has been shown that *PSORS2* is due to gain of function mutations in the caspase recruitment domain family member 14 (*CARD14*) [96] using linkage analysis, targeted and exome capture in combination with NGS. On the basis of these findings, further work has uncovered rare missense variations in *CARD14* linked to PV using a large case-control study [95]. *CARD14* encodes a NF-κB activator within the skin epidermis. The mutations identified lie within the coiled-coiled domain of *CARD14* and result in enhanced NF-κB activity compared with wild-type CARD14 [95].

Generalised pustular psoriasis (GPP) can present with an acute, widespread and life-threatening eruption associated with fever and leukocytosis. It has long been considered a variant of PV. Mutations in *IL36RN*, which encodes the IL36 receptor antagonist and abrogates downstream activation of NF-κB signaling, have been shown to underlie GPP in consanguineous pedigrees of North African origin [103]. This mutation results in enhanced production of IL1, IL6, and IL8 inflammatory cytokines, which may contribute to the profound systemic inflammatory response seen clinically in these patients [103]. Similar recessive mutations in *IL36RN* have not been observed in patients with PV alone [104]. Genetic studies suggest that in fact, PV and GPP are etiologically distinct clinical entities, which consequently have important therapeutic implications [105].

7. Atopic Dermatitis (Eczema)

AD is a chronic inflammatory skin disease characterised by disturbed skin barrier function and dry, itchy skin. Its prevalence worldwide is increasing and in some countries affects almost 20% of children [106]. Like PV, concordance is observed in twin studies with rates of 0.72–0.86 in monozygotic and 0.21–0.23 in dizygotic twin pairs [107]. A complex interplay between environmental, genetic and immunological factors, as for many common disorders, all contribute to susceptibility and severity.

The filaggrin story is central to our understanding of AD and ichthyosis vulgaris (IV). It exemplifies how the study of a monogenic disorder can translate to a complex trait disease. In 2006, null mutations in the filaggrin gene *FLG* were first identified in Irish families with IV, which often causes dry, scaly skin and is also a strong genetic risk factor for AD [108]. Histological evidence for the possible lack of filaggrin in IV dates back to 1985 [109] however these preliminary studies were hindered by the daunting size and repetitive nature of *FLG*, particularly exon 3. The McLean group developed a successful strategy to analyse this locus with the use of long range PCR to amplify exon 3 in combination with short specific PCRs to amplify remaining overlapping fragments that were then used to reconstruct the repetitive sequence [108]. Further research has identified significant associations of *FLG* mutations with atopic asthma, allergic rhinitis and peanut allergy [110], as well as early onset and increased severity of AD [111]. These studies have been reproduced in a variety of geographical populations, including European, Japanese, Taiwanese, Chinese, and Korean [112–114]. Indeed, the correlation between *FLG* mutations and AD is considered one of the most robust examples of genotype-phenotype relationship in complex trait disease with an odds ratio of up to 13.4 [115].

Filaggrin plays a key role in epidermal barrier function. Briefly, its degradation products act as "natural moisturising factors" in the skin and assist the formation of a flattened granular cell layer upon keratinocyte terminal differentiation [116]. Studies describing murine models of filaggrin haploinsufficiency have shown skin barrier impairment and enhanced sensitisation to percutaneous allergens [117,118]. The significant effect of *FLG* mutations on AD risk highlights the role of impaired skin barrier function in the pathogenesis of atopic diseases. Filaggrin replacement therapies could prove significant in the management of AD. Recently, Otsuka *et al.* (2014) [119] identified a novel compound JTC801, with potential therapeutic applicability. This has been shown to increase expression of filaggrin in both human and murine keratinocytes and, when administered orally, it can hinder the development of AD-like inflammation in the NC/nga AD mouse model [119].

Although the AD spotlight has focused largely on filaggrin, several other genes have been implicated in the pathogenesis of this disorder. To date, a total of 19 genome-wide significant ($p < 5 \times 10^{-8}$) susceptibility loci have been identified through GWAS [120]. The first GWAS data was published in 2009 and included 939 cases and 975 controls in addition to 270 complete nuclear families with two affected siblings [121]. It identified a novel susceptibility locus in 11q13.5, located 38 kb downstream of *C11orf30*. The peak association was observed 68 kb upstream of the leucine rich repeat containing 32 gene (*LRRC32*) which has been shown to be expressed in activated human regulatory T cells [122]. Carriers have a risk of developing AD that is 1.47 times that of controls [121]. A 2011 Meta analysis of GWAS for AD included 5606 cases and 20565 controls and an additional 5419 cases and 19833 controls in a validation study [114]. Three novel risk loci reached genome-wide significance: rs479844 upstream of ovo-like zinc finger 1 (*OVOL1*), rs2164983 near actin-like 9 (*ACTL9*) and rs2897442 in kinesin family member 3A (*KIF3A*). They also confirmed association with the *FLG* locus. *OVOL1* disruption in mice leads to keratinocyte hyperproliferation and hair shaft abnormalities [93]. It is thought to play a role in regulating epidermal proliferation and loricrin expression, impairing premature terminal differentiation [123]. *KIF3A* associated SNPs map within a cluster of cytokine and immune mediated genes including Th2 cytokine genes: *IL13* and *IL4*. These cytokines have been implicated in other autoimmune and inflammatory diseases including PV [124], Crohn's Disease [125] and asthma [125]. Increased levels of Th2 cytokines such as these have been reported in AD as well as greater levels of mRNA expression in acute skin lesions compared with unaffected skin in patients [126–128]. These GWAS findings highlight the role of skin barrier function (*FLG*), epidermal proliferation and differentiation (*OVOL1*) and the adaptive immune system response (*IL13-RAD50*, *LRRC32*) in the pathophysiology of AD.

Despite these promising discoveries, less than 20% of disease variance has been explained [129]. The phenomenon of "missing heritability" has been observed across other complex diseases and suggests that unmapped common and rare variants with small effect size in GWAS as well as genetic interactions may contribute to the remaining heritability [129]. Epigenetic studies focusing on the contribution of DNA and chromatin methylation may also explain the role that they play in the formation and progression of complex diseases by regulating gene expression [130]. Future work integrating GWAS and epigenetic data may provide insights into our understanding of complex trait disease. In summary, GWAS data reinforces the concept that multiple low risk variants are most likely to contribute to AD and PV, but that larger sample sizes may be necessary to identify them.

8. Conclusions

The post-Human Genome Project era has seen remarkable advances in our understanding of genes underlying both rare and common skin disease. Such insights have proved significant beyond the field of dermatology because of shared mechanisms of disease for example, PV and inflammatory bowel disease. The wider relevance of skin disease is highlighted by the fact that skin is frequently a marker of internal disease. For example, mutations in *ADAM17* not only cause inflammatory skin and bowel disease but increased susceptibility to infection and cardiomyopathy [131]. Similarly, the study of tylosis with oesophageal cancer, an autosomal dominant cancer syndrome that presents with skin thickening of the palms and soles, has brought to light the role of the inactive rhomboid family member iRHOM2 in cancer pathophysiology [132] and wound healing [133]. This also highlights that mechanistic studies are facilitated by the relative ease with which patient material can be obtained by skin biopsy to derive cell lines for functional studies.

Skin disease is particularly remarkable for its intragenic heterogeneity, for example distinct dominant and recessive mutations in the desmosomal Desmoplakin gene *DSP* can result in a spectrum of disease phenotypes ranging from arrhythmogenic right ventricular cardiomyopathy (ARVC) and striate palmoplantar keratoderma to palmoplantar keratoderma with woolly hair and ARVC (reviewed in [134]).

GWAS, WES and whole genome sequencing (WGS) involving increasingly larger cohorts of ethnically diverse populations may also identify additional low and high penetrance variants that contribute to phenotypic variability. WGS is becoming increasingly affordable and offers scope to become the most cost-effective method for genetic diagnostics. In parallel, advances in bioinformatics and statistics are necessary to analyse the vast quantity of data generated by these studies, and distinguish significant findings. We may also see a move towards re-classification of skin diseases and malignancies based on genome sequence and subsequently, a targeted therapeutic approach to optimise treatment outcome.

Acknowledgments

TM is funded by a Medical Research Council Clinical Fellowship and CAS is funded by a Barts and The London Charity project grant (awarded to DPK).

Author Contributions

Thiviyani Maruthappu, Claire A. Scott and David P. Kelsell wrote the paper.

References

1. Sudbrak, R.; Brown, J.; Dobson-Stone, C.; Carter, S.; Ramser, J.; White, J.; Healy, E.; Dissanayake, M.; Larregue, M.; Perrussel, M.; *et al.* Hailey-hailey disease is caused by mutations in ATP2C1 encoding a novel Ca^{2+} pump. *Hum. Mol. Genet.* **2000**, *9*, 1131–1140.
2. Chavanas, S.; Bodemer, C.; Rochat, A.; Hamel-Teillac, D.; Ali, M.; Irvine, A.D.; Bonafe, J.L.; Wilkinson, J.; Taieb, A.; Barrandon, Y.; *et al.* Mutations in spink5, encoding a serine protease inhibitor, cause netherton syndrome. *Nat. Genet.* **2000**, *25*, 141–142.
3. Sakuntabhai, A.; Ruiz-Perez, V.; Carter, S.; Jacobsen, N.; Burge, S.; Monk, S.; Smith, M.; Munro, C.S.; O'Donovan, M.; Craddock, N.; *et al.* Mutations in ATP2A2, encoding a Ca^{2+} pump, cause darier disease. *Nat. Genet.* **1999**, *21*, 271–277.
4. Miyamura, Y.; Suzuki, T.; Kono, M.; Inagaki, K.; Ito, S.; Suzuki, N.; Tomita, Y. Mutations of the RNA-specific adenosine deaminase gene (DSRAD) are involved in dyschromatosis symmetrica hereditaria. *Am. J. Hum. Genet.* **2003**, *73*, 693–699.

5. Irvine, A.D.; McLean, W.H. The molecular genetics of the genodermatoses: Progress to date and future directions. *Br. J. Dermatol.* **2003**, *148*, 1–13.
6. Mardis, E.R. Next-generation DNA sequencing methods. *Ann. Rev. Genomics Hum. Genet.* **2008**, *9*, 387–402.
7. Ainsworth, C. Immunology: A many layered thing. *Nature* **2012**, *492*, S52–S54.
8. Choate, K.A.; Lu, Y.; Zhou, J.; Choi, M.; Elias, P.M.; Farhi, A.; Nelson-Williams, C.; Crumrine, D.; Williams, M.L.; Nopper, A.J.; *et al.* Mitotic recombination in patients with ichthyosis causes reversion of dominant mutations in KRT10. *Science* **2010**, *330*, 94–97.
9. Kelsell, D.P.; Norgett, E.E.; Unsworth, H.; Teh, M.T.; Cullup, T.; Mein, C.A.; Dopping-Hepenstal, P.J.; Dale, B.A.; Tadini, G.; Fleckman, P.; *et al.* Mutations in ABCA12 underlie the severe congenital skin disease harlequin ichthyosis. *Am. J. Hum. Genet.* **2005**, *76*, 794–803.
10. Oji, V.; Tadini, G.; Akiyama, M.; Blanchet Bardon, C.; Bodemer, C.; Bourrat, E.; Coudiere, P.; DiGiovanna, J.J.; Elias, P.; Fischer, J.; *et al.* Revised nomenclature and classification of inherited ichthyoses: Results of the first ichthyosis consensus conference in soreze 2009. *J. Am. Acad. Dermatol.* **2010**, *63*, 607–641.
11. Rajpopat, S.; Moss, C.; Mellerio, J.; Vahlquist, A.; Ganemo, A.; Hellstrom-Pigg, M.; Ilchyshyn, A.; Burrows, N.; Lestringant, G.; Taylor, A.; *et al.* Harlequin ichthyosis: A review of clinical and molecular findings in 45 cases. *Arch. Dermatol.* **2011**, *147*, 681–686.
12. Dean, M.; Rzhetsky, A.; Allikmets, R. The human ATP-binding cassette (ABC) transporter superfamily. *Genome Res.* **2001**, *11*, 1156–1166.
13. Lefevre, C.; Audebert, S.; Jobard, F.; Bouadjar, B.; Lakhdar, H.; Boughdene-Stambouli, O.; Blanchet-Bardon, C.; Heilig, R.; Foglio, M.; Weissenbach, J.; *et al.* Mutations in the transporter ABCA12 are associated with lamellar ichthyosis type 2. *Hum. Mol. Genet.* **2003**, *12*, 2369–2378.
14. Akiyama, M.; Sugiyama-Nakagiri, Y.; Sakai, K.; McMillan, J.R.; Goto, M.; Arita, K.; Tsuji-Abe, Y.; Tabata, N.; Matsuoka, K.; Sasaki, R.; *et al.* Mutations in lipid transporter ABCA12 in harlequin ichthyosis and functional recovery by corrective gene transfer. *J. Clin. Investig.* **2005**, *115*, 1777–1784.
15. Natsuga, K.; Akiyama, M.; Kato, N.; Sakai, K.; Sugiyama-Nakagiri, Y.; Nishimura, M.; Hata, H.; Abe, M.; Arita, K.; Tsuji-Abe, Y.; *et al.* Novel ABCA12 mutations identified in two cases of non-bullous congenital ichthyosiform erythroderma associated with multiple skin malignant neoplasia. *J. Investig. Dermatol.* **2007**, *127*, 2669–2673.
16. Sakai, K.; Akiyama, M.; Yanagi, T.; McMillan, J.R.; Suzuki, T.; Tsukamoto, K.; Sugiyama, H.; Hatano, Y.; Hayashitani, M.; Takamori, K.; *et al.* ABCA12 is a major causative gene for non-bullous congenital ichthyosiform erythroderma. *J. Investig. Dermatol.* **2009**, *129*, 2306–2309.
17. Akiyama, M. ABCA12 mutations and autosomal recessive congenital ichthyosis: A review of genotype/phenotype correlations and of pathogenetic concepts. *Hum. Mutat.* **2010**, *31*, 1090–1096.
18. Thomas, A.C.; Cullup, T.; Norgett, E.E.; Hill, T.; Barton, S.; Dale, B.A.; Sprecher, E.; Sheridan, E.; Taylor, A.E.; Wilroy, R.S.; *et al.* ABCA12 is the major harlequin ichthyosis gene. *J. Investig. Dermatol.* **2006**, *126*, 2408–2413.
19. Thomas, A.C.; Sinclair, C.; Mahmud, N.; Cullup, T.; Mellerio, J.E.; Harper, J.; Dale, B.A.; Turc-Carel, C.; Hohl, D.; McGrath, J.A.; *et al.* Novel and recurring ABCA12 mutations associated with harlequin ichthyosis: Implications for prenatal diagnosis. *Br. J. Dermatol.* **2008**, *158*, 611–613.

20. Akiyama, M.; Sakai, K.; Sugiyama-Nakagiri, Y.; Yamanaka, Y.; McMillan, J.R.; Sawamura, D.; Niizeki, H.; Miyagawa, S.; Shimizu, H. Compound heterozygous mutations including a *de novo* missense mutation in ABCA12 led to a case of harlequin ichthyosis with moderate clinical severity. *J. Investig. Dermatol.* **2006**, *126*, 1518–1523.
21. Umemoto, H.; Akiyama, M.; Yanagi, T.; Sakai, K.; Aoyama, Y.; Oizumi, A.; Suga, Y.; Kitagawa, Y.; Shimizu, H. New insight into genotype/phenotype correlations in ABCA12 mutations in harlequin ichthyosis. *J. Dermatol. Sci.* **2011**, *61*, 136–139.
22. Scott, C.A.; Plagnol, V.; Nitoiu, D.; Bland, P.J.; Blaydon, D.C.; Chronnell, C.M.; Poon, D.S.; Bourn, D.; Gardos, L.; Csaszar, A.; *et al.* Targeted sequence capture and high-throughput sequencing in the molecular diagnosis of ichthyosis and other skin diseases. *J. Investig. Dermatol.* **2013**, *133*, 573–576.
23. Sakai, K.; Akiyama, M.; Sugiyama-Nakagiri, Y.; McMillan, J.R.; Sawamura, D.; Shimizu, H. Localization of ABCA12 from Golgi apparatus to lamellar granules in human upper epidermal keratinocytes. *Exp. Dermatol.* **2007**, *16*, 920–926.
24. Dale, B.A.; Holbrook, K.A.; Fleckman, P.; Kimball, J.R.; Brumbaugh, S.; Sybert, V.P. Heterogeneity in harlequin ichthyosis, an inborn error of epidermal keratinization: Variable morphology and structural protein expression and a defect in lamellar granules. *J. Investig. Dermatol.* **1990**, *94*, 6–18.
25. Milner, M.E.; O'Guin, W.M.; Holbrook, K.A.; Dale, B.A. Abnormal lamellar granules in harlequin ichthyosis. *J. Investig. Dermatol.* **1992**, *99*, 824–829.
26. Thomas, A.C.; Tattersall, D.; Norgett, E.E.; O'Toole, E.A.; Kelsell, D.P. Premature terminal differentiation and a reduction in specific proteases associated with loss of ABCA12 in harlequin ichthyosis. *Am. J. Pathol.* **2009**, *174*, 970–978.
27. Mitsutake, S.; Suzuki, C.; Akiyama, M.; Tsuji, K.; Yanagi, T.; Shimizu, H.; Igarashi, Y. ABCA12 dysfunction causes a disorder in glucosylceramide accumulation during keratinocyte differentiation. *J. Dermatol. Sci.* **2010**, *60*, 128–129.
28. Yanagi, T.; Akiyama, M.; Nishihara, H.; Ishikawa, J.; Sakai, K.; Miyamura, Y.; Naoe, A.; Kitahara, T.; Tanaka, S.; Shimizu, H. Self-improvement of keratinocyte differentiation defects during skin maturation in ABCA12-deficient harlequin ichthyosis model mice. *Am. J. Pathol.* **2010**, *177*, 106–118.
29. Zuo, Y.; Zhuang, D.Z.; Han, R.; Isaac, G.; Tobin, J.J.; McKee, M.; Welti, R.; Brissette, J.L.; Fitzgerald, M.L.; Freeman, M.W. ABCA12 maintains the epidermal lipid permeability barrier by facilitating formation of ceramide linoleic esters. *J. Biol. Chem.* **2008**, *283*, 36624–36635.
30. Smyth, I.; Hacking, D.F.; Hilton, A.A.; Mukhamedova, N.; Meikle, P.J.; Ellis, S.; Satterley, K.; Collinge, J.E.; de Graaf, C.A.; Bahlo, M.; *et al.* A mouse model of harlequin ichthyosis delineates a key role for ABCA12 in lipid homeostasis. *PLoS Genet.* **2008**, *4*, e1000192.
31. Li, Q.; Frank, M.; Akiyama, M.; Shimizu, H.; Ho, S.Y.; Thisse, C.; Thisse, B.; Sprecher, E.; Uitto, J. Abca12-mediated lipid transport and Snap29-dependent trafficking of lamellar granules are crucial for epidermal morphogenesis in a zebrafish model of ichthyosis. *Dis. Model. Mech.* **2011**, *4*, 777–785.
32. Scott, C.A.; Rajpopat, S.; Di, W.L. Harlequin ichthyosis: ABCA12 mutations underlie defective

lipid transport, reduced protease regulation and skin-barrier dysfunction. *Cell Tissue Res.* **2013**, *351*, 281–288.

33. Akiyama, M.; Suzumori, K.; Shimizu, H. Prenatal diagnosis of harlequin ichthyosis by the examination of keratinized hair canals and amniotic fluid cells at 19 weeks' estimated gestational age. *Prenat. Diagn.* **1999**, *19*, 167–171.
34. Vohra, N.; Rochelson, B.; Smith-Levitin, M. Three-dimensional sonographic findings in congenital (harlequin) ichthyosis. *J. Ultrasound Med.* **2003**, *22*, 737–739.
35. Akiyama, M. Harlequin ichthyosis and other autosomal recessive congenital ichthyoses: The underlying genetic defects and pathomechanisms. *J. Dermatol. Sci.* **2006**, *42*, 83–89.
36. Blaydon, D.C.; Nitoiu, D.; Eckl, K.M.; Cabral, R.M.; Bland, P.; Hausser, I.; van Heel, D.A.; Rajpopat, S.; Fischer, J.; Oji, V.; *et al.* Mutations in CSTA, encoding cystatin a, underlie exfoliative ichthyosis and reveal a role for this protease inhibitor in cell-cell adhesion. *Am. J. Hum. Genet.* **2011**, *89*, 564–571.
37. Hatsell, S.J.; Stevens, H.; Jackson, A.P.; Kelsell, D.P.; Zvulunov, A. An autosomal recessive exfoliative ichthyosis with linkage to chromosome 12q13. *Br. J. Dermatol.* **2003**, *149*, 174–180.
38. Krunic, A.L.; Stone, K.L.; Simpson, M.A.; McGrath, J.A. Acral peeling skin syndrome resulting from a homozygous nonsense mutation in the CSTA gene encoding cystatin A. *Pediatr. Dermatol.* **2013**, *30*, e87–e88.
39. Turk, V.; Bode, W. The cystatins: Protein inhibitors of cysteine proteinases. *FEBS Lett.* **1991**, *285*, 213–219.
40. Steven, A.C.; Steinert, P.M. Protein composition of cornified cell envelopes of epidermal keratinocytes. *J. Cell Sci.* **1994**, *107*, 693–700.
41. Palungwachira, P.; Kakuta, M.; Yamazaki, M.; Yaguchi, H.; Tsuboi, R.; Takamori, K.; Ogawa, H. Immunohistochemical localization of cathepsin l and cystatin A in normal skin and skin tumors. *J. Dermatol.* **2002**, *29*, 573–579.
42. Kato, T.; Takai, T.; Mitsuishi, K.; Okumura, K.; Ogawa, H. Cystatin a inhibits IL-8 production by keratinocytes stimulated with Der p 1 and Der f 1: Biochemical skin barrier against mite cysteine proteases. *J. Allergy Clin. Immunol.* **2005**, *116*, 169–176.
43. Li, C.; Chen, L.; Wang, J.; Zhang, L.; Tang, P.; Zhai, S.; Guo, W.; Yu, N.; Zhao, L.; Liu, M.; *et al.* Expression and clinical significance of cathepsin B and stefin a in laryngeal cancer. *Oncol. Rep.* **2011**, *26*, 869–875.
44. Strojan, P.; Budihna, M.; Smid, L.; Svetic, B.; Vrhovec, I.; Skrk, J. Cathepsin B and L and stefin A and B levels as serum tumor markers in squamous cell carcinoma of the head and neck. *Neoplasma* **2001**, *48*, 66–71.
45. Parker, B.S.; Ciocca, D.R.; Bidwell, B.N.; Gago, F.E.; Fanelli, M.A.; George, J.; Slavin, J.L.; Moller, A.; Steel, R.; Pouliot, N.; *et al.* Primary tumour expression of the cysteine cathepsin inhibitor stefin A inhibits distant metastasis in breast cancer. *J. Pathol.* **2008**, *214*, 337–346.
46. Bilodeau, M.; MacRae, T.; Gaboury, L.; Laverdure, J.P.; Hardy, M.P.; Mayotte, N.; Paradis, V.; Harton, S.; Perreault, C.; Sauvageau, G. Analysis of blood stem cell activity and cystatin gene expression in a mouse model presenting a chromosomal deletion encompassing Csta and Stfa2l1. *PLoS One* **2009**, *4*, e7500.

47. Lin, Z.; Chen, Q.; Lee, M.; Cao, X.; Zhang, J.; Ma, D.; Chen, L.; Hu, X.; Wang, H.; Wang, X.; *et al.* Exome sequencing reveals mutations in TRPV3 as a cause of olmsted syndrome. *Am. J. Hum. Genet.* **2012**, *90*, 558–564.
48. Haghighi, A.; Scott, C.A.; Poon, D.S.; Yaghoobi, R.; Saleh-Gohari, N.; Plagnol, V.; Kelsell, D.P. A missense mutation in the MBTPS2 gene underlies the X-linked form of olmsted syndrome. *J. Investig. Dermatol.* **2013**, *133*, 571–573.
49. Cambiaghi, S.; Tadini, G.; Barbareschi, M.; Caputo, R. Olmsted syndrome in twins. *Arch. Dermatol.* **1995**, *131*, 738–739.
50. Larregue, M.; Callot, V.; Kanitakis, J.; Suau, A.M.; Foret, M. Olmsted syndrome: Report of two new cases and literature review. *J. Dermatol.* **2000**, *27*, 557–568.
51. Giehl, K.A.; Eckstein, G.N.; Pasternack, S.M.; Praetzel-Wunder, S.; Ruzicka, T.; Lichtner, P.; Seidl, K.; Rogers, M.*; Graf, E.; Langbein,* L.**;** *et al.* Nonsense mutations in AAGAB cause punctate palmoplantar keratoderma type buschke-fischer-brauer. *Am. J. Hum. Genet.* **2012**, *91*, 754–759.
52. Pohler, E.; Mamai, O.; Hirst, J.; Zamiri, M.; Horn, H.; Nomura, T.; Irvine, A.D.; Moran, B.; Wilson, N.J.; Smith, F.J.; *et al.* Haploinsufficiency for AAGAB causes clinically heterogeneous forms of punctate palmoplantar keratoderma. *Nat. Genet.* **2012**, *44*, 1272–1276.
53. Kono, M.; Sugiura, K.; Suganuma, M.; Hayashi, M.; Takama, H.; Suzuki, T.; Matsunaga, K.; Tomita, Y.; Akiyama, M. Whole-exome sequencing identifies ADAM10 mutations as a cause of reticulate acropigmentation of kitamura, a clinical entity distinct from dowling-degos disease. *Hum. Mol. Genet.* **2013**, *22*, 3524–3533.
54. Blaydon, D.C.; Lind, L.K.; Plagnol, V.; Linton, K.J.; Smith, F.J.; Wilson, N.J.; McLean, W.H.; Munro, C.S.; South, A.P.; Leigh, I.M.; *et al.* Mutations in *AQP5*, encoding a water-channel protein, cause autosomal-dominant diffuse nonepidermolytic palmoplantar keratoderma. *Am. J. Hum. Genet.* **2013**, *93*, 330–335.
55. Eytan, O.; Morice-Picard, F.; Sarig, O.; Ezzedine, K.; Isakov, O.; Li, Q.; Ishida-Yamamoto, A.; Shomron, N.; Goldsmith, T.; Fuchs-Telem, D.; *et al.* Cole disease results from mutations in ENPP1. *Am. J. Hum. Genet.* **2013**, *93*, 752–757.
56. McGrath, J.A.; Stone, K.L.; Begum, R.; Simpson, M.A.; Dopping-Hepenstal, P.J.; Liu, L.; McMillan, J.R.; South, A.P.; Pourreyron, C.; McLean, W.H.; *et al.* Germline mutation in EXPH5 implicates the Rab27B effector protein Slac2-b in inherited skin fragility. *Am. J. Hum. Genet.* **2012**, *91*, 1115–1121.
57. Lin, Z.; Chen, Q.; Shi, L.; Lee, M.; Giehl, K.A.; Tang, Z.; Wang, H.; Zhang, J.; Yin, J.; Wu, L.; *et al.* Loss-of-function mutations in HOXC13 cause pure hair and nail ectodermal dysplasia. *Am. J. Hum. Genet.* **2012**, *91*, 906–911.
58. Ramot, Y.; Molho-Pessach, V.; Meir, T.; Alper-Pinus, R.; Siam, I.; Tams, S.; Babay, S.; Zlotogorski, A. Mutation in KANK2, encoding a sequestering protein for steroid receptor coactivators, causes keratoderma and woolly hair. *J. Med. Genet.* **2014**, *51*, 388–394.
59. Li, M.; Cheng, R.; Liang, J.; Yan, H.; Zhang, H.; Yang, L.; Li, C.; Jiao, Q.; Lu, Z.; He, J.; *et al.* Mutations in POFUT1, encoding protein *O*-fucosyltransferase 1, cause generalized dowling-degos disease. *Am. J. Hum. Genet.* **2013**, *92*, 895–903.

60. Basmanav, F.B.; Oprisoreanu, A.M.; Pasternack, S.M.; Thiele, H.; Fritz, G.; Wenzel, J.; Grosser, L.; Wehner, M.; Wolf, S.; Fagerberg, C.; *et al.* Mutations in POGLUT1, encoding protein O-glucosyltransferase 1, cause autosomal-dominant dowling-degos disease. *Am. J. Hum. Genet.* **2014**, *94*, 135–143.
61. Kubo, A.; Shiohama, A.; Sasaki, T.; Nakabayashi, K.; Kawasaki, H.; Atsugi, T.; Sato, S.; Shimizu, A.; Mikami, S.; Tanizaki, H.; *et al.* Mutations in SERPINB7, encoding a member of the serine protease inhibitor superfamily, cause nagashima-type palmoplantar keratosis. *Am. J. Hum. Genet.* **2013**, *93*, 945–956.
62. Eytan, O.; Fuchs-Telem, D.; Mevorach, B.; Indelman, M.; Bergman, R.; Sarig, O.; Goldberg, I.; Adir, N.; Sprecher, E. Olmsted syndrome caused by a homozygous recessive mutation in TRPV3. *J. Investig. Dermatol.* **2014**, *134*, 1752–1754.
63. Smith, G.D.; Gunthorpe, M.J.; Kelsell, R.E.; Hayes, P.D.; Reilly, P.; Facer, P.; Wright, J.E.; Jerman, J.C.; Walhin, J.P.; Ooi, L.; *et al.* TRPV3 is a temperature-sensitive vanilloid receptor-like protein. *Nature* **2002**, *418*, 186–190.
64. Peier, A.M.; Reeve, A.J.; Andersson, D.A.; Moqrich, A.; Earley, T.J.; Hergarden, A.C.; Story, G.M.; Colley, S.; Hogenesch, J.B.; McIntyre, P.; *et al.* A heat-sensitive TRP channel expressed in keratinocytes. *Science* **2002**, *296*, 2046–2049.
65. Xu, H.; Ramsey, I.S.; Kotecha, S.A.; Moran, M.M.; Chong, J.A.; Lawson, D.; Ge, P.; Lilly, J.; Silos-Santiago, I.; Xie, Y.; *et al.* TRPV3 is a calcium-permeable temperature-sensitive cation channel. *Nature* **2002**, *418*, 181–186.
66. Asakawa, M.; Yoshioka, T.; Matsutani, T.; Hikita, I.; Suzuki, M.; Oshima, I.; Tsukahara, K.; Arimura, A.; Horikawa, T.; Hirasawa, T.; *et al.* Association of a mutation in TRPV3 with defective hair growth in rodents. *J. Investig. Dermatol.* **2006**, *126*, 2664–2672.
67. Cheng, X.; Jin, J.; Hu, L.; Shen, D.; Dong, X.P.; Samie, M.A.; Knoff, J.; Eisinger, B.; Liu, M.L.; Huang, S.M.; *et al.* Trp channel regulates EGFR signaling in hair morphogenesis and skin barrier formation. *Cell* **2010**, *141*, 331–343.
68. Xiao, R.; Tian, J.; Tang, J.; Zhu, M.X. The TRPV3 mutation associated with the hairless phenotype in rodents is constitutively active. *Cell Calcium* **2008**, *43*, 334–343.
69. Lai-Cheong, J.E.; Sethuraman, G.; Ramam, M.; Stone, K.; Simpson, M.A.; McGrath, J.A. Recurrent heterozygous missense mutation, p.Gly573ser, in the TRPV3 gene in an Indian boy with sporadic olmsted syndrome. *Br. J. Dermatol.* **2012**, *167*, 440–442.
70. Duchatelet, S.; Guibbal, L.; de Veer, S.; Fraitag, S.; Nitschke, P.; Zarhrate, M.; Bodemer, C.; Hovnanian, A. Olmsted syndrome with erythromelalgia caused by recessive TRPV3 mutations. *Br. J. Dermatol.* **2014**, doi:10.1111/bjd.12951.
71. Duchatelet, S.; Pruvost, S.; de Veer, S.; Fraitag, S.; Nitschke, P.; Bole-Feysot, C.; Bodemer, C.; Hovnanian, A. A new TRPV3 missense mutation in a patient with olmsted syndrome and erythromelalgia. *JAMA Dermatol.* **2014**, *150*, 303–306.
72. Yaghoobi, R.; Omidian, M.; Sina, N.; Abtahian, S.A.; Panahi-Bazaz, M.R. Olmsted syndrome in an Iranian family: Report of two new cases. *Arch. Iran. Med.* **2007**, *10*, 246–249.
73. Oeffner, F.; Fischer, G.; Happle, R.; Konig, A.; Betz, R.C.; Bornholdt, D.; Neidel, U.; Boente Mdel, C.; Redler, S.; Romero-Gomez, J.; *et al.* IFAP syndrome is caused by deficiency in

MBTPS2, an intramembrane zinc metalloprotease essential for cholesterol homeostasis and ER stress response. *Am. J. Hum. Genet.* **2009**, *84*, 459–467.

74. Ding, Y.G.; Wang, J.Y.; Qiao, J.J.; Mao, X.H.; Cai, S.Q. A novel mutation in MBTPS2 causes ichthyosis follicularis, alopecia and photophobia (IFAP) syndrome in a chinese family. *Br. J. Dermatol.* **2010**, *163*, 886–889.
75. Tang, L.; Liang, J.; Wang, W.; Yu, L.; Yao, Z. A novel mutation in MBTPS2 causes a broad phenotypic spectrum of ichthyosis follicularis, atrichia, and photophobia syndrome in a large chinese family. *J. Am. Acad. Dermatol.* **2011**, *64*, 716–722.
76. Naiki, M.; Mizuno, S.; Yamada, K.; Yamada, Y.; Kimura, R.; Oshiro, M.; Okamoto, N.; Makita, Y.; Seishima, M.; Wakamatsu, N. MBTPS2 mutation causes bresek/bresheck syndrome. *Am. J. Med. Genet. Part A* **2012**, *158A*, 97–102.
77. Aten, E.; Brasz, L.C.; Bornholdt, D.; Hooijkaas, I.B.; Porteous, M.E.; Sybert, V.P.; Vermeer, M.H.; Vossen, R.H.; van der Wielen, M.J.; Bakker, E.; *et al.* Keratosis follicularis spinulosa decalvans is caused by mutations in MBTPS2. *Hum. Mutat.* **2010**, *31*, 1125–1133.
78. Sakai, J.; Nohturfft, A.; Goldstein, J.L.; Brown, M.S. Cleavage of sterol regulatory element-binding proteins (srebps) at site-1 requires interaction with SREBP cleavage-activating protein. Evidence from *in vivo* competition studies. *J. Biol. Chem.* **1998**, *273*, 5785–5793.
79. Rawson, R.B.; Zelenski, N.G.; Nijhawan, D.; Ye, J.; Sakai, J.; Hasan, M.T.; Chang, T.Y.; Brown, M.S.; Goldstein, J.L. Complementation cloning of S2P, a gene encoding a putative metalloprotease required for intramembrane cleavage of SREBPs. *Mol. Cell* **1997**, *1*, 47–57.
80. Ye, J.; Rawson, R.B.; Komuro, R.; Chen, X.; Dave, U.P.; Prywes, R.; Brown, M.S.; Goldstein, J.L. ER stress induces cleavage of membrane-bound ATF6 by the same proteases that process SREBPs. *Mol. Cell* **2000**, *6*, 1355–1364.
81. Bisgaard, H.; Simpson, A.; Palmer, C.N.; Bonnelykke, K.; McLean, I.; Mukhopadhyay, S.; Pipper, C.B.; Halkjaer, L.B.; Lipworth, B.; Hankinson, J.; *et al.* Gene-environment interaction in the onset of eczema in infancy: Filaggrin loss-of-function mutations enhanced by neonatal cat exposure. *PLoS Med.* **2008**, *5*, e131.
82. Enamandram, M.; Kimball, A.B. Psoriasis epidemiology: The interplay of genes and the environment. *J. Investig. Dermatol.* **2013**, *133*, 287–289.
83. Carroll, C.L.; Balkrishnan, R.; Feldman, S.R.; Fleischer, A.B., Jr.; Manuel, J.C. The burden of atopic dermatitis: Impact on the patient, family, and society. *Pediatr. Dermatol.* **2005**, *22*, 192–199.
84. Baker, C.S.; Foley, P.A.; Braue, A. Psoriasis uncovered—Measuring burden of disease impact in a survey of australians with psoriasis. *Australas. J. Dermatol.* **2013**, *54*, 1–6.
85. Stranger, B.E.; Stahl, E.A.; Raj, T. Progress and promise of genome-wide association studies for human complex trait genetics. *Genetics* **2011**, *187*, 367–383.
86. Weinstein, G.D.; Frost, P. Abnormal cell proliferation in psoriasis. *J. Investig. Dermatol.* **1968**, *50*, 254–259.
87. Wuepper, K.D.; Coulter, S.N.; Haberman, A. Psoriasis vulgaris: A genetic approach. *J. Investig. Dermatol.* **1990**, *95*, 2S–4S.
88. Capon, F.; Barker, J.N. The quest for psoriasis susceptibility genes in the postgenome-wide association studies era: Charting the road ahead. *Br. J. Dermatol.* **2012**, *166*, 1173–1175.

89. Capon, F.; Burden, A.D.; Trembath, R.C.; Barker, J.N. Psoriasis and other complex trait dermatoses: From loci to functional pathways. *J. Investig. Dermatol.* **2012**, *132*, 915–922.
90. Nair, R.P.; Stuart, P.; Henseler, T.; Jenisch, S.; Chia, N.V.; Westphal, E.; Schork, N.J.; Kim, J.; Lim, H.W.; Christophers, E.; *et al.* Localization of psoriasis-susceptibility locus PSORS1 to a 60-kb interval telomeric to HLA-C. *Am. J. Hum. Genet.* **2000**, *66*, 1833–1844.
91. Capon, F.; Munro, M.; Barker, J.; Trembath, R. Searching for the major histocompatibility complex psoriasis susceptibility gene. *J. Investig. Dermatol.* **2002**, *118*, 745–751.
92. Nair, R.P.; Stuart, P.E.; Nistor, I.; Hiremagalore, R.; Chia, N.V.; Jenisch, S.; Weichenthal, M.; Abecasis, G.R.; Lim, H.W.; Christophers, E.; *et al.* Sequence and haplotype analysis supports HLA-C as the psoriasis susceptibility 1 gene. *Am. J. Hum. Genet.* **2006**, *78*, 827–851.
93. Nair, M.; Teng, A.; Bilanchone, V.; Agrawal, A.; Li, B.; Dai, X. Ovol1 regulates the growth arrest of embryonic epidermal progenitor cells and represses c-myc transcription. *J. Cell Biol.* **2006**, *173*, 253–264.
94. Nair, R.P.; Duffin, K.C.; Helms, C.; Ding, J.; Stuart, P.E.; Goldgar, D.; Gudjonsson, J.E.; Li, Y.; Tejasvi, T.; Feng, B.J.; *et al.* Genome-wide scan reveals association of psoriasis with IL-23 and NF-kappab pathways. *Nat. Genet.* **2009**, *41*, 199–204.
95. Jordan, C.T.; Cao, L.; Roberson, E.D.; Duan, S.; Helms, C.A.; Nair, R.P.; Duffin, K.C.; Stuart, P.E.; Goldgar, D.; Hayashi, G.; *et al.* Rare and common variants in CARD14, encoding an epidermal regulator of NF-kappab, in psoriasis. *Am. J. Hum. Genet.* **2012**, *90*, 796–808.
96. Jordan, C.T.; Cao, L.; Roberson, E.D.; Pierson, K.C.; Yang, C.F.; Joyce, C.E.; Ryan, C.; Duan, S.; Helms, C.A.; Liu, Y.; *et al.* PSORS2 is due to mutations in CARD14. *Am. J. Hum. Genet.* **2012**, *90*, 784–795.
97. Di Cesare, A.; di Meglio, P.; Nestle, F.O. The IL-23/Th17 axis in the immunopathogenesis of psoriasis. *J. Investig. Dermatol.* **2009**, *129*, 1339–1350.
98. Weaver, C.T.; Hatton, R.D.; Mangan, P.R.; Harrington, L.E. IL-17 family cytokines and the expanding diversity of effector T cell lineages. *Ann. Rev. Immunol.* **2007**, *25*, 821–852.
99. Zheng, Y.; Danilenko, D.M.; Valdez, P.; Kasman, I.; Eastham-Anderson, J.; Wu, J.; Ouyang, W. Interleukin-22, a T(h)17 cytokine, mediates IL-23-induced dermal inflammation and acanthosis. *Nature* **2007**, *445*, 648–651.
100. Krueger, G.G.; Langley, R.G.; Leonardi, C.; Yeilding, N.; Guzzo, C.; Wang, Y.; Dooley, L.T.; Lebwohl, M.; Group, C.P.S. A human interleukin-12/23 monoclonal antibody for the treatment of psoriasis. *N. Engl. J. Med.* **2007**, *356*, 580–592.
101. Suarez-Farinas, M.; Shah, K.R.; Haider, A.S.; Krueger, J.G.; Lowes, M.A. Personalized medicine in psoriasis: Developing a genomic classifier to predict histological response to alefacept. *BMC Dermatol.* **2010**, *10*, doi:10.1186/1471-5945-10-1.
102. Tomfohrde, J.; Silverman, A.; Barnes, R.; Fernandez-Vina, M.A.; Young, M.; Lory, D.; Morris, L.; Wuepper, K.D.; Stastny, P.; Menter, A.; *et al.* Gene for familial psoriasis susceptibility mapped to the distal end of human chromosome 17q. *Science* **1994**, *264*, 1141–1145.
103. Marrakchi, S.; Guigue, P.; Renshaw, B.R.; Puel, A.; Pei, X.Y.; Fraitag, S.; Zribi, J.; Bal, E.; Cluzeau, C.; Chrabieh, M.; *et al.* Interleukin-36-receptor antagonist deficiency and generalized pustular psoriasis. *N. Engl. J. Med.* **2011**, *365*, 620–628.

104. Berki, D.M.; Mahil, S.K.; Burden, A.D.; Trembath, R.C.; Smith, C.H.; Capon, F.; Barker, J.N. Loss of IL36RN function does not confer susceptibility to psoriasis vulgaris. *J. Investig. Dermatol.* **2014**, *134*, 271–273.
105. Capon, F. IL36RN mutations in generalized pustular psoriasis: Just the tip of the iceberg? *J. Investig. Dermatol.* **2013**, *133*, 2503–2504.
106. Flohr, C.; Mann, J. New insights into the epidemiology of childhood atopic dermatitis. *Allergy* **2014**, *69*, 3–16.
107. Larsen, F.S.; Holm, N.V.; Henningsen, K. Atopic dermatitis: A genetic-epidemiologic study in a population-based twin sample. *J. Am. Acad. Dermatol.* **1986**, *15*, 487–494.
108. Smith, F.J.; Irvine, A.D.; Terron-Kwiatkowski, A.; Sandilands, A.; Campbell, L.E.; Zhao, Y.; Liao, H.; Evans, A.T.; Goudie, D.R.; Lewis-Jones, S.; *et al.* Loss-of-function mutations in the gene encoding filaggrin cause ichthyosis vulgaris. *Nat. Genet.* **2006**, *38*, 337–342.
109. Sybert, V.P.; Dale, B.A.; Holbrook, K.A. Ichthyosis vulgaris: Identification of a defect in synthesis of filaggrin correlated with an absence of keratohyaline granules. *J. Investig. Dermatol.* **1985**, *84*, 191–194.
110. Brown, S.J.; Asai, Y.; Cordell, H.J.; Campbell, L.E.; Zhao, Y.; Liao, H.; Northstone, K.; Henderson, J.; Alizadehfar, R.; Ben-Shoshan, M.; *et al.* Loss-of-function variants in the filaggrin gene are a significant risk factor for peanut allergy. *J. Allergy Clin. Immunol.* **2011**, *127*, 661–667.
111. Brown, S.J.; Relton, C.L.; Liao, H.; Zhao, Y.; Sandilands, A.; McLean, W.H.; Cordell, H.J.; Reynolds, N.J. Filaggrin haploinsufficiency is highly penetrant and is associated with increased severity of eczema: Further delineation of the skin phenotype in a prospective epidemiological study of 792 school children. *Br. J. Dermatol.* **2009**, *161*, 884–889.
112. Hirota, T.; Takahashi, A.; Kubo, M.; Tsunoda, T.; Tomita, K.; Sakashita, M.; Yamada, T.; Fujieda, S.; Tanaka, S.; Doi, S.; *et al.* Genome-wide association study identifies eight new susceptibility loci for atopic dermatitis in the Japanese population. *Nat. Genet.* **2012**, *44*, 1222–1226.
113. Sun, L.D.; Xiao, F.L.; Li, Y.; Zhou, W.M.; Tang, H.Y.; Tang, X.F.; Zhang, H.; Schaarschmidt, H.; Zuo, X.B.; Foelster-Holst, R.; *et al.* Genome-wide association study identifies two new susceptibility loci for atopic dermatitis in the Chinese han population. *Nat. Genet.* **2011**, *43*, 690–694.
114. Paternoster, L.; Standl, M.; Chen, C.M.; Ramasamy, A.; Bonnelykke, K.; Duijts, L.; Ferreira, M.A.; Alves, A.C.; Thyssen, J.P.; Albrecht, E.; *et al.* Meta-analysis of genome-wide association studies identifies three new risk loci for atopic dermatitis. *Nat. Genet.* **2012**, *44*, 187–192.
115. Palmer, C.N.; Irvine, A.D.; Terron-Kwiatkowski, A.; Zhao, Y.; Liao, H.; Lee, S.P.; Goudie, D.R.; Sandilands, A.; Campbell, L.E.; Smith, F.J.; *et al.* Common loss-of-function variants of the epidermal barrier protein filaggrin are a major predisposing factor for atopic dermatitis. *Nat. Genet.* **2006**, *38*, 441–446.
116. Sandilands, A.; Sutherland, C.; Irvine, A.D.; McLean, W.H. Filaggrin in the frontline: Role in skin barrier function and disease. *J. Cell Sci.* **2009**, *122*, 1285–1294.
117. Fallon, P.G.; Sasaki, T.; Sandilands, A.; Campbell, L.E.; Saunders, S.P.; Mangan, N.E.; Callanan, J.J.; Kawasaki, H.; Shiohama, A.; Kubo, A.; *et al.* A homozygous frameshift mutation in the mouse flg gene facilitates enhanced percutaneous allergen priming. *Nat. Genet.* **2009**, *41*, 602–608.

118. Oyoshi, M.K.; Murphy, G.F.; Geha, R.S. Filaggrin-deficient mice exhibit Th17-dominated skin inflammation and permissiveness to epicutaneous sensitization with protein antigen. *J. Allergy Clin. Immunol.* **2009**, *124*, 485–493
119. Otsuka, A.; Doi, H.; Egawa, G.; Maekawa, A.; Fujita, T.; Nakamizo, S.; Nakashima, C.; Nakajima, S.; Watanabe, T.; Miyachi, Y.; *et al.* Possible new therapeutic strategy to regulate atopic dermatitis through upregulating filaggrin expression. *J. Allergy Clin. Immunol.* **2014**, *133*, 139–146.
120. Tamari, M.; Hirota, T. Genome-wide association studies of atopic dermatitis. *J. Dermatol.* **2014**, *41*, 213–220.
121. Esparza-Gordillo, J.; Weidinger, S.; Folster-Holst, R.; Bauerfeind, A.; Ruschendorf, F.; Patone, G.; Rohde, K.; Marenholz, I.; Schulz, F.; Kerscher, T.; *et al.* A common variant on chromosome 11q13 is associated with atopic dermatitis. *Nat. Genet.* **2009**, *41*, 596–601.
122. Wang, R.; Wan, Q.; Kozhaya, L.; Fujii, H.; Unutmaz, D. Identification of a regulatory T cell specific cell surface molecule that mediates suppressive signals and induces Foxp3 expression. *PLoS One* **2008**, *3*, e2705.
123. Buschke, S.; Stark, H.J.; Cerezo, A.; Pratzel-Wunder, S.; Boehnke, K.; Kollar, J.; Langbein, L.; Heldin, C.H.; Boukamp, P. A decisive function of transforming growth factor-beta/smad signaling in tissue morphogenesis and differentiation of human HaCat keratinocytes. *Mol. Biol. Cell* **2011**, *22*, 782–794.
124. Chang, M.; Li, Y.; Yan, C.; Callis-Duffin, K.P.; Matsunami, N.; Garcia, V.E.; Cargill, M.; Civello, D.; Bui, N.; Catanese, J.J.; *et al.* Variants in the 5q31 cytokine gene cluster are associated with psoriasis. *Genes Immun.* **2008**, *9*, 176–181.
125. Li, Y.; Chang, M.; Schrodi, S.J.; Callis-Duffin, K.P.; Matsunami, N.; Civello, D.; Bui, N.; Catanese, J.J.; Leppert, M.F.; Krueger, G.G.; *et al.* The 5q31 variants associated with psoriasis and crohn's disease are distinct. *Hum. Mol. Genet.* **2008**, *17*, 2978–2985.
126. Leung, D.Y. New insights into atopic dermatitis: Role of skin barrier and immune dysregulation. *Allergol. Int.* **2013**, *62*, 151–161.
127. Hamid, Q.; Naseer, T.; Minshall, E.M.; Song, Y.L.; Boguniewicz, M.; Leung, D.Y. *In vivo* expression of IL-12 and IL-13 in atopic dermatitis. *J. Allergy Clin. Immunol.* **1996**, *98*, 225–231.
128. Hamid, Q.; Boguniewicz, M.; Leung, D.Y. Differential *in situ* cytokine gene expression in acute *versus* chronic atopic dermatitis. *J. Clin. Investig.* **1994**, *94*, 870–876.
129. Maher, B. Personal genomes: The case of the missing heritability. *Nature* **2008**, *456*, 18–21.
130. Gudjonsson, J.E.; Krueger, G. A role for epigenetics in psoriasis: Methylated cytosine-guanine sites differentiate lesional from nonlesional skin and from normal skin. *J. Investig. Dermatol.* **2012**, *132*, 506–508.
131. Blaydon, D.C.; Biancheri, P.; Di, W.L.; Plagnol, V.; Cabral, R.M.; Brooke, M.A.; van Heel, D.A.; Ruschendorf, F.; Toynbee, M.; Walne, A.; *et al.* Inflammatory skin and bowel disease linked to ADAM17 deletion. *N. Engl. J. Med.* **2011**, *365*, 1502–1508.
132. Blaydon, D.C.; Etheridge, S.L.; Risk, J.M.; Hennies, H.C.; Gay, L.J.; Carroll, R.; Plagnol, V.; McRonald, F.E.; Stevens, H.P.; Spurr, N.K.; *et al.* RHBDF2 mutations are associated with tylosis, a familial esophageal cancer syndrome. *Am. J. Hum. Genet.* **2012**, *90*, 340–346.

133. Brooke, M.A.; Etheridge, S.L.; Kaplan, N.; Simpson, C.; O'Toole, E.A.; Ishida-Yamamoto, A.; Marches, O.; Getsios, S.; Kelsell, D.P. iRHOM*2*-dependent regulation of ADAM17 in cutaneous disease and epidermal barrier function. *Hum. Mol. Genet.* **2014**, *23*, 4064–4076.
134. Brooke, M.A.; Nitoiu, D.; Kelsell, D.P. Cell-cell connectivity: Desmosomes and disease. *J. Pathol.* **2012**, *226*, 158–171.

7

A Balanced Look at the Implications of Genomic (and Other "Omics") Testing for Disease Diagnosis and Clinical Care

Scott D. Boyd [1,*], Stephen J. Galli [1,2], Iris Schrijver [1,3], James L. Zehnder [1,4], Euan A. Ashley [4] and Jason D. Merker [1]

1 Department of Pathology, Stanford University, Stanford, CA 94305, USA;
E-Mails: sgalli@stanford.edu (S.J.G.); ischrijver@stanfordmed.org (I.S.); zehnder@stanford.edu (J.L.Z.); jdmerker@stanford.edu (J.D.M.)

2 Department of Microbiology and Immunology, Stanford University, Stanford, CA 94305, USA

3 Department of Pediatrics, Stanford University, Stanford, CA 94305, USA

4 Department of Medicine, Stanford University, Stanford, CA 94305, USA;
E-Mail: euan@stanford.edu

* Author to whom correspondence should be addressed; E-Mail: sboyd1@stanford.edu;

Abstract: The tremendous increase in DNA sequencing capacity arising from the commercialization of "next generation" instruments has opened the door to innumerable routes of investigation in basic and translational medical science. It enables very large data sets to be gathered, whose interpretation and conversion into useful knowledge is only beginning. A challenge for modern healthcare systems and academic medical centers is to apply these new methods for the diagnosis of disease and the management of patient care without unnecessary delay, but also with appropriate evaluation of the quality of data and interpretation, as well as the clinical value of the insights gained. Most critically, the standards applied for evaluating these new laboratory data and ensuring that the results and their significance are clearly communicated to patients and their caregivers should be at least as rigorous as those applied to other kinds of medical tests. Here, we present an overview of conceptual and practical issues to be considered in planning for the integration of genomic methods or, in principle, any other type of "omics" testing into clinical care.

Keywords: personalized medicine; clinical genomics; regulation; quality assurance

1. Introduction

Improvements in DNA sequencing technology in the past decade represent one of the most significant technological achievements in recent history, with far-reaching implications for medicine and society. Most human diseases have at least some genetic factors that contribute to their incidence, or course, either related to the germline genome inherited from an individual's parents or the somatic genetic changes that can lead to the development of malignancies. The ability to read the DNA from an individual's cells with next generation sequencing (NGS) should therefore offer insights relevant for medical care. However, there is a significant gap between our current ability to acquire sequence data and the ultimate goal of extracting all of the useful medical genetic knowledge from the sequences. In particular, societal expectations and ethical considerations require that any correlations between DNA sequences and predictions of disease risk, prognosis or optimal treatment choice should be held to higher standards of evidence than those that are typically applied in the peer review process for publication of a research article. In this overview, we initially highlight areas of recent progress and promise in clinical genomic testing (including whole genome sequence analysis, as well as analysis of selected fractions of the genome, such as the protein coding exome, or large panels of genes of clinical interest) and discuss these new approaches in the context of medical laboratory testing and the current regulatory framework governing such tests in the United States. The potential benefits of clinical genomic testing are tremendous, but devising appropriate systems for quality assurance, data sharing and validation, incorporation into clinical trials and cost-benefit analysis of this new diagnostic area will be an ongoing effort in the coming years.

2. The Promise of Genomic Methods

In the past decade, the quantity of DNA sequencing that can be performed per dollar spent has increased by several orders of magnitude, as a result of technological innovations enabling highly parallelized simultaneous sequencing of millions of spatially-separated template molecules, with optical or electronic readout of the sequencing reaction as it occurs [1,2]. With the latest generation of sequencing instruments, the cost for a whole human genome sequence with 30× coverage may approach $1,000 [3]. Although such estimates notably exclude the costs of interpreting the data, it is clear that genome sequencing is now within the range of costs for many other diagnostic methods, such as radiologic imaging studies or full evaluation of tissue biopsies by anatomic pathologists. As a result of painstaking earlier studies of the genetics of inherited diseases and the genetic changes that are found in cancer cells, there are already many known gene mutations whose significance in relation to particular diseases is described. For many of the most well-established variants, there are single-gene diagnostic sequencing tests already available, either from private companies or specialized, typically academic, diagnostic laboratories. The CDC estimates that genetic tests for use in the clinical setting have been developed for approximately 2,000 diseases [4]. The critical difference between current genome sequencing capabilities and these earlier test methods is that a significantly greater amount of data (whether genome, exome or gene panels) now can be gathered as readily and easily as the sequence from a single gene. Many medical centers and companies are hurrying to stake claims as the preferred destination for testing and interpretation of genomic sequence data for clinical purposes, as this methodology has begun to be adopted and standardized.

Already, there are multiple published medical success stories using these methods. Sequencing of genomes or exomes (which includes the protein coding portions of the genome) for the diagnosis of patients with heritable syndromic disorders has resulted in a number of exciting case reports and studies of patients in whom likely causative mutations have been discovered, and in some cases, such findings have guided successful clinical treatment decisions [5–7]. Systematic efforts in the NIH Undiagnosed Diseases Program to apply genomic methods to arrive at diagnoses for patients with unusual or mysterious clinical presentations, particularly for cases where family history suggests a possible genetic cause, have also yielded new causative mutations and discoveries in human biology [8]. Likewise, clinical laboratories have used exome sequencing since 2011 to evaluate patients with suspected genetic disorders and have identified a molecular diagnostic yield of approximately 25% [9].

Sequencing of cancer genomes has revealed many new recurrent mutations that may contribute to the development of particular cancers, and has revealed new candidates for targeted therapies [10]. These new molecular insights into cancer are already beginning to influence the ways that cancers are classified and treated [11]. Similarly, sequencing of fetal DNA from the plasma of a pregnant woman to screen for trisomies 18 and 21 demonstrates improved performance relative to prior aneuploidy screening methods [12]. Studies of viruses, bacteria and other microbes via sequencing of their small genomes is revolutionizing epidemiological tracking of infectious diseases and the rapid detection of new and emerging pathogens [13]. The most challenging area of all, predicting the risks of diseases in healthy human beings, has also show promise in particular disease categories, such as prediction of cancer risk in women having germline mutations in the *BRCA1* or *BRCA2* genes. More broadly, it is likely that only a fraction of potential disease-associated variants have been identified at present, and the cooperative or competitive relationships between the effects of different sequence variants are only beginning to be described [14]. Many common diseases (e.g., diabetes, schizophrenia and autism) do not appear to be the result of simple sequence alterations that could be easily diagnosed by DNA sequencing. In these diseases, there are many genes associated with increased risk, but the conditions may be caused by combinations of these genes acting together in association with environmental factors. In addition, while it is now possible to identify numerous variants in cancer genomes, the biological significance of many variants is unknown, and their annotation is not standardized. The number of clinically actionable variants, at present, is small.

However, even given the above limitations, these advances herald the increasing importance that genome sequencing and related methods are likely to play in the diagnosis and management of diseases across all medical specialties. We are currently in a transition period in which methods initially applied in research settings and limited small clinical studies need to be adapted for application to large numbers of patients. With that transition comes a requirement for increased standardization, reliability and monitoring of experimental steps, as well as agreed-upon standards for data analysis, storage, clinical interpretation and communication with patients and/or their guardians.

3. Research Experiments, Clinical Testing and Genomic Testing

3.1. Research Assays and Methods

The experimental methods used in medical testing typically originate in research laboratories in universities, government or private research institutes or corporations. However, there is a substantial difference between the assay performance characteristics needed for use in the published scientific literature, compared to those used for clinical diagnosis and guiding the treatment of patients. Authors using a new experimental method as part of a published peer-reviewed research study must convince the scientific reviewers selected by the journal that the assay is a valid method of measurement and has been appropriately applied to the research topic in question, but often, reviewers are not experts in all aspects of the experimental methodology and data analysis approaches used. The expectation in a research setting is that efforts by other researchers to replicate the results in question and to use the methods described will eventually test the reported results and reveal any limitations or errors. This approach has been the basis for essentially all scientific advances, despite the fact that very few published papers are entirely free from errors or only partially correct conclusions, and some are entirely false [15].

Recent studies have highlighted common problems with experimental design and statistical analysis that contribute disproportionately to preventable errors in the scientific literature, especially in attempting to identify genetic contributions to diseases [16,17]. These include choices in experimental design that can introduce biases favoring the discovery of apparently significant effects, even in randomized trials, where patient selection, problems with randomization, lack of blinded data analysis and changes in the plan for data analysis once a trial is underway can all have an impact. In case-control or retrospective studies, the potential for mistaken conclusions is even greater. Studies using large data sets are particularly vulnerable to errors from over-fitting a model to the data or insufficiently accounting for multiple hypothesis testing, especially if independent validation data sets are not used to test the robustness of initial conclusions. Other well-known factors that can distort the scientific literature are publication biases in favor of positive results and the competitive social and economic factors that disproportionately reward scientists publishing papers reporting apparently highly novel findings in prominent journals, while imposing minimal penalties for prior publications later found to be partly or completely in error.

Several suggested improvements to the research methodology have been proposed, including advance registration of a wider range of clinical studies, better documentation of experimental protocols, results and data analysis approaches, more consistent involvement of statisticians and other experts in study design and analysis, full transparency and availability of experimental data and computer scripts used in data analysis and greater attention paid to research reproducibility in the professional evaluation of scientists [16]. In practice, an excessively regulated research environment would probably serve to stifle and limit some of the creative and perhaps poorly planned, but ultimately serendipitous efforts in science that can lead to unexpected insights, so a compromise between higher standards for clinical trial research and continued freedom of inquiry (subject to ethical and safety review by institutional review boards) in more exploratory areas of medical science would probably best serve the public and the ultimate goals of funding providers.

3.2. Clinical Tests

Clinical laboratory testing in the United States is regulated and subjected to greater methodological scrutiny than basic or clinical research. All clinical laboratory testing done for purposes of patient care (as opposed to research) must be performed in a CLIA (Clinical Laboratory Improvement Amendments of 1988)-certified laboratory. CLIA established key laboratory quality standards to ensure that test performance consistently meets patient care needs. CLIA certification may be achieved through the Centers for Medicare and Medicaid Services (CMS), which administers CLIA laboratory certification, or through CMS-approved accrediting organizations (e.g., the College of American Pathologists (CAP), or The Joint Commission). CLIA requirements are stratified according to the complexity of testing performed, with genomic testing generally falling into the highest level of complexity. Such high-complexity testing laboratories must meet specified quality standards, including those related to personnel qualifications and responsibilities, proficiency testing, facilities, general laboratory systems and quality management, as well as preanalytic, analytic and postanalytic systems. (CLIA Brochure, ICN #006270, May 2013). Laboratory compliance with CLIA regulations is evaluated by biennial on-site inspections. Depending on the nature of the tests performed, additional requirements may need to be satisfied for such laboratories, including those of the AABB (formerly, the American Association of Blood Banks) (Bethesda, MD, USA), FDA (U.S. Food and Drug Administration, Silver Spring, MD, USA), ASHI (the American Society for Histocompatibility and Immunogenetics, Mt. Laurel, NJ, USA), FAA (Federal Aviation Administration, Washington, DC, USA) and state agencies. High-complexity testing must be done by or subject to the oversight of laboratory professionals with an advanced degree and with appropriate credentials in laboratory medicine.

New types of laboratory tests, including most genomic tests, are not available in the form of FDA approved/cleared test kits, but rather are typically created within academic or commercial clinical laboratories as laboratory-developed tests (LDTs). Such development follows a strictly prescribed process of test validation prior to clinical implementation, which is when the test becomes orderable [18,19]. Recently, one high-throughput sequencing instrument and reagent kit system has received FDA clearance for use in clinical testing. In a recent position statement, the Association for Molecular Pathology (AMP) has introduced the term, laboratory-developed procedure, which is defined as "a professional service that encompasses and integrates the design, development, validation, verification, and quality systems used in laboratory testing and interpretative reporting in the context of clinical care" and which much more accurately reflects the highly complex nature of molecular laboratory testing, as well as the central contribution of highly trained and qualified laboratory professionals to the patient care process. AMP also concluded that CMS can ensure the effective oversight and validation of most molecular genetic laboratory tests [20].

3.3. Genomic Testing

The nature of genomic testing, in which large data sets of DNA sequences can be conveniently gathered, but where only a fraction of the overall data can be interpreted at present, places these data sets at the interface between research and conventional clinical testing. Interpretation of genome sequence data is not unique in requiring sophisticated understanding, both of the methods used to gather the data, and the medical literature and body of prior investigation, to arrive at accurate conclusions. For example, histologic diagnosis of cancers in modern pathology practice depends on years of training in recognizing the visual characteristics of aberrant cell populations in tissue sections and selecting appropriate confirmatory tests. The history of revised and improved tumor classification systems in modern oncology reflects the increased understanding gained over decades by studying tumors with new experimental methods, revisiting prior data and correlating the features of each cancer with its response to treatment. Genomic sequence data are vast and complex in other ways, in that the data are generated as lists of nucleotide identities that require computational tools and comparison to sequences in reference databases for analysis, before the clinical significance of sequence variants can begin to be assessed.

Despite amazing progress in the past decade, the technologies and analytical approaches for sequencing and interpreting genomes still have significant blind spots, such as the greater difficulty of detecting structural variants including insertions, deletions, inversions, and trinucleotide repeat expansions, compared to single nucleotide variants in sequences [21]. Reference databases of sequence variants also contain many artifactual annotations, such as 'variants' that are actually sequences derived from pseudogenes similar to the gene in question. It is likely to take many years to resolve such ambiguities or errors in prior gene sequencing work and to ensure that all sequence variants can be correctly annotated for medical applications. Judging from earlier work in human genetics, the interpretation of each patient's genome sequence will require a personalized approach that takes into account the population group from which that person is derived. Even for the most extensively studied genes, such as the cystic fibrosis transmembrane conductance regulator (*CFTR*), the databases of mutations and their significance are quite limited, because functional studies have not been performed for most mutations and because mutations in population groups that have not been studied as extensively as European-derived groups are not well characterized [22].

In the near future, we should not necessarily expect that any given patient's genome sequence, particularly if they are currently healthy, will reveal sequence features requiring any sort of response or clinical guidance beyond those that would already be provided by a physician in a routine checkup, such as advice about healthy diet, exercise, vaccinations and safety topics. A recent study of the potential and current limitations of whole genome sequence interpretation for clinical use in 12 healthy individuals highlighted the relatively small effects associated with most known disease-associated sequence variants, but also revealed that each individual had at least one gene variant from the Clinical Pharmacogenomics Implementation Consortium list of variants that can affect responses to drug therapies [14]. In addition, two of the 12 subjects studied received actionable information that could affect their future health, including one subject who underwent prophylactic surgery. Other studies have underlined the relatively small effect sizes of genetic variants for the most common and serious diseases, such as cardiovascular disease, but rare and highly deleterious variants can also be identified, such as those causing familial hypercholesterolemia, which would warrant immediate medical intervention, such as statin therapy [23,24]. Recently, standards of evidence for concluding that gene variants are associated with diseases have been proposed, taking into account the large amount of data gathered in genome or exome studies and the potential for false discoveries associated with such large numbers of observations of a sample; for example, in exome studies, $p = 5 \times 10^{-7}$ has been suggested as one threshold for claiming significance [25,26].

4. Quality Assurance in the Genome Sequencing Era

Molecular diagnostic laboratories have provided innovative testing since the emergence of diagnostic methods applied to DNA or RNA molecules several decades ago. Clinical molecular laboratories are experienced in validating new methods for single gene testing and data analysis. In one sense, genomic testing is "just another" such innovation. However, it could be argued that the scale of data now obtainable represents a challenge for analysis that is not merely incremental, but is qualitatively different, as human beings cannot manually go through the sequences and interpret them visually in a practical amount of time. Instead, computational methods and bioinformatics tools are required to help carry out the analysis. This represents a significant break with prior traditions of medical training in all specialties, where, with rare exceptions, computer science and bioinformatics were not learned by the generations of physicians who are currently in practice and in positions of authority. Physicians currently in training, particularly in laboratory medicine, now have the opportunity to learn and help to develop these new methods before they enter practice, and currently practicing physicians must at least learn how the results of genomic testing should be applied in their patient care decisions [27,28]. In spite of these challenges, the fundamental features required for reliable clinical laboratory tests based on genome sequencing are the same as those for other kinds of tests and are, in our view, compatible with current regulatory frameworks for ensuring the quality of medical laboratory testing. Some of the most critical initiatives underway in this area are as follows:

(a) Guidelines for Clinical NGS Implementation

Initial laboratory guidelines for clinical diagnostic NGS testing have been established and published by several laboratory professional organizations. These include initiatives by professional organizations, such as the Association for Molecular Pathology (AMP) [29] and the American College of Medical Genetics and Genomics (ACMG) [30,31], as well as those by entities, such as the Clinical Laboratory Standards Institute (CLSI) and the Division of Laboratory Science and Standards at the Centers for Disease Control (CDC) [32]. These efforts are expected eventually to result in consistent recommendations for the clinical validation process of NGS testing, as well as for performance metrics and genomic reference materials for clinical use.

(b) Checks and Balances: The College of American Pathologists Checklist for NGS Testing

The College of American Pathologists (CAP), a CMS-approved accrediting organization, has recently developed a new set of checklist requirements that are specific to NGS, which advances greater standardization in clinical NGS testing. CAP checklists are available to subscribing laboratories and cover key aspects of laboratory function: policies, procedures and pre-analytical, analytical and post-analytical aspects of clinical testing. There is a customized checklist for every section of a clinical laboratory, as well as a general checklist that applies to all sections. During a laboratory inspection, CAP inspectors use these checklists in their evaluation process, to assess whether laboratories follow regulations and practice guidelines and operate at a quality level that is worthy of CAP accreditation and CLIA certification. The NGS section of the molecular checklist contains a set of requirements for both the analytical wet bench processes, as well as for the various bio-informatics steps required for data analysis and annotation. Even though these requirements are not rigidly prescriptive, they highlight key points that must be considered when documenting the reliability and usefulness of clinical NGS testing methods. Many medical centers are within the second year or third of carrying out such testing and therefore have undergone inspection of their genomic or next-generation DNA sequencing assays by CAP or other groups that carry out inspections of CLIA-certified laboratories. Feedback from inspectors and participant laboratories will be very useful for identifying the areas in which checklists need to be revised or made more detailed, explicit or prescriptive, as well as for highlighting the more difficult or uncertain areas in genome sequence data gathering, interpretation and reporting. Some evaluation of the thoroughness of inspections in this new area and the knowledge and qualifications of the inspectors selected for this task will also be warranted as part of the laboratory medicine profession's due diligence in incorporating these new testing methods into mainstream clinical testing.

(c) Assay Validation Requirements

The assay validation conducted before a test is offered clinically documents that a test is consistently and accurately detecting what it claims to be able to identify. CLIA regulations (Code of Federal Regulations § 493.1253 (b) (2)) stipulate that certain core analytical characteristics must be assessed and documented. These include accuracy, precision, analytical sensitivity, analytical specificity, reportable range, reference intervals (normal values) and any other performance characteristic required for test performance (e.g., carryover, dilutions and calculations). The same parameters should be applied to NGS testing, which ranges in scope from single genes or mutation panels to genome sequencing. Limitations of sequence library generation and interpretation, including poorer quality analysis of repetitive sequence regions, less reliable detection or inability to detect certain categories of variation (e.g., insertions, deletions, and other structural variants), inadequately covered regions and similar problems should be reflected and noted in descriptions of the testing method. During the validation process, every single step of NGS must be evaluated, including sample library preparation, clonal fragment amplification, sequencing and all steps of the analysis.

A key need for NGS assay development and validation is the availability of well-characterized "gold-standard" reference materials. Fortunately, there are several public efforts and commercial products that are beginning to meet this need. As an example, the National Institute of Standards and Technology (NIST), with the Genome in a Bottle Consortium, has developed well-characterized single genome reference material for SNVs and small insertions and deletions [33]. Continued support of these and related efforts are needed to generate additional reference genomes and other reference materials for additional applications (e.g., somatic variants).

(d) Interpretation and Reporting of NGS Results

The ACMG has previously published recommendations for the interpretation and reporting of sequence variations for heritable disease, and updates to these recommendations that include interpretation and reporting of NGS-derived sequence changes are expected to be released soon [34]. Additional recommendations will likely be required for other applications (e.g., somatic mutation testing in cancer, pharmacogenetic variation). Despite the advances in sequencing technology, many of the key principles of interpretation still apply. SNP databases and disease-related collections of sequence variants are immensely helpful in variant annotation and interpretation, but there are significant issues that prevent them from being reliably used for clinical diagnostic purposes. Many population databases contain individuals that have developed or will develop disease, and many of the disease-specific databases include benign variants. This underscores the importance of centralized efforts to generate clinical-grade variant databases, such as ClinVar [35].

The final formal interpretation of NGS results, their official posting into the patient's medical record and their translation into clinical care by the physicians responsible for doing that requires interdisciplinary collaboration, whereby pathologists, geneticists and other laboratory professionals become even more directly involved with others in the healthcare team, in order to ensure accurate diagnostic information for individual patients in the context of their disease phenotype.

(e) Proficiency Testing for NGS Assays

Apart from the creation of NGS-specific checklist items, the CAP is in the process of developing NGS proficiency testing products, which are expected to become available in the near future. No other NGS proficiency testing is available in the U.S. from CMS or a CMS-approved accrediting organization. However, laboratories are required to participate in proficiency testing at least twice per year, and this requirement is currently met by alternative assessment. The purpose of such proficiency testing is to be a central quality assessment tool that is an integral component of laboratory inspections and regulatory requirements. To this end, laboratories commonly perform a blinded proficiency testing exchange with other laboratories.

Assessing the Utility of Genomic Information in Clinical Patient Care

We anticipate that NGS technologies will continuously improve in their ability to detect sequence changes and will increase their overall accuracy and ease of use. The increasing capabilities and enhancements to these instruments will facilitate the clinical use of genomic data. The information that is returned to the patient, however, reaches further than the technical and interpretational aspects alone and includes the perceived and, therefore, subjective value of the information. For individual patients, therefore, there is the aspect of personal usefulness (value from the patient's perspective), as well as clinical utility, which constitutes the net health benefits or the balance of benefit *versus* harm. Clinical utility is a complex metric that includes a variety of aspects, such as the patient population tested, the clinical manifestations of the finding and the rationale for testing. Currently, some clinical questions that are explored with NGS can be addressed with considerable confidence, whereas others reach beyond established knowledge. This includes an assessment of the pathogenicity of some variants detected by NGS. Full disclosure of the level of confidence in the clinical meaning of confirmed results, careful patient selection and informed consenting, with a clear understanding of the context in which NGS testing is sought, and genetic counseling before and after testing are important quality measures for the clinical use of NGS testing.

5. Physician Education and Training

It seems certain that the incorporation of genomic methods into clinical patient care will significantly change some aspects of the practice of medicine, affecting not only those who are performing and interpreting genomic testing, but virtually all healthcare professionals. Efforts to integrate genomic approaches, or approaches to analyze transcriptomes, proteomes, metabolomes, microbiomes, *etc*., into clinical care need to be paralleled by the education of our medical students, residents, clinical fellows and faculty to provide a fund of knowledge and an understanding of the possibilities, strengths and limitations of these approaches when they are translated from research to the bedside. This can be accomplished through core efforts in medical school curriculum design, residency and fellowship training programs, as well as in the form of informal and formal continuing medical education (CME) for practicing physicians. Virtually every medical specialty will need to incorporate genomics aspects, as they pertain to that specialty, into their education. A concerted educational effort in medical schools will be critical to ensure the appropriate application of genomic testing, and resources are beginning to become available from professional and medical specialty organizations (for example, the Training Residents in Genomics Working Group, [36]).

Recently, several pathology residency programs have introduced curriculum changes to include more genomic medicine teaching [27,28,37]. Admittedly, in the face of all the other knowledge that residents must acquire in the course of their training, these initial efforts will not produce pathologists who are equally expert in bioinformatics, histology and the wide scope of other laboratory testing methods, but they do represent a first step toward systematic training in the interpretation of large DNA sequence data sets.

6. Ethical and Privacy Considerations

Quality of Patient Information and Informed Consent

Informed consent is the keystone of the ethical treatment of patients in clinical care settings. Patients must have the opportunity to learn about the benefits and limitations that may be associated with genomic testing and consider whether they wish to proceed with such tests. Especially because genomic methods (and other types of "omics" testing) can, in principle, provide magnitudes more information than those traditionally derived, the informed consent process can be more challenging. The level of that challenge depends on the scope of testing and is very different for single gene tests compared to the large net that can be cast with methods that determine the entire exome or genome of the patient. In the latter scenario, incidental findings may disclose medical or personal issues that the patient was not aware of and that were not part of the reason for the current care episode, but that may impact overall health and medical care. The process of returning incidental findings, especially when they may have direct medical ramifications, is an area of active discussion in the medical genetics community [38,39].

The ACMG has issued a policy statement emphasizing the need for informed consent and the content of such a process prior to exome or genome sequencing for germline conditions [40]. Additional recommendations by the ACMG addressing incidental findings have been released and updated [39]. The initial release of these recommendations gave rise to controversy about the informed consent process, and these issues are being considered by multiple additional medical specialties. Until community consensus is reached, individual institutions need to consider how to evaluate and communicate incidental findings of known significance, as well as those genome changes that are of, as yet, unknown clinical significance. What is reported, therefore, needs to be established upfront and clearly communicated to and meaningfully discussed with the patient or his or her guardian, so that expectations are accurate. An informed consent process includes patient education and counseling prior to test ordering, with information addressing the results that will be included in the patient's report, as well as a discussion of the limits of genomic testing and the interpretation of sequence variants. In addition, as with any other testing, patients need to be able to rely on the privacy and confidentiality of their data.

7. Guiding Principles for Clinical Genomic and Other "Omic" Testing

Any medical center or healthcare organization seeking to incorporate genomic or other "omic" testing into its system of patient care would be well served by: (1) ensuring that the testing is performed in a manner that is fully in accord with relevant legal and regulatory requirements and by personnel with the appropriate training and credentials to perform and interpret such testing; and (2) involving representatives of medical specialties, researchers in genetics, statisticians, computer scientists, bioethicists and hospital administrators in the planning, implementation and integration of this new kind of testing into clinical decision-making. Some key principles we recommend for consideration are listed in Box 1 and are described below:

Box 1. Guiding principles for clinical genomic and other "omic" testing:

(a) Clinical laboratory testing is an integral component of patient care and is held to different standards than research testing not used to guide clinical care.

(b) Clinical genomic testing requires extra effort to be dedicated to designing the informed consent and patient education processes.

(c) Education of physicians and other care-givers about genomic testing methods will be critical for appropriate use and maximal patient benefit.

(d) The use of less-extensively validated genomic testing approaches for clinical care ordinarily should progress in a graded manner from use in "innovative care" settings, followed by use in clinical research settings, before being added to "standard" clinical laboratory testing.

(e) Individual and institutional conflicts of interest in clinical genomic testing must be identified and managed.

(f) These guiding principles also apply to efforts to introduce other clinical "omics" testing into clinical care (such as transcriptomes, proteomes, metabolomes and microbiomes).

(g) Clinical genomic and other "omic" data and methodologies should, to the greatest extent possible, be shared openly with the wider medical and research communities, to accelerate the pace of medical discovery and to increase the quality and reproducibility of clinical genomic data analysis.

(a) Clinical Laboratory Testing Is an Integral Component of Patient Care and Is Held to Different Standards than Research Testing Not Used to Guide Clinical Care

Clinical laboratory testing, regardless of the assay methodology or test complexity, is done to guide patient care decisions, including making diagnoses, counseling patients regarding their prognosis or their future risk of developing disease, guiding management of the patient's condition and making recommendations about reproductive or life style choices. Multidisciplinary committees of clinician specialists and clinical laboratory geneticists, guided by recommendations from medical specialty organizations, as well as other sources of information, may be best able to decide which new genomic tests or applications are sufficiently well-supported by evidence in the scientific literature to be adapted for clinical use.

Any implementation of clinical genomic testing must, of course, comply fully with all relevant laws and regulations governing laboratory tests and, in the United States of America, meet the standards of the professional bodies, such as the College of American Pathologists (CAP) and/or ASHI (the American Society for Histocompatibility and Immunogenetics) that, together with The Joint Commission, have been deemed the status to inspect clinical laboratories on behalf of the Centers for Medicare and Medicaid Services (CMS) to ensure that requirements for CLIA certification are met and all medical tests, including genomic tests, are being carried out responsibly.

(b) Clinical Genomic Testing Requires Extra Effort to be Dedicated to Designing the Informed Consent and Patient Education Processes

Patients must be able to obtain sufficient information about the potential value, future implications and limitations of genomic testing so as to be able to give informed consent if they choose to "opt-in" to the use of such tests for their care. Patient education about genomic test results will help to ensure that any subsequent clinical decision-making is carried out as an informed collaborative process between the patient and their physician. In many cases, this may require additional time to be spent by hospital personnel with the patient to ensure that they understand what is measured and what is interpreted from genomic tests. It is likely that the development of additional educational resources for patients will be necessary for this process. As with almost any other clinical interaction with patients, the use of genomic testing should be done only on the basis of a patient decision to "opt-in", rather than as a default pathway from which patients need to "opt-out".

(c) Education of Physicians and Other Care-Givers about Genomic Testing Methods will be Critical for Appropriate Use and Maximal Patient Benefit

Medical centers and healthcare organizations should consider establishing a set of resources, including a service staffed by individuals with training in molecular genetic pathology, medical genetics and genetic counseling, to educate and advise medical personnel about the proper selection of genetic tests and the appropriate interpretation of their results. The need for such a resource has been highlighted by a recent study performed at a large reference laboratory, which documented a strikingly high rate of inappropriate selection of genetic tests (e.g., ordering the incorrect test, ordering tests that were not needed or ordering suboptimal tests given the clinical question being asked). Approximately 26% of all requests for complex genetic testing for heritable disease were changed following review [41]. These misorders result in unnecessary costs to the healthcare system and, more importantly, may result in significant clinical consequences (e.g., failure or delays in receiving necessary testing, receiving incidental or secondary findings that were not requested or desired).

If clinical findings indicate that genomic tests could be helpful for the care of a particular patient, integrated "tumor board"-style meetings are particularly important in evaluating whether genomic sequencing methods should be applied for the care of that individual patient and for discussing the results and implications of the new genomic data for that patient's care, particularly in challenging cases. It is still an open question as to how the cost of physician and other professional time and effort will be compensated for such diagnostic conferences, but the trend toward health system payments based on patient outcomes rather than the sheer volume of clinical work performed in a patient's care may be compatible with such new efforts, if genomic testing contributes significantly to optimal diagnostic and management decisions and the cost-effectiveness of caring for individuals and populations in coming years.

(d) The Use of Less-Extensively Validated Genomic Testing Approaches for Clinical Care Ordinarily Should Progress in a Graded Manner from Use in "Innovative Care" Settings, Followed by Use in Clinical Research Settings, before being Added to "Standard" Clinical Laboratory Testing

It is likely that a range of different approaches or applications of "genomic" testing will continue to be proposed by physician scientists and other medical investigators, spanning a wide range of different kinds of measurements and interpretations, with widely-varying levels of evidence for their actual clinical utility in different clinical contexts. There are preexisting good models for incorporating innovative clinical methods into practice, and these can be applied to the evaluation of genomic tests supported by various levels of prior evidence. For applications of "genomic" technology that measure already well-established genetic variants with clear clinical significance, typically by replacing older Sanger DNA sequencing assays, rapid incorporation into clinically and analytically validated molecular genetic pathology testing in the CLIA-certified clinical laboratories is advisable. The results of such sequence interpretation are applied for clinical decision-making in the same manner as equivalent results obtained using prior testing methods.

When the clinical value of genomic testing is not well established, but few or no other adequate diagnostic testing options exist, then, as with other types of "innovative clinical care" adopted by medical centers, the application of these tools on very limited numbers of patients for specific purposes at the discretion of the clinician can be considered. If genomic testing will be applied systematically on multiple subjects, without established evidence of clinical utility, it should be carried out in the context of a clinical research study with the accompanying human subject protections and regulations associated with this activity. This would, of course, include obtaining informed consent from the patient for the research, following an explicit and detailed discussion of the limitations of the novel test as a basis for making clinical decisions or healthcare recommendations, the kinds of unexpected (and in some cases, unwanted by the patient) test results that could be reported to the patient and their physician and additional confirmatory testing that would be required before making clinical decisions based on the results of the novel test.

As genomic testing and interpretation methods are validated by ongoing clinical research studies and evidence accumulates for the clinical utility of particular approaches in a given clinical context, some testing and interpretation methods would become sufficiently mature to join the list of "standard" laboratory tests that can be ordered for individual patients by clinicians without the additional safeguards and consultations described above. To the extent that patient informed consent permits, the data and interpreted results of genomic tests should be stored in databases that will permit additional research and discovery to proceed and derive additional clinical insights and knowledge from the testing process.

(e) Individual and Institutional Conflicts of Interest in Clinical Genomic Testing must be Identified and Managed

Conflicts of interest are a serious concern for physicians and others responsible for patient care, and the current period of great discovery and commercial interest in clinical genomics has presented opportunities for physician-scientists and others to become involved in the commercialization of new genomic testing or interpretation methods. All physicians and healthcare institutions must be vigilant to ensure that such potential conflicts of interest do not lead to inappropriate decisions about the kinds of testing approaches to pursue or not to pursue. Individual conflicts of interest, where a faculty member or physician has a financial stake in a private company and/or intellectual property related to genomic testing and analysis methods, are similar to those that apply to the use of other medical technologies, pharmaceuticals and devices. Institutional conflicts of interest are those where the healthcare organization or those directing it could influence decisions about which kinds of diagnostic testing would be used for the care of patients, either to encourage the use of particular test methods, instruments, analytical systems or outsourcing to particular genomic testing to particular companies, or else the avoidance of particular tests or companies. Information about any such potential conflicts should be publicly available, as well as scrutinized and managed within the organization, to ensure and document the propriety and ethical behavior of all participants.

(f) These Guiding Principles also Apply to Efforts to Introduce Other Clinical "Omics" Testing into Clinical Care (Such as Transcriptomes, Proteomes, Metabolomes and Microbiomes).

The improvement in NGS-based sequencing methods over the past five years has resulted in dramatic decreases in the cost per base of DNA sequence. Initially, these technologies revolutionized research endeavors, but clinical laboratories rapidly adopted these technologies. Other related complex testing using NGS or other technologies in the research setting shows significant potential clinical utility. These methods and technologies include RNA sequencing (complementary DNA sequencing), proteomics, metabolomics and metagenomics, as well as functional genomic studies. We suggest that these guiding principles for genomic testing can also be applied to the incorporation of other complex clinical laboratory testing into patient care when sufficient evidence is available to support clinical utility.

(g) Clinical Genomic and Other "Omic" Data and Methodologies Should, to the Greatest Extent Possible, be Shared Openly with the Wider Medical and Research Communities, to Accelerate the Pace of Medical Discovery and to Increase the Quality and Reproducibility of Clinical Genomic Data Analysis

With many technological advances, there are opportunities for private enrichment, as well as the creation of new public resources. The balance between these two components can shape the pace of adoption and the ultimate impact of the technology; the history of the development of the Internet and the role of "open source" contributions to it show the key impact that communities with some element of altruism or public spiritedness can have on the success of a technology. The recent U.S. Supreme Court ruling in *AMP v. Myriad Genetics, Inc.* [42], which determined that human genes themselves are not able to be patented, and the preceding ruling in *Mayo Collaborative Services v. Prometheus Laboratories, Inc.* [43], which found that correlations between measured analytes and medical interpretations of such data do not qualify as patentable inventions, may have somewhat decreased the likelihood that private companies will attempt to use litigation to deter testing for particular human gene mutations [44]. These rulings may increase the likelihood that individual companies may try to amass, and keep in private hands, human genetic information and clinical interpretations as trade secrets. In spite of this possibility, there is now an opportunity for medical centers and other healthcare institutions to cooperate in sharing data, interpretations and analysis methods, to speed the identification of correlations between genomic sequences and disease risks, prognosis and treatment outcomes. Currently, the initial frameworks for such data sharing and coordination efforts are promising, but medical organizations and, particularly, their patients will benefit from further joint activity in the public domain that advances clinical genomics [45] and that can serve as a counter-balance to siloed, competitive, inward-looking efforts (whether in academic or commercial settings).

8. Conclusions

We have entered a new era of clinical testing, in which genetic data and other types of "omics" data are much more easily obtained, but the challenges of their interpretation will likely continue for many years. A balance between efficient adoption of new genomic tests and careful consideration of the reliability and clinical value of the results derived from genomic sequence data is needed as these methods become more widely disseminated and utilized within healthcare systems. There are key differences between the quality standards for assays used in research experimentation compared to those that must be maintained for clinical testing, and these standards are under active development and refinement by laboratory professional organizations, as well as associations focusing on particular clinical conditions or specialties. We have outlined seven key principles for consideration in implementing clinical genomic testing, encompassing the technical, as well as the human elements that must be engaged and coordinated to enable optimal utilization of this new form of clinical care. Above all, we feel that this period of intense exploration and discovery in human genetics represents a major opportunity for cooperative and transparent work between different areas of laboratory and clinical medicine, for the ultimate benefit of the patients.

Author Contributions

All authors contributed to planning, writing and editing the article.

References

1. Boyd, S.D. Diagnostic applications of high-throughput DNA sequencing. *Annu. Rev. Pathol.* **2013**, *8*, 381–410.
2. Mardis, E.R. A decade's perspective on DNA sequencing technology. *Nature* **2011**, *470*, 198–203.
3. Sheridan, C. Illumina claims $1,000 genome win. *Nat. Biotechnol.* **2014**, *32*, 115.
4. Centers for disease control and prevention. Aavailable online: http://www.cdc.gov/genomics/gtesting/ (accessed on 16 July 2014).
5. Ng, S.B.; Buckingham, K.J.; Lee, C.; Bigham, A.W.; Tabor, H.K.; Dent, K.M.; Huff, C.D.; Shannon, P.T.; Jabs, E.W.; Nickerson, D.A.; *et al.* Exome sequencing identifies the cause of a mendelian disorder. *Nat. Genet.* **2010**, *42*, 30–35.
6. Worthey, E.A.; Mayer, A.N.; Syverson, G.D.; Helbling, D.; Bonacci, B.B.; Decker, B.; Serpe, J.M.; Dasu, T.; Tschannen, M.R.; Veith, R.L.; Basehore, M.J.; *et al.* Making a definitive diagnosis: Successful clinical application of whole exome sequencing in a child with intractable inflammatory bowel disease. *Genet. Med.* **2011**, *13*, 255–262.
7. Lupski, J.R.; Reid, J.G.; Gonzaga-Jauregui, C.; Rio Deiros, D.; Chen, D.C.; Nazareth, L.; Bainbridge, M.; Dinh, H.; Jing, C.; Wheeler, D.A.; *et al.* Whole-genome sequencing in a patient with Charcot-Marie-Tooth neuropathy. *N. Engl. J. Med.* **2010**, *362*, 1181–1191.
8. Gahl, W.A.; Tifft, C.J. The NIH undiagnosed diseases program: Lessons learned. *JAMA* **2011**, *305*, 1904–1905.
9. Yang, Y.; Muzny, D.M.; Reid, J.G.; Bainbridge, M.N.; Willis, A.; Ward, P.A.; Braxton, A.; Beuten, J.; Xia, F.; Niu, Z.; *et al.* Clinical whole-exome sequencing for the diagnosis of mendelian disorders. *N. Engl. J. Med.* **2013**, *369*, 1502–1511.
10. Weinstein, J.N.; Collisson, E.A.; Mills, G.B.; Shaw, K.R.; Ozenberger, B.A.; Ellrott, K.; Shmulevich, I.; Sander, C.; Stuart, J.M. The cancer genome atlas pan-cancer analysis project. *Nat. Genet.* **2013**, *45*, 1113–1120.
11. Verhaak, R.G.; Hoadley, K.A.; Purdom, E.; Wang, V.; Qi, Y.; Wilkerson, M.D.; Miller, C.R.; Ding, L.; Golub, T.; Mesirov, J.P.; *et al.* Integrated genomic analysis identifies clinically relevant subtypes of glioblastoma characterized by abnormalities in PDGFRA, IDH1, EGFR, and NF1. *Cancer Cell* **2010**, *17*, 98–110.
12. Bianchi, D.W.; Parker, R.L.; Wentworth, J.; Madankumar, R.; Saffer, C.; Das, A.F.; Craig, J.A.; Chudova, D.I.; Devers, P.L.; Jones, K.W.; *et al.* DNA sequencing versus standard prenatal aneuploidy screening. *N. Engl. J. Med.* **2014**, *370*, 799–808.
13. Cruz-Rivera, M.; Forbi, J.C.; Yamasaki, L.H.; Vazquez-Chacon, C.A.; Martinez-Guarneros, A.; Carpio-Pedroza, J.C.; Escobar-Gutierrez, A.; Ruiz-Tovar, K.; Fonseca-Coronado, S.; Vaughan, G. Molecular epidemiology of viral diseases in the era of next generation sequencing. *J. Clin. Virol.* **2013**, *57*, 378–380.
14. Dewey, F.E.; Grove, M.E.; Pan, C.; Goldstein, B.A.; Bernstein, J.A.; Chaib, H.; Merker, J.D.;

Goldfeder, R.L.; Enns, G.M.; David, S.P.; *et al*. Clinical interpretation and implications of whole-genome sequencing. *JAMA* **2014**, *311*, 1035–1045.
15. Ioannidis, J.P. Why most published research findings are false. *PLoS Med.* **2005**, *2*, e124.
16. Ioannidis, J.P.; Greenland, S.; Hlatky, M.A.; Khoury, M.J.; Macleod, M.R.; Moher, D.; Schulz, K.F.; Tibshirani, R. Increasing value and reducing waste in research design, conduct, and analysis. *Lancet* **2014**, *383*, 166–175.
17. Ioannidis, J.P.; Tarone, R.; McLaughlin, J.K. The false-positive to false-negative ratio in epidemiologic studies. *Epidemiology* **2011**, *22*, 450–456.
18. Halling, K.C.; Schrijver, I.; Persons, D.L. Test verification and validation for molecular diagnostic assays. *Arch. Pathol. Lab. Med.* **2012**, *136*, 11–13.
19. Jennings, L.; van Deerlin, V.M.; Gulley, M.L. Recommended principles and practices for validating clinical molecular pathology tests. *Arch. Pathol. Lab. Med.* **2009**, *133*, 743–755.
20. Ferreira-Gonzalez, A.; Emmadi, R.; Day, S.P.; Klees, R.F.; Leib, J.R.; Lyon, E.; Nowak, J.A.; Pratt, V.M.; Williams, M.S.; Klein, R.D. Revisiting oversight and regulation of molecular-based laboratory-developed tests: A position statement of the association for molecular pathology. *J. Mol. Diagn.* **2014**, *16*, 3–6.
21. Vandeweyer, G.; Kooy, R.F. Detection and interpretation of genomic structural variation in health and disease. *Expert Rev. Mol. Diagn.* **2013**, *13*, 61–82.
22. Rohlfs, E.M.; Zhou, Z.; Heim, R.A.; Nagan, N.; Rosenblum, L.S.; Flynn, K.; Scholl, T.; Akmaev, V.R.; Sirko-Osadsa, D.A.; Allitto, B.A.; *et al*. Cystic fibrosis carrier testing in an ethnically diverse US population. *Clin. Chem.* **2011**, *57*, 841–848.
23. Paynter, N.P.; Chasman, D.I.; Pare, G.; Buring, J.E.; Cook, N.R.; Miletich, J.P.; Ridker, P.M. Association between a literature-based genetic risk score and cardiovascular events in women. *JAMA* **2010**, *303*, 631–637.
24. Palomaki, G.E.; Melillo, S.; Neveux, L.; Douglas, M.P.; Dotson, W.D.; Janssens, A.C.; Balkite, E.A.; Bradley, L.A. Use of genomic profiling to assess risk for cardiovascular disease and identify individualized prevention strategies—A targeted evidence-based review. *Genet. Med.* **2010**, *12*, 772–784.
25. Do, R.; Kathiresan, S.; Abecasis, G.R. Exome sequencing and complex disease: practical aspects of rare variant association studies. *Hum. Mol. Genet.* **2012**, *21*, R1–R9.
26. MacArthur, D.G.; Manolio, T.A.; Dimmock, D.P.; Rehm, H.L.; Shendure, J.; Abecasis, G.R.; Adams, D.R.; Altman, R.B.; Antonarakis, S.E.; Ashley, E.A.; *et al*. Guidelines for investigating causality of sequence variants in human disease. *Nature* **2014**, *508*, 469–476.
27. Schrijver, I.; Natkunam, Y.; Galli, S.; Boyd, S.D. Integration of genomic medicine into pathology residency training: The stanford open curriculum. *J. Mol. Diagn.* **2013**, *15*, 141–148.
28. Haspel, R.L.; Olsen, R.J.; Berry, A.; Hill, C.E.; Pfeifer, J.D.; Schrijver, I.; Kaul, K.L. Progress and potential: Training in genomic pathology. *Arch. Pathol. Lab. Med.* **2014**, *138*, 498–504.
29. Schrijver, I.; Aziz, N.; Farkas, D.H.; Furtado, M.; Gonzalez, A.F.; Greiner, T.C.; Grody, W.W.; Hambuch, T.; Kalman, L.; Kant, J.A.; *et al*. Opportunities and challenges associated with clinical diagnostic genome sequencing: A report of the association for molecular pathology. *J. Mol. Diagn.* **2012**, *14*, 525–540.

30. Alford, R.L.; Arnos, K.S.; Fox, M.; Lin, J.W.; Palmer, C.G.; Pandya, A.; Rehm, H.L.; Robin, N.H.; Scott, D.A.; Yoshinaga-Itano, C. American college of medical genetics and genomics guideline for the clinical evaluation and etiologic diagnosis of hearing loss. *Genet. Med.* **2014**, *16*, 347–355.
31. Rehm, H.L.; Bale, S.J.; Bayrak-Toydemir, P.; Berg, J.S.; Brown, K.K.; Deignan, J.L.; Friez, M.J.; Funke, B.H.; Hegde, M.R.; Lyon, E. ACMG clinical laboratory standards for next-generation sequencing. *Genet. Med.* **2013**, *15*, 733–747.
32. Gargis, A.S.; Kalman, L.; Berry, M.W.; Bick, D.P.; Dimmock, D.P.; Hambuch, T.; Lu, F.; Lyon, E.; Voelkerding, K.V.; Zehnbauer, B.A.; *et al.* Assuring the quality of next-generation sequencing in clinical laboratory practice. *Nat. Biotechnol.* **2012**, *30*, 1033–1036.
33. Zook, J.M.; Chapman, B.; Wang, J.; Mittelman, D.; Hofmann, O.; Hide, W.; Salit, M. Integrating human sequence data sets provides a resource of benchmark SNP and indel genotype calls. *Nat. Biotechnol.* **2014**, *32*, 246–251.
34. Richards, C.S.; Bale, S.; Bellissimo, D.B.; Das, S.; Grody, W.W.; Hegde, M.R.; Lyon, E.; Ward, B.E. ACMG recommendations for standards for interpretation and reporting of sequence variations: Revisions 2007. *Genet. Med.* **2008**, *10*, 294–300.
35. Landrum, M.J.; Lee, J.M.; Riley, G.R.; Jang, W.; Rubinstein, W.S.; Church, D.M.; Maglott, D.R. Clinvar: Public archive of relationships among sequence variation and human phenotype. *Nucleic. Acids. Res.* **2014**, *42*, D980–985.
36. Training Residents in Genomics Working Group. Avaliable online: http://www.ascp.org/trig (accessed on 14 July 2014).
37. Haspel, R.L.; Arnaout, R.; Briere, L.; Kantarci, S.; Marchand, K.; Tonellato, P.; Connolly, J.; Boguski, M.S.; Saffitz, J.E. A call to action: Training pathology residents in genomics and personalized medicine. *Am. J. Clin. Pathol.* **2010**, *133*, 832–834.
38. Bennette, C.S.; Trinidad, S.B.; Fullerton, S.M.; Patrick, D.; Amendola, L.; Burke, W.; Hisama, F.M.; Jarvik, G.P.; Regier, D.A.; Veenstra, D.L. Return of incidental findings in genomic medicine: Measuring what patients value—Development of an instrument to measure preferences for information from next-generation testing (IMPRINT). *Genet. Med.* **2013**, *15*, 873–881.
39. Green, R.C.; Berg, J.S.; Grody, W.W.; Kalia, S.S.; Korf, B.R.; Martin, C.L.; McGuire, A.L.; Nussbaum, R.L.; O'Daniel, J.M.; Ormond, K.E.; *et al.* ACMG recommendations for reporting of incidental findings in clinical exome and genome sequencing. *Genet. Med.* **2013**, *15*, 565–574.
40. ACMG Board of Directors. Points to consider for informed consent for genome/exome sequencing. *Genet. Med.* **2013**, *15*, 748–749.
41. Miller, C.E.; Krautscheid, P.; Baldwin, E.E.; Tvrdik, T.; Openshaw, A.S.; Hart, K.; Lagrave, D. Genetic counselor review of genetic test orders in a reference laboratory reduces unnecessary testing. *Am. J. Med. Genet. A* **2014**, *164*, 1094–1101.
42. Association for Molecular Pathology V. Myriad Genetics. 569 US, Supreme Court, 12–398, 2013.
43. Mayo Collaborative Services V. Prometheus Laboratories. Inc., 566 US, Supreme Court, 10–1150, 2012.
44. Klein, R.D. AMP V. Myriad: The supreme court gives a win to personalized medicine. *J. Mol.*

Diagn. **2013**, *15*, 731–732.

45. National Research Council (US) Committee on a Framework for Developing a New Taxonomy of Disease. *Toward Precision Medicine: Building a Knowledge Network for Biomedical Research and a New Taxonomy of Disease*; National Academies Press: Washington, DC, USA, 2011.

8

From Genotype to Functional Phenotype: Unraveling the Metabolomic Features of Colorectal Cancer

Oliver F. Bathe [1,2,*] **and Farshad Farshidfar** [1]

[1] Department of Surgery, Tom Baker Cancer Center, University of Calgary, 1331 29th St NW, Calgary, AB T2N 4N2, Canada; E-Mail: farshidf@ucalgary.ca

[2] Department of Oncology, Tom Baker Cancer Center, University of Calgary, 1331 29th St NW, Calgary, AB T2N 4N2, Canada

* Author to whom correspondence should be addressed; E-Mail: bathe@ucalgary.ca;

Abstract: Much effort in recent years has been expended in defining the genomic and epigenetic alterations that characterize colorectal adenocarcinoma and its subtypes. However, little is known about the functional ramifications related to various subtypes. Metabolomics, the study of small molecule intermediates in disease, provides a snapshot of the functional phenotype of colorectal cancer. Data, thus far, have characterized some of the metabolic perturbations that accompany colorectal cancer. However, further studies will be required to identify biologically meaningful metabolic subsets, including those corresponding to specific genetic aberrations. Moreover, further studies are necessary to distinguish changes due to tumor and the host response to tumor.

Keywords: metabolomics; colorectal cancer; functional genomics

---

1. Introduction

Colorectal cancer (CRC) is the second leading cause of cancer death in the Western world. CRC invades locally to involve successive layers of the colon or rectum, then spreads via the lymphatic system to regional lymph nodes and/or metastasizes hematogenously to involve distant organs, such as the liver or lungs. Distant metastatic disease is present in about 25% of individuals. The features of the tumor that describe this behavior are reflected in the TNM (Tumor, Lymph Node, Metastasis) staging classification of tumor, where higher degrees of disease portend a worse prognosis. Other clinical and pathologic features that are well known to reflect biological behavior include the presence of obstruction or perforation; degree of differentiation; and presence of lymph or vascular invasion.

In recent years, it has become apparent that CRC has variable biologic behavior, which is not always reflected by its clinicopathological features. Rather, its growth properties, its tendency to metastasize, and its susceptibility to treatments are a function of its molecular properties. For example, colorectal cancers with lymph node involvement more frequently have *BRAF* and *KRAS* mutations [1,2], high levels of CCR7 [3], low levels of thymidylate gene expression [4], as well as *p16INK4A* promoter methylation [1]. The recognition that the molecular properties of CRC dictate biological behavior has led various investigators to subclassify CRC based on its "molecular signature" at the transcriptomic level [5–8]. Even more recently, coordinated efforts have been made to obtain highly detailed analysis of the CRC genome, as well as the downstream transcriptomic and epigenomic events [9]. As a result of these efforts, it is now feasible to refine the classification of CRC, based on molecular pathogenesis.

2. Molecular Subclassification of CRC: From Genomics to Phenotype

Most investigators classify sporadic CRC according to molecular pathway leading to its pathogenesis (Figure 1): chromosomal instability (CIN), microsatellite instability (MSI) and CpG island methylator phenotype (CIMP). Others have proposed a molecular classification system in which groups of CRC are defined according only to MSI and CIMP status in conjunction with clinicopathological features (Figure 2) [10,11]. More recently, based on data generated from The Cancer Genome Atlas Project, CRC has been designated as hypermutated or non-hypermutated, based on mutation rates (Figure 3).

CRC due to CIN represents 80%–85% of sporadic cases. There are imbalances in chromosome number and loss of heterozygosity, as well as accumulation of mutations in tumor suppressor genes and oncogenes that activate pathways critical for CRC initiation and progression. In that pathway, CRC arises from adenomas.

Tumors with a high degree of MSI (MSI-H; about 15% of CRC) are characterized by frequent microsatellite length mutations. MSI occurs due to deficiencies in the mismatch repair (MMR) system, which recognizes and repairs nucleotide mismatches. Most sporadic MSI-H CRCs are caused by hypermethylation (epigenetic silencing) of the mismatch-repair gene *MLH1*. This silencing typically occurs in tumors of the CIMP phenotype. There is substantial overlap between MSI-H cancers and cancers containing a high degree of CIMP (Figure 2).

CIMP represents a specific type of epigenomic stability that is characterized by widespread promoter CpG island methylation and epigenetic gene silencing including tumor suppressor genes. CRCs with a high degree of CIMP are associated with older age, female gender, proximal tumor location, poor tumor differentiation, *BRAF* mutation, wild-type TP53, and high levels of global DNA methylation [12–14]. CIMP tumors have a distinct mRNA expression profile [9]. CIMP is also significantly associated with mucinous or signet ring cell morphology, as well as a marked peritumoral lymphocytic reaction, features that are also associated with MSI-H tumors [15,16]. CIMP and MSI-H tumors are thought to arise via the serrated adenoma pathway (*i.e.*, derived from sessile serrated adenomas, with progressive dysplasia) [16,17].

Figure 1. Classification of CRC by pathogenic pathway. The classical pathway involves a progressive accumulation of mutations due to chromosomal instability as an adenoma develops into adenocarcinoma. Serrated polyps, which are thought to develop from hyperplastic polyps, are generated due to microsatellite instability and/or high levels of CpG island methylation. The adenocarcinomas that emerge from that pathway have distinct clinical and pathological features.

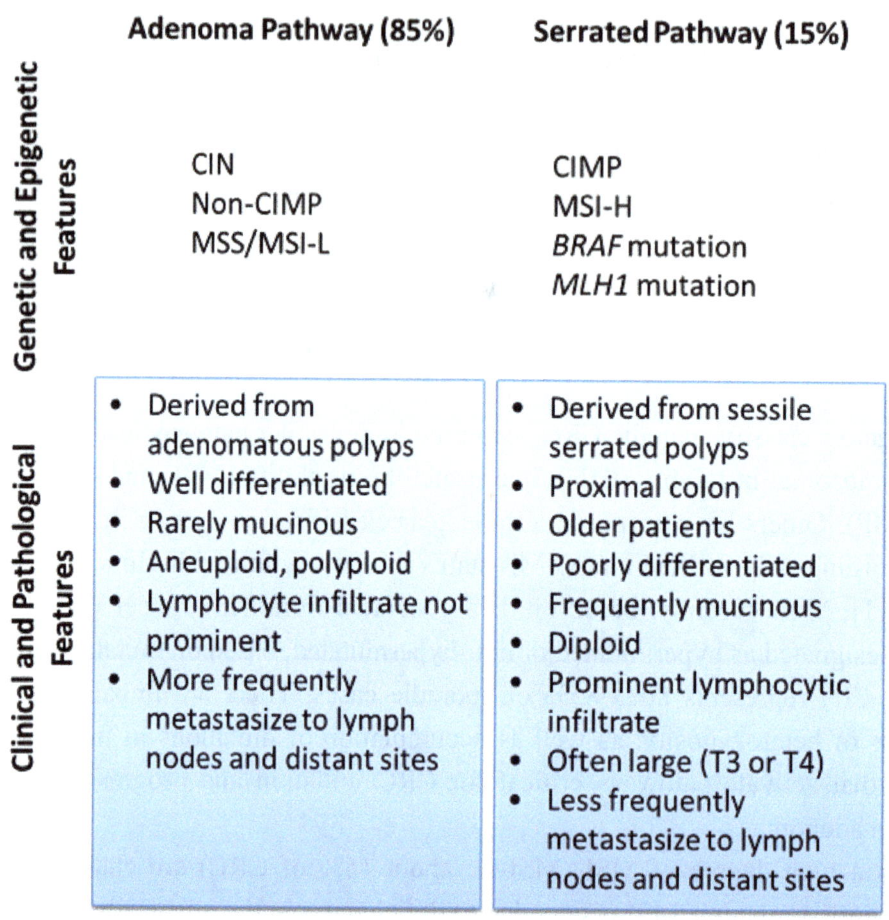

Figure 2. Relationship of CIMP expression phenotype and microsatellite instability. There is significant overlap between MSI-H tumors and CIMP-H tumors, although microsatellite stable (MSS) tumors and tumors with a low level of MSI (MSI-L) may have high levels of CpG island methylation.

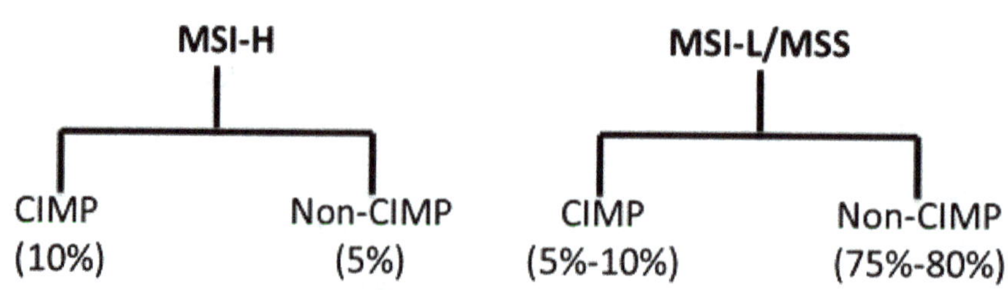

Figure 3. CRC subgroups identified by analysis of molecular data compiled from 224 tumors analyzed by The Cancer Genome Atlas Project [9]. Hypermutated CRC is highly enriched for hypermethylation, CIMP expression phenotype and *BRAF* mutations; it most frequently occurs in the proximal colon.

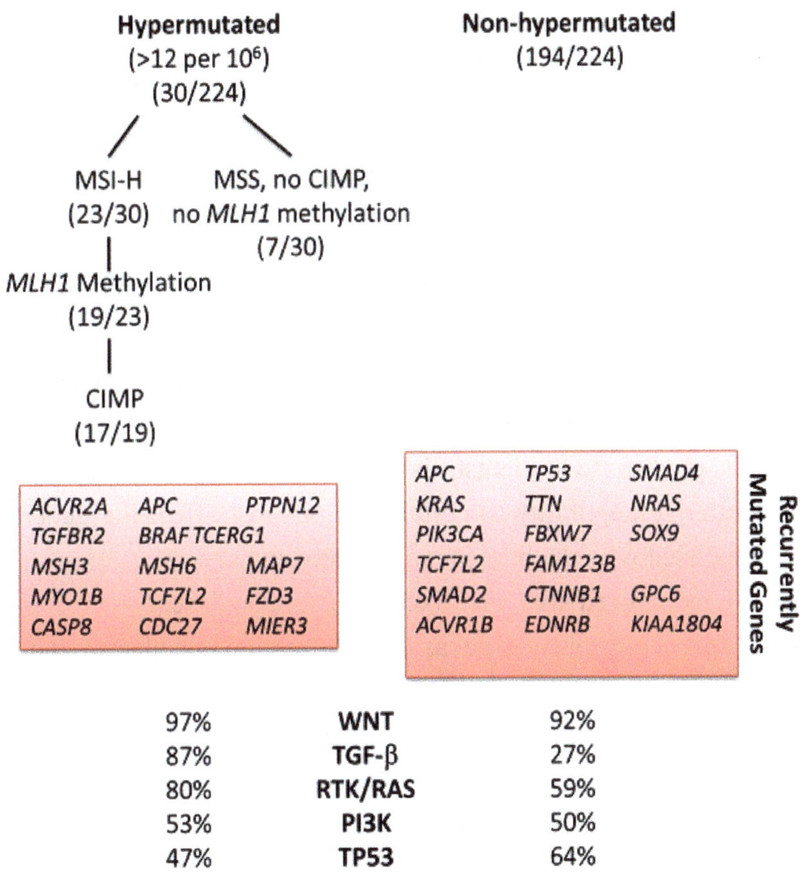

Importantly, the genomic and epigenomic subclass of CRC has clinical significance (and, therefore, biological significance). CIN is associated with a worse prognosis independent of stage and type of therapy [18]. MSI-H tumors have a better prognosis than microsatellite stable (MSS) CRC [19,20]. Hypermethylation is more common in cancers of the proximal colon, and most hypermutated CRCs originate in the proximal colon [9]. The effect of CIMP on prognosis is controversial, and analysis has been complicated by the high degree of overlap with MSI. MSI-H tumors containing a high level of CIMP are less prone to lymph node or distant metastasis [21]. However, in tumors that are not MSI-H, CIMP appears to be associated with worse survivals [22,23]. There are also numerous studies on patient outcomes as a function of individual molecular events (*BRAF* mutation, *KRAS* mutation, *TP53* mutation, *etc.*). However, given the numerous possible interactions with other genetic and molecular events, the findings from such studies should be interpreted with caution. As an example of this, among MSI-H cancers, *TGFBR2* mutations are associated with better survival [24], and this association with improved survival is even more pronounced in the presence of *BAX* mutations [25]. Finally, there is evidence that there is a link between genomic and epigenomic subclass and chemosensitivity [26]. Together, these observations demonstrate that the molecular features of CRC have biological consequences that translate to clinically significant outcomes. To this end, subclassification schemes will provide a good initial framework on which to study the disease.

However, subclassification based on genomic and epigenomic features has limitations, particularly when applying that information to an individual. That is, given the large number of combinations of molecular aberrations that are possible for any subclass of CRC, it is clear that there is substantial heterogeneity even within each subclass—even at the genomic level, let alone at the transcriptomic and proteomic levels. For example, while *BRAF* mutation frequency is particularly high in CIMP-H tumors, it may also occur (albeit infrequently) in CIMP-low tumors [10]. Indeed, there are no known mutations that are pathognomonic for any particular subtype of CRC. Therefore, if therapeutic strategies and decisions are to be derived for any individual, more work is required to define the biological phenotype by studying downstream pathophysiologies.

3. Functional Genomics: Defining the Biological Impact of the CRC Genome

While cataloging the molecular features of CRC subtypes is important, what is most relevant from a clinical standpoint is how particular molecular features translate to differences in tumor biology. Which molecular features predispose to metastasis to various sites in the body, enhance tumor growth, induce angiogenesis, confer susceptibility to certain drugs, most affect the general health of the host (the patient), and are associated with the best (or worst) prognosis?

In general, it is well recognized that the fixed structure and sequence comprising the genome does not closely predict function or phenotype, as the information encoded by the genome is subject to modifications through a multitude of mechanisms and downstream events (Figure 4). There are multiple examples relevant to CRC. Hypermethylation of promoter-associated CpG islands is frequently seen in CRC and leads to transcriptional silencing; hypomethylation of CpG islands outside of promoter regions is also a hallmark of CRC [27]. Long non-coding RNAs that overlap the 5' and 3' termini of genes may regulate the function of one or more genes [28–30]. MicroRNAs (miRNA) function to regulate expression at the post-transcriptional level, and numerous miRNAs are seen to be dysregulated in CRC [31–35]. Pseudogenes further complicate the interpretation of genomic and transcriptomic information. Pseudogenes resemble real genes, but contain premature stop codons and mutations that preclude their translation into functional proteins. Pseudogene-derived transcripts may act as a decoy to functionally significant miRNAs. For example, transcripts corresponding to the pseudogenes *PTENP1* and *KRAS1P* act as a decoy for miRNAs targeting *PTEN* and *KRAS* [36,37], which are known to be important in CRC. Protein translation is further regulated by a number of mechanisms that may vary in CRC [38,39]. Post-translational modifications, which may be altered in CRC, and differential expression of protein isoforms can further affect tumor biology or the host response to CRC [40].

In CRC, the genes and pathways that are particularly important in the initiation and progression of CRC include the WNT, MAPK, phosphatidylinositol-3-kinase (PI3K), TGF-β and p53 signaling pathways [9]. What is becoming increasingly clear is that there is a multitude of genetic and molecular events that can lead to the dysregulation of these signaling pathways in CRC. Moreover, the degree of dysregulation in each of these pathways may vary considerably between individuals, and the biological consequences of those variations are not currently predictable.

Figure 4. CRC is derived from an accumulation of genomic and epigenomic alterations. Alterations at the transcriptional level also occur due to the influence of regulatory RNAs (e.g., long noncoding RNA, miRNA, pseudogenes). Post-translational regulation and post-translational modifications further contribute to functional perturbations in tumor cells. These sequential and synchronous events contribute to the phenotype. Phenotype can further modify the genotype as well as any of the downstream events. According to this model, the metabolome represents the closest molecular representation of phenotype.

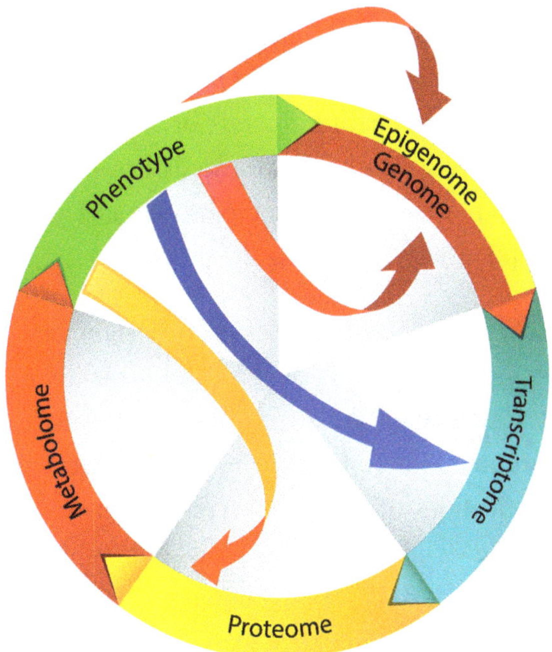

One example of how diverse (but related) mechanisms could lead to the same biological manifestation is in activation of the insulin-like growth factor (IGF)-PI3K pathway. It has recently been identified that about 7% of CRCs have an amplification of insulin-like growth factor 2 (*IGF2*) [9]. IGF2 overexpression may play an important role in the promotion of CRC [41,42]. In 15% of tumors without *IGF2* amplification, *IGF2* gene expression was also overexpressed. *IGF2* amplification or overexpression was associated with genomic events known to activate the PI3K pathway. We and others have also observed that CRC is often associated with high levels of expression of IGF1R (the receptor for IGF2) [43]. The *IRS2* gene, which encodes a protein that links IGF1R with PI3K, is frequently gained in CRC, which may also enhance the activity of this pathway [9]. Activation of the IGF1R-PI3K pathway is associated with multiple functional alterations in tumor cells, including changes in lipid metabolism and inflammatory events. Given the diversity of mechanisms by which this pathway (and any pathway) could be affected, it would be impractical to dissect all of the individual molecular events contributing to the function of that pathway. Rather, it may be more practical to dissect the phenotype so that therapeutic efforts could be directed at that particular imbalance.

Phenotype is therefore a product of preliminary molecular instructions (the genome) amended by a number of sequential and synchronous molecular events at the transcriptional and translational levels. Genomic information is, therefore, least reflective of phenotype and, as more downstream events are taken into account (especially in sum), molecular features more closely predict phenotype. Using this as a framework, it would be expected that proteomic and metabolomic profiles most closely reflect the phenotype and functional state of a cell (Figure 4). As an extension of this functional genomic model, phenotype can further modify any of the preceding molecular processes (including genotype, in cells susceptible to mutation), by affecting the conditions of the intracellular or extracellular microenvironment.

4. Metabolism: A Terminal Function Reflecting Phenotype

While it is possible to identify and measure some of the terminal elements of the proteome, protein function is significantly modified by the abundance of other proteins that may not be simultaneously measured. For example, soluble receptors and binding proteins may compete with functional receptors; ligand function can be modified by competing ligands or inhibitory proteins; and protein fragments can have significant biological effects. Since our knowledge of protein function is far from complete and since it is not yet possible to measure every single protein and protein fragment, one cannot make accurate inferences on phenotype just by evaluating the proteome.

Perhaps the best reflection of tumor phenotype (besides measurement of specific biological functions) is the metabolome. Any biological function is dependent on metabolic functions. Any alterations in metabolism and bioenergetics will alter the efficiency of downstream biological functions. Any perturbations in metabolism are identifiable by simultaneously measuring the abundance of co-related metabolites. Importantly, modifications of any of the constituents of the metabolome are measurable. Constituents of the metabolome—similar to proteins—are subject to some modifications such hydroxylation or amination. Derivatization of functionally known metabolites is traceable by mass spectrometry methods and flux analysis. Moreover, because of the co-relationship of various metabolites, it is possible to extrapolate functional consequences of a metabolomic state despite such modifications.

Metabolic phenotype is dynamic; it is not static or even stable. There are a number of sources for the dynamic nature of the phenotype in tumor. First, while we often consider a tumor to be of a particular genotype, it must be appreciated that any given tumor is actually comprised of cells with substantial molecular heterogeneity [44]. Second, tumor phenotype is sculpted by the host immune response, where susceptible cells are eliminated and resistant cells survive the immune response, immunoediting [45,46]. Third, treatments, such as chemotherapy, alter phenotype by selective mechanisms, based on the pharmacology of any particular pharmacologic agent [44,47,48]. Fourth, the metabolic and inflammatory milieu within the tumor microenvironment may affect the function and phenotype of tumor cells, irrespective of genotype [49,50]. Finally, these same microenvironmental factors can predispose tumor cells to further mutational events [51]. The metabolic state of any tumor is, therefore, a product of all of these ever-changing influences at any given time.

4.1. Disordered Metabolism Is a Hallmark of Cancer

It has recently been recognized that one of the hallmarks of a cancer cell is the reprogramming of energy metabolism [52]. Cancer consists of rapidly proliferating cells, and large amounts of adenosine triphosphate (ATP) and substrates are required to support this proliferation. Sufficient ATP production is possible due to adaptations in metabolic pathways, including processes of carbohydrate, protein, lipid and nucleotide synthesis, which sustain the high metabolic demand.

The classic example of metabolic reprogramming is the Warburg Effect [53]. In normal cells, in the presence of sufficient oxygen, glucose is processed through oxidative phosphorylation, which is the most efficient means of generating ATP. Glycolysis only becomes a primary means to metabolize glucose in hypoxic conditions. However, in cancer cells, glycolysis is the dominant pathway for glucose metabolism, and this is independent of oxygen supply. The advantage to the tumor cell is that this is a much more rapid means of ATP production, which is necessary to support rapid cellular proliferation. The increased glucose processing in cancer cells forms the basis of ^{18}F-fluorodeoxyglucose positron emission tomography (FDG-PET), which is used to detect and monitor tumors including CRC [54,55].

Tumor cells also have other characteristic features of metabolic reprogramming, each of which function to support the rapidly expanding biomass within tumor. Glutamine uptake is enhanced to replenish the tricarboxylic acid (TCA) cycle; glutaminolysis also contributes to the production of acetyl-coenzyme A for subsequent lipid biosynthesis; and there is increased fatty acid and lipid synthesis, which sustains synthesis of cell membranes and lipid derivatives.

The altered metabolism of malignancy influences other hallmark functions of cancer such as proliferation, apoptosis and inflammation. A high rate of cell proliferation is supported by high levels of ATP and substrate. Defective mitochondrial morphology and function can affect susceptibility to apoptosis [56,57]. The metabolic milieu within tumor can produce an immunosuppressive microenvironment. For example, tryptophan depletion due to overexpression of indoleamine 2,3-dioxygenase can suppress T cell responses against tumor [58–60]. Finally, alterations in fatty acid metabolism can result in a proinflammatory state, which is known to deleteriously affect tumor biology in CRC [61–63]. Therefore, not only is altered metabolism in itself a hallmark of cancer; it also supports the other disordered functions that characterize malignancy.

4.2. Genomic and Molecular Events Influencing Metabolism in CRC

While the genetic and molecular pathogenesis of CRC is being delineated in ever greater detail, the effects of the genetic and transcriptional events that characterize CRC are relatively poorly understood. That genotype affects metabolism is apparent in the features of FDG-PET, which varies with *KRAS* mutation status and HIF-1 expression [55]. Table 1 summarizes some of the known metabolic effects of genetic and epigenetic features that frequently accompany CRC. In a number of instances, important proteins can be dysregulated as a result of modulatory events at multiple levels (Figure 5).

KRAS mutations frequently accompany CRC. Tumors with KRAS mutations express high levels of GLUT1 (glucose transporter-1), providing the ability for enhanced glucose uptake and glycolysis, enabling survival in low glucose conditions [64]. KRAS protein expression and activation can be further modified by a number of mechanisms, including post-translational modification (Figure 5).

Figure 5. Examples of predicted metabolic effects of dysregulated proteins associated with CRC. The proteins are dysregulated as a consequence of modulatory molecular events at multiple levels. Alterations in metabolic function (and ultimately phenotype) result from the closely connected functional networks' response to these upstream signals. (GDP: guanine diphosphate; GTP: guanine triphosphate; GEF: guanine nucleotide-exchange factor; GAP: GTPase-activating protein; PPP: pentose phosphate pathway) [65–68].

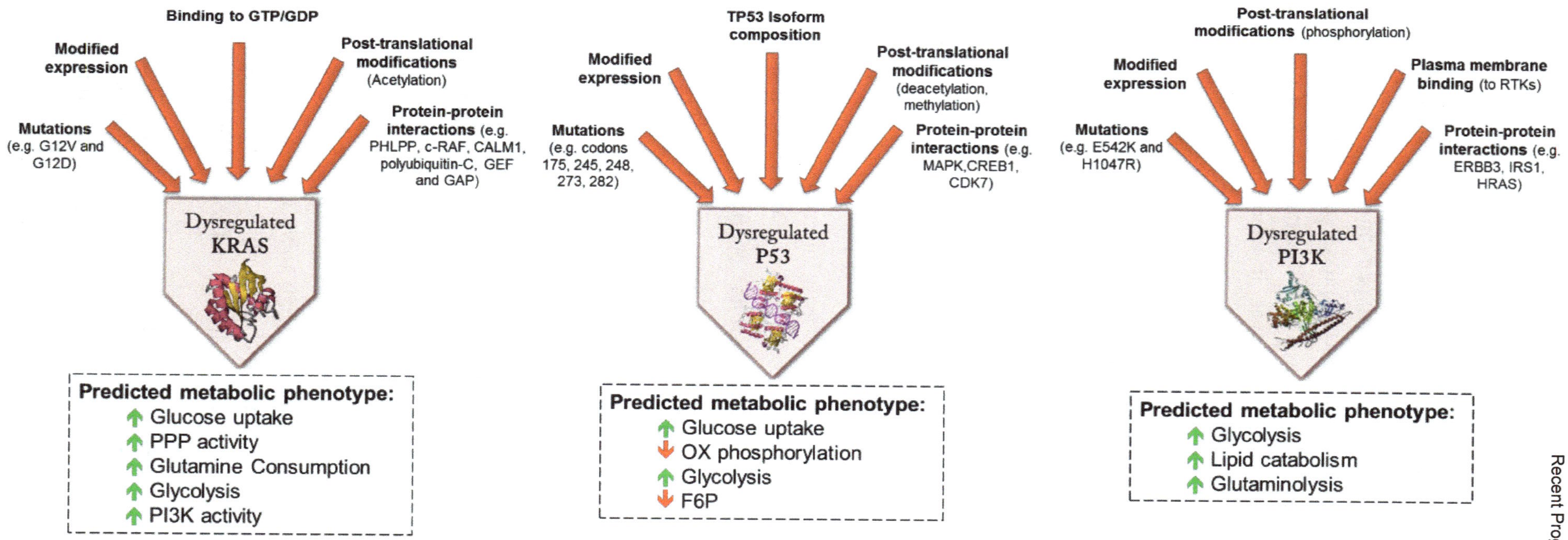

Table 1. Examples of genetic and epigenetic alterations in colorectal cancer (CRC) that have known or potential metabolic consequences.

Gene	Protein Product	Mechanism of Change in Function	Metabolic Effect
TGFBR2	TGF-beta receptor type-2	Inactivating mutation, overexpression	Activation of MAPK/ERK and TGF-β-SMAD pathway; inactivation leads to increased proliferation and decreased apoptosis
TP53	Tumor protein p53	Inactivating mutation or SNP in tumor suppressor	Inhibition of glucose transporters, inhibition of insulin receptor, activation of TCA cycle and oxidative phosphorylation
KRAS	GTPase kras	Activating mutation	Increased glucose uptake. Increased glycolysis, activation of PI3K pathway
PI3KCA	Phosphatidylinositol-4,5-bisphosphate 3-kinase, catalytic subunit alpha	Activating mutation	Increased lipid metabolism, growth-factor independence, increased glycolysis and glutaminolysis
SMAD4	Mothers against decapentaplegic homolog 4	Inactivating mutation	TGF-β signaling
TCF7L2	Transcription factor 7-like 2	Activating mutation	Increased Wnt signaling, increased glycolysis and lactate production
SMAD2	Mothers against decapentaplegic homolog 2	Inactivating mutation in tumor suppressor	TGF-β signaling
CTNNB1	Catenin beta-1	Activating mutation	Wnt signaling pathway
SOX9	SRY (sex determining region Y)-box 9	Mutation or Overexpression of transcription factor	Wnt signaling pathway, inactivation of insulin signaling, anti-proliferation
SOX9	SRY (sex determining region Y)-box 9	Mutation or Overexpression of transcription factor	Wnt signaling pathway, inactivation of insulin signaling, anti-proliferation
ACVR1B	Activin receptor type-1B	Mutation, Overexpression	Activation of TGF-β signaling
EDNRB	Endothelin B receptor	Mutation, hypermethylation, Overexpression	Response to peptide hormonal stimuli
FASN	Fatty acid synthase	Overexpression	Production of fatty acids from Acetyl-CoA
PTGS2 (COX2)	Prostaglandin G/H synthase 2	Overexpression	Modulated by HIF-2α, inducing TGF-β pathway
E-Cadherin (CDH1)	Cadherin 1, type 1, E-cadherin	Mutation, Overexpression	Activates Wnt signaling and lipid metabolism pathway
CDKN2A (p16-INK4a)	Cyclin dependent kinase inhibitor 2A	Mutation, deletion, Methylation	Leads to mitochondrial dysfunction and impaired phosphorylative oxidation, increased glycolysis

Table 1. *Cont.*

Gene	Protein Product	Mechanism of Change in Function	Metabolic Effect
THBS1/TSP1	Thrombospondin 1	Methylation	Regulator of TGF-β signaling, increased inflammation in adipose tissue
SDH	Succinate dehydrogenase complex, subunit B, iron sulfur	Underexpression (mechanism unclear)	Enzyme for TCA cycle, phosphorylative oxidation activity, decreased glucose uptake
PTEN	Phosphatase and tensin homolog	Inactivating mutation in tumor suppressor	Suppressor of PI3K/Akt pathway. Inactivation leads to increased glycolysis, lipogenesis and glycogenesis.
HIF-1α	Hypoxia-inducible factor 1-alpha	Overexpression and molecular stabilization	Activates glycolysis, deactivates TCA cycle and phosphorylative oxidation

HIF-1 overexpression also frequently accompanies CRC [69]. HIF-1 transcription factor activates numerous target genes (reviewed in [70]). Not only is HIF-1 transcription factor a pivotal regulator of oxygen homeostasis; it also encourages glycolysis, contributes to the metabolism of nucleotides and iron, and exerts additional effects on cellular bioenergetics through its mitogenic effects. HIF-1 regulates genes involved in glucose metabolism, encouraging conversion of glucose to lactate. Specifically, HIF-1 increases expression of glucose transporters (GLUT1 and GLUT3) [71] and increases transcription of hexokinase-2 [71]. Direct activation of pyruvate dehydrogenase kinase 1 (PDK1) by HIF-1 leads to inactivation of pyruvate dehydrogenase (PDH), a key enzyme in TCA cycle. This inhibition of mitochondrial biogenesis results in shunting of pyruvate to lactate [72]. HIF-1 also alters the expression of pyruvate kinase (PK), the enzyme that catalyzes the last step of glycolysis. In a number of cancers, including CRC [73], the M2 isoform of (normally found in embryonic tissue) is the predominant PK isoform, and this is encouraged by HIF-1. PKM2 enhances aerobic glycolysis, which represents a selective growth advantage to tumor cells [74].

TP53 mutations frequently occur in CRC. TP53 inhibits transcription of glucose transporters (GLUT1 and GLUT4) and the insulin receptor, reducing glucose uptake by cells. Glucose uptake is further inhibited by activation of NF-κB, which inhibits GLUT3 expression. TP53 also activates TP53-inducible glycolysis and apoptosis regulator (TIGAR), which catalyzes the conversion of fructose-2,6-bisphosphate to fructose-6-phosphate, an important substrate for the pentose phosphate pathway [75]. Accumulation of fructose-6-phosphate also inhibits glycolysis and encourages gluconeogenesis. TIGAR also activates γH2AX complex, which plays an important role in histone methylation of many genes, some of which may have metabolic functions such as *PTEN* [76]. Finally, TP53 activates transcription of *SCO2* (cytochrome C oxidase assembly gene), which encourages oxidative phosphorylation through regulation of complex IV of the mitochondrial respiratory chain [75]. Loss of TP53 function, which results from *TP53* mutations in CRC, therefore, results in enhanced cellular glucose uptake, accelerated glycolysis, as well as reduced oxidative phosphorylation.

At the protein level, p53 isoform composition, post-translational deacetylation and interactions with other proteins may further modify p53 activity (Figure 5).

In CRC, *PTEN* inactivation occurs through mixed genetic and epigenetic mechanisms [77], and it is also controlled by post-translational modifications [78]. *PTEN* is a tumor suppressor gene that functions to antagonize the PI3K/AKT/mammalian target of rapamycin (mTOR) pathway. *PTEN* loss of function results in deregulation of PI3K signaling, leading to constitutively activated AKT. Constitutive activation of AKT results in metabolic changes that are characteristic of the Warburg effect [79], as well as lipogenesis [80].

Alterations in TGF-β signaling are frequent in CRC. Binding of TGF-β initiates downstream signaling involving phosphorylation of SMADs. *TGFBR2* mutations are among the recurrently mutated genes in hypermutated CRCs [9,81]. Mutations in *SMAD2*, *SMAD3*, and *SMAD4* occur primarily in tumors of mucinous histology [82]. These alterations typically result in suppression of the TGF-β antiproliferative effect [83,84].

4.3. Metabolomic Studies Related to CRC

Recently, the field of metabolomics has emerged as a means to more comprehensively study the contribution of tumor on the overall metabolic milieu. The metabolome can be evaluated in a multiplexed fashion using two primary technologies: nuclear magnetic resonance (NMR) spectroscopy and mass spectrometry (MS). Currently no single analytical technology is capable of detecting all metabolites in a biological sample and, since metabolites detectable in one analytical modality are not necessarily detectable in the other, these modalities are complimentary.

There has been considerable interest in the metabolomic analysis of CRC. Several groups have shown that metabolomic profiling of colon mucosa could distinguish between normal and malignant tissues using high resolution magic angle spinning (HR-MAS) ^{1}H-NMR spectroscopy, as well as gas chromatography-mass spectrometry (GC-MS) [85–87]. In general, metabolites associated with the TCA cycle were found to be lower in malignant tissues; and intermediates of the urea cycle, purines, pyrimidines, amino acids, and choline containing compounds were more abundant, consistent with the higher metabolic requirements of rapidly dividing cells.

Fecal water extracts have also been submitted to metabolomic analysis. Monleon *et al.* reported that patients with CRC had low fecal concentrations of short chain fatty acids such as acetate and butyrate and higher levels of proline and cysteine [88]. While there may be some diagnostic utility in this observation, further study is required to understand the origin of these alterations. CRC-associated changes in fecal metabolites may be related to differences in gut mucosa, malabsorption of certain nutrients, or alterations in the gut microflora; any of these differences may represent factors that predispose to CRC or that occur secondary to CRC. In an attempt to delineate the contribution of intestinal bacteria, Weir and coworkers simultaneously assessed stool metabolome (by GC-MS) and gut microbiome [89]. Indeed, in patients with CRC, butyrate-producing species were under-represented and a mucin-degrading species (*Akkermansia muciniphila*) was present at a higher level.

A number of reports have appeared describing the serum metabolome associated with CRC [90–96]. Subtle differences in disease-associated metabolomic changes may reflect population-based differences based on dietary, environmental and genetic factors, although the contribution of these factors to results are currently difficult to identify because of the diversity of analytical platforms used in each of the studies. Qiu *et al.* compared 64 Chinese patients with CRC to healthy controls; metabolomic profiles were determined by GC-MS and liquid chromatography-mass spectrometry (LC-MS) [90]. A distinct metabolomic signature for CRC was identified. In a follow-up replication study, similar findings were observed, which demonstrated alterations in the TCA cycle, urea cycle, glutamine metabolism, and gut flora metabolism [96]. Similar efforts have emerged from Japan, using GC-MS [94]. CRC is associated with changes in the composition of serum fatty acid profile (evaluated by GC-MS) [91]. The serum amino acid profile as identified by electrospray tandem mass spectrometry differs from normal controls [93]. The biological meaning of alterations in fatty acid and amino acid profile will require further interrogation. Stage of CRC as well as site of metastasis may affect metabolomic profile [95]. In addition, there is evidence that the serum metabolomic profile may be related to survival, although it is unclear whether it is of prognostic or predictive significance [97]. In all, these studies demonstrate the feasibility of using metabolomics biomarkers to diagnose CRC, and also suggest that it may be possible to identify subgroups based on metabolomic profile.

Efforts at describing the metabolomic changes that occur in various tissues and biofluids in CRC have largely been descriptive in nature. Much more work will be required to understand the biological implications of any of the metabolomic alterations that typify CRC. Diet, environment and genetic background can represent confounding factors. In metabolomic studies using blood, urine, and stool, it is difficult to determine whether metabolic perturbations are derived from tumor, host factors (including response to tumor), or gut microbiome. To complicate the analysis further, it is possible that some of the metabolic perturbations are not actually a result of CRC, but rather reflect the metabolic state of an individual who is predisposed to CRC. Therefore, to understand the biological basis of metabolomic studies, more detailed experiments and bioinformatic analyses will be required.

5. Linking Genotypic Subsets with Functional Subsets of CRC

As we improve our understanding of the metabolic changes associated with CRC, it will be important to dissect the genetic, epigenetic and transcriptional events that accompany these changes. Using the multiplexed information derived from metabolomic studies, bioinformatic approaches have been described to identify pathways that are putatively involved in the metabolic derangements of CRC [98–100]. This is facilitated by the fact that many metabolites behave in a collinear fashion due to their relationships in metabolic processes. Such an approach generates hypotheses, facilitating a more focused interrogation based on experiments. Indeed, even the limited information derived from a proteomic screen can be used to seed a bioinformatic search for functional networks at the protein and transcriptional levels [101].

Alternatively, multiplexed genomic, transcriptomic, proteomic, and metabolomic data sets can be generated in parallel to catalogue relationships in genotype and phenotype. Integrating "omics" information has been attempted by several groups [100,102–105]. The Cancer Genome Atlas (TCGA) Project represents a large-scale effort at synchronously cataloguing genome sequence, DNA copy number, promoter methylome, and transcriptome for a number of tumor types, including CRC [9]. The tools to analyze these parallel "omics" data sets are evolving quickly. For example, the cBio Cancer Genomics Portal [106] is an online open-access application that enables public access to the multidimensional raw data derived from TCGA, as well as evaluation for possible molecular relationships [107]. The University of California, Santa Cruz (UCSC) Interaction Browser [108] facilitates the simultaneous visualization and analysis of multiple "omics" data sets, enabling integration of biological networks at multiple levels [105]. So far, metabolomic data have not been a large part of such large-scale efforts, but the means to perform such research is now available. The Subpathway-GM method for identifying important metabolic subpathways using genomic and metabolomic data sets represents one example of an effort to integrate metabolomics with upstream molecular events [100].

While integrative analysis of multiple "omics" data sets is attractive, there are some limitations. Even alterations in a single metabolic enzyme or any other molecular event may affect multiple pathways, and such effects can be modified by consequential molecular events. Therefore, such an approach will prove challenging from a bioinformatic perspective. Importantly, opposing molecular events may appear at different biological levels (for example, at the genomic and transcriptional levels), and the net effect of those events cannot be predicted without the final phenotypical and functional information. Finally, it is important to keep in mind that bioinformatic analyses are largely hypothesis-generating, and specific experiments remain an essential means to accentuate our biological knowledge.

To more efficiently derive an understanding of the linkages between genome, epigenome, transcriptome and metabolome, targeted analyses of specific events in large data sets will be required. Moreover, to derive mechanistic insight, discreet experimental systems must be utilized. This represents an important direction in the field, and publications taking this approach are beginning to appear. One such recent effort involved the analysis of 376 surgical samples of CRC and adjacent normal colon from US and Chinese patients, using GC-MS [109]. Fifteen metabolites were found to be differentially abundant in tumor and normal colonic epithelium. Investigators observed several metabolic variations to support proliferation; findings consistent with the Warburg effect (glycolysis and aerobic fermentation); and activation of the pentose phosphate pathway (providing substrates for nucleic acid and fatty acid synthesis). A subsequent targeted transcriptomic analysis was performed based on the pattern of metabolites seen to be differentially abundant in CRC tissue. Fatty acid synthase (FASN) and stearoyl-CoA desaturase-1 (SCD1) were among the most highly upregulated transcripts. Thus, using a combination of bioinformatic interrogation and experimental work, transcriptional and metabolic linkages could be identified.

Manna and coworkers analyzed the urine metabolome of *Apc*$^{Min/+}$ mice as well as mice with azoxymethane (AOM) induced tumors by LC-MS [110]. In mice with colon-specific disruption of *APC*, urinary excretion of amino acid metabolites (e.g., glutamine, proline, Nα-acetyl lysine) and nucleic acid metabolites (e.g., xanthosine, inosine, xanthine, cytidine, deoxyuridine, thymidine) increased progressively during tumorigenesis. Similar changes occurred in mice with AOM-induced colorectal tumors, although there were some differences in individual metabolites in this model. In *Apc*$^{Min/+}$ mice, these metabolomic changes were associated with expression of key genes involved in related pathways. For example, there was overexpression of a number of genes involved in amino acid metabolism, urea cycle and polyamine metabolism. The interconnectivity of these events suggested that the pathogenesis of CRC involved a coordinated reprogramming of metabolic pathways during tumorigenesis.

Finally, Tessem *et al.* utilized HR-MAS ^{1}H-NMR spectroscopy to determine differences in CRC tissue between MSI-H tumors and MSS tumors [87]. The metabolomic profiles were easily distinguished. MSI-H tumors were characterized by higher levels of lactate, glycine, taurine, creatine, and choline; myo-inositol and glucose were decreased. Interestingly, there were also differences seen in adjacent normal colon between MSI-H tumors and MSS tumors. The biological significance of these findings might become apparent with a more comprehensive analysis of the metabolic perturbations seen in each of these CRC subtypes.

Further studies will clearly be required to connect genomic, epigenomic, transcriptomic, and proteomic events to alterations in metabolism in CRC. A combined approach including bioinformatics and targeted experimental analysis appears to be quite constructive. It is possible, since metabolism is a terminal event preceding function, that new phenotypical subtypes can be identified, which may aid in individualizing systemic therapy for CRC.

6. Metabolomics as a Means to Discover Novel Therapeutic Targets

Genotype is known to affect sensitivity to treatment. For example, tumors containing KRAS or NRAS mutations are resistant to EGFR inhibitors [111,112]. Tumors that are MSI-H are resistant to fluoropyrimidines [26]. While genotype has some clinical utility as a predictive biomarker, it does not predict with any certainty whether an individual tumor will respond to any antineoplastic agent, for coincidental mutations and molecular alterations may additionally influence chemosensitivity.

Information derived from metabolomic studies may identify improved ways of targeting the disordered metabolism seen in CRC, which is particularly critical for those tumors that are resistant to other agents. Defining the metabolic phenotype may also enhance therapeutic efforts in individuals and in subgroups. For example, one CRC variant may have dysregulation of specific metabolic pathways that provide it with a growth advantage; other variants may have different metabolic disorders that could be targeted. Indeed, in CRC several metabolic variations have been described [87,109].

The recognition that disordered metabolism is a hallmark of cancer has spurred some interest in therapies targeting metabolism. Cytotoxic drugs, such as fluoropyrimidines, target metabolism and they have been used in practice for years. Because tumors frequently have disordered mitochondrial function, drugs have been developed that affect mitochondrial function [113]. Interventions influencing the disordered carbohydrate metabolism that characterizes most cancers are also attracting interest. For example, oral hypoglycemics used to treat diabetes are being investigated, and retrospective studies have demonstrated reductions in cancer-related mortality in diabetics taking metformin [114,115]. Metformin inhibits the mTOR pathway. Interestingly, metformin is toxic to cancer stem cells [116]. In breast cancer patients, metformin is associated with higher response rates to cytotoxic chemotherapy [117]. A clinical trial is in progress assessing the role of metformin in colorectal adenoma formation [118]. Other mTOR inhibitors are also being tested in CRC [119–121], as are other drugs targeting the insulin-like growth factor pathway [122,123]. These are only some of the examples of pharmacologic agents that target specific metabolic processes that are being evaluated for cancer.

Genotype may aid in identifying the metabolic disorders that are likely to be contained in any particular tumor. For example, tumors with *KRAS* mutations are known to have typical metabolic derangements. *KRAS* transformed fibroblasts lose their proliferative ability with glutamine deprivation [124]. In preclinical models, targeting metabolic enzymes to disrupt glucose metabolism is effective in the treatment of tumors driven by *KRAS* [64,125]. Other common genotypes may similarly be treatable by a specific metabolism-targeted therapy.

The main challenge in designing agents that target metabolism will be to avoid toxicity related to targeting metabolic pathways in normal proliferating cells. Therefore, it will be vital to identify pathways that are redundant in normal cells but absent in cancer cells. Identification of such a therapeutic window may be facilitated by comprehensive analysis of the metabolome in cancer cells and normal cells.

7. Conclusions

It has become apparent that cataloguing the static structural and sequence alterations in the CRC genome merely represents a start to understanding the biology of CRC. It is essential to develop a greater understanding of the dynamic functional perturbations that accompany the genomic changes that characterize CRC and its subtypes, including the multitude of changes in the transcriptome, the proteome, and the metabolome. Moreover, the interactions of each of these elements that comprise each of these downstream molecular events must be dissected. Understanding the functional (or phenotypic) implications of each genotype is imperative to the clinician for a number of reasons. The biological behavior of subsets of CRC can be defined for the purpose of prognostication; and therapies targeting specific biological events can be better engineered. Metabolomics allows a comprehensive analysis of some of the most fundamental biological processes that typify CRC and its subtypes, and is perhaps the closest molecular representation of phenotype currently available.

Author Contributions

Oliver F. Bathe and Farshad Farshidfar both conceived of the paper, researched relevant literature, and wrote the manuscript.

References

1. Miranda, E.; Destro, A.; Malesci, A.; Balladore, E.; Bianchi, P.; Baryshnikova, E.; Franchi, G.; Morenghi, E.; Laghi, L.; Gennari, L.; *et al.* Genetic and epigenetic changes in primary metastatic and nonmetastatic colorectal cancer. *Br. J. Cancer* **2006**, *95*, 1101–1107.
2. Oliveira, C.; Velho, S.; Moutinho, C.; Ferreira, A.; Preto, A.; Domingo, E.; Capelinha, A.F.; Duval, A.; Hamelin, R.; Machado, J.C.; *et al.* KRAS and BRAF oncogenic mutations in MSS colorectal carcinoma progression. *Oncogene* **2007**, *26*, 158–163.
3. Gunther, K.; Leier, J.; Henning, G.; Dimmler, A.; Weissbach, R.; Hohenberger, W.; Forster, R. Prediction of lymph node metastasis in colorectal carcinoma by expression of chemokine receptor CCR7. *Int. J. Cancer* **2005**, *116*, 726–733.

4. Artinyan, A.; Essani, R.; Lake, J.; Kaiser, A.M.; Vukasin, P.; Danenberg, P.; Danenberg, K.; Haile, R.; Beart, R.W., Jr. Molecular predictors of lymph node metastasis in colon cancer: Increased risk with decreased thymidylate synthase expression. *J. Gastrointest. Surg.* **2005**, *9*, 1216–1221.
5. Lin, Y.M.; Furukawa, Y.; Tsunoda, T.; Yue, C.T.; Yang, K.C.; Nakamura, Y. Molecular diagnosis of colorectal tumors by expression profiles of 50 genes expressed differentially in adenomas and carcinomas. *Oncogene* **2002**, *21*, 4120–4128.
6. Arango, D.; Wilson, A.J.; Shi, Q.; Corner, G.A.; Aranes, M.J.; Nicholas, C.; Lesser, M.; Mariadason, J.M.; Augenlicht, L.H. Molecular mechanisms of action and prediction of response to oxaliplatin in colorectal cancer cells. *Br. J. Cancer* **2004**, *91*, 1931–1946.
7. Mariadason, J.M.; Arango, D.; Shi, Q.; Wilson, A.J.; Corner, G.A.; Nicholas, C.; Aranes, M.J.; Lesser, M.; Schwartz, E.L.; Augenlicht, L.H. Gene expression profiling-based prediction of response of colon carcinoma cells to 5-fluorouracil and camptothecin. *Cancer Res.* **2003**, *63*, 8791–8812.
8. Li, M.; Lin, Y.M.; Hasegawa, S.; Shimokawa, T.; Murata, K.; Kameyama, M.; Ishikawa, O.; Katagiri, T.; Tsunoda, T.; Nakamura, Y.; *et al.* Genes associated with liver metastasis of colon cancer, identified by genome-wide cDNA microarray. *Int. J. Oncol.* **2004**, *24*, 305–312.
9. The Cancer Genome Atlas Network. Comprehensive molecular characterization of human colon and rectal cancer. *Nature* **2012**, *487*, 330–337.
10. Ogino, S.; Goel, A. Molecular classification and correlates in colorectal cancer. *J. Mol. Diagn.* **2008**, *10*, 13–27.
11. Jass, J.R. Classification of colorectal cancer based on correlation of clinical, morphological and molecular features. *Histopathology* **2007**, *50*, 113–130.
12. Ogino, S.; Cantor, M.; Kawasaki, T.; Brahmandam, M.; Kirkner, G.J.; Weisenberger, D.J.; Campan, M.; Laird, P.W.; Loda, M.; Fuchs, C.S. Cpg island methylator phenotype (CIMP) of colorectal cancer is best characterised by quantitative DNA methylation analysis and prospective cohort studies. *Gut* **2006**, *55*, 1000–1006.
13. Samowitz, W.S.; Albertsen, H.; Herrick, J.; Levin, T.R.; Sweeney, C.; Murtaugh, M.A.; Wolff, R.K.; Slattery, M.L. Evaluation of a large, population-based sample supports a cpg island methylator phenotype in colon cancer. *Gastroenterology* **2005**, *129*, 837–845.
14. Weisenberger, D.J.; Siegmund, K.D.; Campan, M.; Young, J.; Long, T.I.; Faasse, M.A.; Kang, G.H.; Widschwendter, M.; Weener, D.; Buchanan, D.; *et al.* Cpg island methylator phenotype underlies sporadic microsatellite instability and is tightly associated with braf mutation in colorectal cancer. *Nat. Genet.* **2006**, *38*, 787–793.
15. Ogino, S.; Odze, R.D.; Kawasaki, T.; Brahmandam, M.; Kirkner, G.J.; Laird, P.W.; Loda, M.; Fuchs, C.S. Correlation of pathologic features with CpG island methylator phenotype (CIMP) by quantitative DNA methylation analysis in colorectal carcinoma. *Am. J. Surg. Pathol.* **2006**, *30*, 1175–1183.
16. Soreide, K.; Janssen, E.A.M.; Soiland, H.; Korner, H.; Baak, J.P.A. Microsatellite instability in colorectal cancer. *Br. J. Surg.* **2006**, *93*, 395–406.
17. Grady, W.M.; Carethers, J.M. Genomic and epigenetic instability in colorectal cancer pathogenesis. *Gastroenterology* **2008**, *135*, 1079–1099.

18. Walther, A.; Houlston, R.; Tomlinson, I. Association between chromosomal instability and prognosis in colorectal cancer: A meta-analysis. *Gut* **2008**, *57*, 941–950.
19. Popat, S.; Hubner, R.; Houlston, R.S. Systematic review of microsatellite instability and colorectal cancer prognosis. *J. Clin. Oncol.* **2005**, *23*, 609–618.
20. Guastadisegni, C.; Colafranceschi, M.; Ottini, L.; Dogliotti, E. Microsatellite instability as a marker of prognosis and response to therapy: A meta-analysis of colorectal cancer survival data. *Eur. J. Cancer* **2010**, *46*, 2788–2798.
21. Laghi, L.; Malesci, A. Microsatellite instability and therapeutic consequences in colorectal cancer. *Dig. Dis.* **2012**, *30*, 304–309.
22. Ward, R.L.; Cheong, K.; Ku, S.L.; Meagher, A.; O'Connor, T.; Hawkins, N.J. Adverse prognostic effect of methylation in colorectal cancer is reversed by microsatellite instability. *J. Clin. Oncol.* **2003**, *21*, 3729–3736.
23. Ogino, S.; Meyerhardt, J.A.; Kawasaki, T.; Clark, J.W.; Ryan, D.P.; Kulke, M.H.; Enzinger, P.C.; Wolpin, B.M.; Loda, M.; Fuchs, C.S. CpG island methylation, response to combination chemotherapy, and patient survival in advanced microsatellite stable colorectal carcinoma. *Virchows Arch.* **2007**, *450*, 529–537.
24. Watanabe, T.; Wu, T.T.; Catalano, P.J.; Ueki, T.; Satriano, R.; Haller, D.G.; Benson, A.B., 3rd; Hamilton, S.R. Molecular predictors of survival after adjuvant chemotherapy for colon cancer. *N. Engl. J. Med.* **2001**, *344*, 1196–1206.
25. Jung, B.; Smith, E.J.; Doctolero, R.T.; Gervaz, P.; Alonso, J.C.; Miyai, K.; Keku, T.; Sandler, R.S.; Carethers, J.M. Influence of target gene mutations on survival, stage and histology in sporadic microsatellite unstable colon cancers. *Int. J. Cancer* **2006**, *118*, 2509–2513.
26. Ribic, C.M.; Sargent, D.J.; Moore, M.J.; Thibodeau, S.N.; French, A.J.; Goldberg, R.M.; Hamilton, S.R.; Laurent-Puig, P.; Gryfe, R.; Shepherd, L.E.; *et al.* Tumor microsatellite-instability status as a predictor of benefit from fluorouracil-based adjuvant chemotherapy for colon cancer. *N. Engl. J. Med.* **2003**, *349*, 247–257.
27. Lao, V.V.; Grady, W.M. Epigenetics and colorectal cancer. *Nat. Rev. Gastroenterol. Hepatol.* **2011**, *8*, 686–700.
28. Matsui, M.; Chu, Y.; Zhang, H.; Gagnon, K.T.; Shaikh, S.; Kuchimanchi, S.; Manoharan, M.; Corey, D.R.; Janowski, B.A. Promoter RNA links transcriptional regulation of inflammatory pathway genes. *Nucleic Acids Res.* **2013**, *41*, 10086–10109.
29. Svoboda, M.; Slyskova, J.; Schneiderova, M.; Makovicky, P.; Bielik, L.; Levy, M.; Lipska, L.; Hemmelova, B.; Kala, Z.; Protivankova, M.; *et al.* HOTAIR long non-coding RNA is a negative prognostic factor not only in primary tumors, but also in the blood of colorectal cancer patients. *Carcinogenesis* **2014**, *35*, 1510–1515.
30. Qi, P.; Xu, M.D.; Ni, S.J.; Shen, X.H.; Wei, P.; Huang, D.; Tan, C.; Sheng, W.Q.; Zhou, X.Y.; Du, X. Down-regulation of ncRAN, a long non-coding RNA, contributes to colorectal cancer cell migration and invasion and predicts poor overall survival for colorectal cancer patients. *Mol. Carcinog.* **2014**, doi:10.1002/mc.22137.
31. Chen, T.; Yao, L.Q.; Shi, Q.; Ren, Z.; Ye, L.C.; Xu, J.M.; Zhou, P.H.; Zhong, Y.S. MicroRNA-31 contributes to colorectal cancer development by targeting factor inhibiting HIF-1alpha (FIH-1). *Cancer Biol. Ther.* **2014**, *15*, 516–523.

32. Pichler, M.; Ress, A.L.; Winter, E.; Stiegelbauer, V.; Karbiener, M.; Schwarzenbacher, D.; Scheideler, M.; Ivan, C.; Jahn, S.W.; Kiesslich, T.; *et al.* MIR-200a regulates epithelial to mesenchymal transition-related gene expression and determines prognosis in colorectal cancer patients. *Br. J. Cancer* **2014**, 110, 1614–1621.
33. Cappuzzo, F.; Sacconi, A.; Landi, L.; Ludovini, V.; Biagioni, F.; D'Incecco, A.; Capodanno, A.; Salvini, J.; Corgna, E.; Cupini, S.; *et al.* Microrna signature in metastatic colorectal cancer patients treated with anti-EGFR monoclonal antibodies. *Clin. Colorectal Cancer* **2014**, *13*, 37–45.
34. Nosho, K.; Igarashi, H.; Nojima, M.; Ito, M.; Maruyama, R.; Yoshii, S.; Naito, T.; Sukawa, Y.; Mikami, M.; Sumioka, W.; *et al.* Association of microRNA-31 with BRAF mutation, colorectal cancer survival and serrated pathway. *Carcinogenesis* **2014**, *35*, 776–783.
35. Pizzini, S.; Bisognin, A.; Mandruzzato, S.; Biasiolo, M.; Facciolli, A.; Perilli, L.; Rossi, E.; Esposito, G.; Rugge, M.; Pilati, P.; *et al.* Impact of micrornas on regulatory networks and pathways in human colorectal carcinogenesis and development of metastasis. *BMC Genomics* **2013**, *14*, doi:10.1186/1471-2164-14-589.
36. Poliseno, L.; Salmena, L.; Zhang, J.; Carver, B.; Haveman, W.J.; Pandolfi, P.P. A coding-independent function of gene and pseudogene mrnas regulates tumour biology. *Nature* **2010**, *465*, 1033–1038.
37. Tay, Y.; Kats, L.; Salmena, L.; Weiss, D.; Tan, S.M.; Ala, U.; Karreth, F.; Poliseno, L.; Provero, P.; di Cunto, F.; *et al.* Coding-independent regulation of the tumor suppressor pten by competing endogenous mrnas. *Cell* **2011**, *147*, 344–357.
38. Matassa, D.S.; Amoroso, M.R.; Agliarulo, I.; Maddalena, F.; Sisinni, L.; Paladino, S.; Romano, S.; Romano, M.F.; Sagar, V.; Loreni, F.; *et al.* Translational control in the stress adaptive response of cancer cells: A novel role for the heat shock protein trap1. *Cell Death Dis.* **2013**, *4*, e851.
39. Dixon, D.A. Dysregulated post-transcriptional control of COX-2 gene expression in cancer. *Curr. Pharm. Des.* **2004**, *10*, 635–646.
40. Pedersen, J.W.; Blixt, O.; Bennett, E.P.; Tarp, M.A.; Dar, I.; Mandel, U.; Poulsen, S.S.; Pedersen, A.E.; Rasmussen, S.; Jess, P.; *et al.* Seromic profiling of colorectal cancer patients with novel glycopeptide microarray. *Int. J. Cancer* **2011**, *128*, 1860–1871.
41. Nakagawa, H.; Chadwick, R.B.; Peltomaki, P.; Plass, C.; Nakamura, Y.; de la Chapelle, A. Loss of imprinting of the insulin-like growth factor II gene occurs by biallelic methylation in a core region of H19-associated CTCF-binding sites in colorectal cancer. *Proc. Natl. Acad. Sci. USA* **2001**, *98*, 591–596.
42. Levine, A.J.; Ihenacho, U.; Lee, W.; Figueiredo, J.C.; Vandenberg, D.J.; Edlund, C.K.; Davis, B.D.; Stern, M.C.; Haile, R.W. Genetic variation in insulin pathway genes and distal colorectal adenoma risk. *Int. J. Colorectal Dis.* **2012**, *27*, 1587–1595.
43. Guo, S.T.; Jiang, C.C.; Wang, G.P.; Li, Y.P.; Wang, C.Y.; Guo, X.Y.; Yang, R.H.; Feng, Y.; Wang, F.H.; Tseng, H.Y.; *et al.* Microrna-497 targets insulin-like growth factor 1 receptor and has a tumour suppressive role in human colorectal cancer. *Oncogene* **2013**, *32*, 1910–1920.
44. Janku, F. Tumor heterogeneity in the clinic: Is it a real problem? *Ther. Adv. Med. Oncol.* **2014**, *6*, 43–51.
45. Mittal, D.; Gubin, M.M.; Schreiber, R.D.; Smyth, M.J. New insights into cancer immunoediting and its three component phases-elimination, equilibrium and escape. *Curr. Opin. Immunol.* **2014**, *27C*, 16–25.

46. Bathe, O.; Dalyot-Herman, N.; Malek, T. Therapeutic limitations in tumor-specific CD8$^+$ memory T cell engraftment. *BMC Cancer* **2003**, *3*, doi:10.1186/1471-2407-3-21.
47. Schwitalla, S. Tumor cell plasticity: The challenge to catch a moving target. *J. Gastroenterol.* **2014**, *49*, 618–627.
48. Navin, N.E. Tumor evolution in response to chemotherapy: Phenotype *versus* genotype. *Cell Rep.* **2014**, *6*, 417–419.
49. Wu, D.; Wu, P.; Huang, Q.; Liu, Y.; Ye, J.; Huang, J. Interleukin-17: A promoter in colorectal cancer progression. *Clin. Dev. Immunol.* **2013**, doi:10.1155/2013/436307.
50. Grivennikov, S.I.; Wang, K.; Mucida, D.; Stewart, C.A.; Schnabl, B.; Jauch, D.; Taniguchi, K.; Yu, G.Y.; Osterreicher, C.H.; Hung, K.E.; *et al.* Adenoma-linked barrier defects and microbial products drive IL-23/IL-17-mediated tumour growth. *Nature* **2012**, *491*, 254–258.
51. Pozza, A.; Scarpa, M.; Ruffolo, C.; Polese, L.; Erroi, F.; Bridda, A.; Norberto, L.; Frego, M. Colonic carcinogenesis in ibd: Molecular events. *Ann. Ital. Chir.* **2011**, *82*, 19–28.
52. Hanahan, D.; Weinberg, R.A. Hallmarks of cancer: The next generation. *Cell* **2011**, *144*, 646–674.
53. Warburg, O. On respiratory impairment in cancer cells. *Science* **1956**, *124*, 269–270.
54. Lee, H.S.; Kim, H.O.; Hong, Y.S.; Kim, T.W.; Kim, J.C.; Yu, C.S.; Kim, J.S. Prognostic value of metabolic parameters in patients with synchronous colorectal cancer liver metastasis following curative-intent colorectal and hepatic surgery. *J. Nucl. Med.* **2014**, *55*, 582–589.
55. Miles, K.A.; Ganeshan, B.; Rodriguez-Justo, M.; Goh, V.J.; Ziauddin, Z.; Engledow, A.; Meagher, M.; Endozo, R.; Taylor, S.A.; Halligan, S.; *et al.* Multifunctional imaging signature for V-Ki-RAS2 Kirsten rat sarcoma viral oncogene homolog (KRAS) mutations in colorectal cancer. *J. Nucl. Med.* **2014**, *55*, 386–391.
56. Babbar, M.; Sheikh, M.S. Metabolic stress and disorders related to alterations in mitochondrial fission or fusion. *Mol. Cell. Pharmacol.* **2013**, *5*, 109–133.
57. Grills, C.; Jithesh, P.V.; Blayney, J.; Zhang, S.D.; Fennell, D.A. Gene expression meta-analysis identifies VDAC1 as a predictor of poor outcome in early stage non-small cell lung cancer. *PLoS One* **2011**, *6*, e14635.
58. Uyttenhove, C.; Pilotte, L.; Theate, I.; Stroobant, V.; Colau, D.; Parmentier, N.; Boon, T.; van den Eynde, B.J. Evidence for a tumoral immune resistance mechanism based on tryptophan degradation by indoleamine 2,3-dioxygenase. *Nat. Med.* **2003**, *9*, 1269–1274.
59. Ino, K.; Yamamoto, E.; Shibata, K.; Kajiyama, H.; Yoshida, N.; Terauchi, M.; Nawa, A.; Nagasaka, T.; Takikawa, O.; Kikkawa, F. Inverse correlation between tumoral indoleamine 2,3-dioxygenase expression and tumor-infiltrating lymphocytes in endometrial cancer: Its association with disease progression and survival. *Clin. Cancer Res.* **2008**, *14*, 2310–2317.
60. Sucher, R.; Kurz, K.; Weiss, G.; Margreiter, R.; Fuchs, D.; Brandacher, G. IDO-mediated tryptophan degradation in the pathogenesis of malignant tumor disease. *Int. J. Tryptophan Res.* **2010**, *3*, 113–120.
61. Cai, F.; Dupertuis, Y.M.; Pichard, C. Role of polyunsaturated fatty acids and lipid peroxidation on colorectal cancer risk and treatments. *Curr. Opin. Clin. Nutr. Metab. Care* **2012**, *15*, 99–106.
62. McMillan, D.C.; Crozier, J.E.; Canna, K.; Angerson, W.J.; McArdle, C.S. Evaluation of an inflammation-based prognostic score (GPS) in patients undergoing resection for colon and rectal cancer. *Int. J. Colorectal Dis.* **2007**, *22*, 881–886.

63. Crozier, J.E.; McKee, R.F.; McArdle, C.S.; Angerson, W.J.; Anderson, J.H.; Horgan, P.G.; McMillan, D.C. Preoperative but not postoperative systemic inflammatory response correlates with survival in colorectal cancer. *Br. J. Surg.* **2007**, *94*, 1028–1032.
64. Yun, J.; Rago, C.; Cheong, I.; Pagliarini, R.; Angenendt, P.; Rajagopalan, H.; Schmidt, K.; Willson, J.K.; Markowitz, S.; Zhou, S.; *et al.* Glucose deprivation contributes to the development of KRAS pathway mutations in tumor cells. *Science* **2009**, *325*, 1555–1559.
65. Weinberg, F.; Hamanaka, R.; Wheaton, W.W.; Weinberg, S.; Joseph, J.; Lopez, M.; Kalyanaraman, B.; Mutlu, G.M.; Budinger, G.R.; Chandel, N.S. Mitochondrial metabolism and ROS generation are essential for Kras-mediated tumorigenicity. *Proc. Natl. Acad. Sci. USA* **2010**, *107*, 8788–8793.
66. Chatr-aryamontri, A.; Breitkreutz, B.-J.; Heinicke, S.; Boucher, L.; Winter, A.; Stark, C.; Nixon, J.; Ramage, L.; Kolas, N.; O'Donnell, L.; *et al.* The biogrid interaction database: 2013 Update. *Nucl. Acids Res.* **2013**, *41*, D816–D823.
67. Berman, H.M.; Westbrook, J.; Feng, Z.; Gilliland, G.; Bhat, T.N.; Weissig, H.; Shindyalov, I.N.; Bourne, P.E. The protein data bank. *Nucl. Acids Res.* **2000**, *28*, 235–242.
68. Magrane, M.; Consortium, U. Uniprot knowledgebase: A hub of integrated protein data. *Database* **2011**, *2011*, doi:10.1093/database/bar009.
69. Furlan, D.; Sahnane, N.; Carnevali, I.; Cerutti, R.; Uccella, S.; Bertolini, V.; Chiaravalli, A.M.; Capella, C. Up-regulation and stabilization of HIF-1alpha in colorectal carcinomas. *Surg. Oncol.* **2007**, *16*, S25–S27.
70. Semenza, G.L. Targeting HIF-1 for cancer therapy. *Nat. Rev. Cancer* **2003**, *3*, 721–732.
71. Levine, A.J.; Puzio-Kuter, A.M. The control of the metabolic switch in cancers by oncogenes and tumor suppressor genes. *Science* **2010**, *330*, 1340–1344.
72. Kim, J.W.; Tchernyshyov, I.; Semenza, G.L.; Dang, C.V. HIF-1-mediated expression of pyruvate dehydrogenase kinase: A metabolic switch required for cellular adaptation to hypoxia. *Cell Metab.* **2006**, *3*, 177–185.
73. Bluemlein, K.; Gruning, N.M.; Feichtinger, R.G.; Lehrach, H.; Kofler, B.; Ralser, M. No evidence for a shift in pyruvate kinase PKM1 to PKM2 expression during tumorigenesis. *Oncotarget* **2011**, *2*, 393–400.
74. Christofk, H.R.; Vander Heiden, M.G.; Harris, M.H.; Ramanathan, A.; Gerszten, R.E.; Wei, R.; Fleming, M.D.; Schreiber, S.L.; Cantley, L.C. The M2 splice isoform of pyruvate kinase is important for cancer metabolism and tumour growth. *Nature* **2008**, *452*, 230–233.
75. Maddocks, O.D.; Vousden, K.H. Metabolic regulation by p53. *J. Mol. Med. (Berl.)* **2011**, *89*, 237–245.
76. Sinha, S.; Ghildiyal, R.; Mehta, V.S.; Sen, E. ATM-NFkappab axis-driven tigar regulates sensitivity of glioma cells to radiomimetics in the presence of TNFalpha. *Cell Death Dis.* **2013**, *4*, e615.
77. Molinari, F.; Frattini, M. Functions and regulation of the PTEN gene in colorectal cancer. *Front. Oncol.* **2013**, *3*, e326.
78. Chalhoub, N.; Baker, S.J. PTEN and the PI3-kinase pathway in cancer. *Annu. Rev. Pathol.* **2009**, *4*, 127–150.

79. Elstrom, R.L.; Bauer, D.E.; Buzzai, M.; Karnauskas, R.; Harris, M.H.; Plas, D.R.; Zhuang, H.; Cinalli, R.M.; Alavi, A.; Rudin, C.M.; *et al.* AKT stimulates aerobic glycolysis in cancer cells. *Cancer Res.* **2004**, *64*, 3892–3899.
80. Buzzai, M.; Bauer, D.E.; Jones, R.G.; Deberardinis, R.J.; Hatzivassiliou, G.; Elstrom, R.L.; Thompson, C.B. The glucose dependence of AKT-transformed cells can be reversed by pharmacologic activation of fatty acid beta-oxidation. *Oncogene* **2005**, *24*, 4165–4173.
81. Ogino, S.; Kawasaki, T.; Ogawa, A.; Kirkner, G.J.; Loda, M.; Fuchs, C.S. TGFBR2 mutation is correlated with CpG island methylator phenotype in microsatellite instability-high colorectal cancer. *Hum. Pathol.* **2007**, *38*, 614–620.
82. Fleming, N.I.; Jorissen, R.N.; Mouradov, D.; Christie, M.; Sakthianandeswaren, A.; Palmieri, M.; Day, F.; Li, S.; Tsui, C.; Lipton, L.; *et al.* SMAD2, SMAD3 and SMAD4 mutations in colorectal cancer. *Cancer Res.* **2013**, *73*, 725–735.
83. Bellam, N.; Pasche, B. TGF-beta signaling alterations and colon cancer. *Cancer Treat. Res.* **2010**, *155*, 85–103.
84. Kim, Y.S.; Yi, Y.; Choi, S.G.; Kim, S.J. Development of TGF-beta resistance during malignant progression. *Arch. Pharm. Res.* **1999**, *22*, 1–8.
85. Chan, E.C.; Koh, P.K.; Mal, M.; Cheah, P.Y.; Eu, K.W.; Backshall, A.; Cavill, R.; Nicholson, J.K.; Keun, H.C. Metabolic profiling of human colorectal cancer using high-resolution magic angle spinning nuclear magnetic resonance (HR-MAS NMR) spectroscopy and gas chromatography mass spectrometry (GC/MS). *J. Proteome Res.* **2009**, *8*, 352–361.
86. Denkert, C.; Budczies, J.; Weichert, W.; Wohlgemuth, G.; Scholz, M.; Kind, T.; Niesporek, S.; Noske, A.; Buckendahl, A.; Dietel, M.; *et al.* Metabolite profiling of human colon carcinoma—Deregulation of TCA cycle and amino acid turnover. *Mol. Cancer* **2008**, *7*, doi:10.1186/1476-4598-7-72.
87. Tessem, M.B.; Selnaes, K.M.; Sjursen, W.; Trano, G.; Giskeodegard, G.F.; Bathen, T.F.; Gribbestad, I.S.; Hofsli, E. Discrimination of patients with microsatellite instability colon cancer using 1H HR MAS MR spectroscopy and chemometric analysis. *J. Proteome Res.* **2010**, *9*, 3664–3670.
88. Monleon, D.; Morales, J.M.; Barrasa, A.; Lopez, J.A.; Vazquez, C.; Celda, B. Metabolite profiling of fecal water extracts from human colorectal cancer. *NMR Biomed.* **2009**, *22*, 342–348.
89. Weir, T.L.; Manter, D.K.; Sheflin, A.M.; Barnett, B.A.; Heuberger, A.L.; Ryan, E.P. Stool microbiome and metabolome differences between colorectal cancer patients and healthy adults. *PLoS One* **2013**, *8*, e70803.
90. Qiu, Y.; Cai, G.; Su, M.; Chen, T.; Zheng, X.; Xu, Y.; Ni, Y.; Zhao, A.; Xu, L.X.; Cai, S.; *et al.* Serum metabolite profiling of human colorectal cancer using GC-TOFMS and UPLC-QTOFMS. *J. Proteome Res.* **2009**, *8*, 4844–4850.
91. Kondo, Y.; Nishiumi, S.; Shinohara, M.; Hatano, N.; Ikeda, A.; Yoshie, T.; Kobayashi, T.; Shiomi, Y.; Irino, Y.; Takenawa, T.; *et al.* Serum fatty acid profiling of colorectal cancer by gas chromatography/mass spectrometry. *Biomark. Med.* **2011**, *5*, 451–460.
92. Ludwig, C.; Ward, D.G.; Martin, A.; Viant, M.R.; Ismail, T.; Johnson, P.J.; Wakelam, M.J.; Gunther, U.L. Fast targeted multidimensional NMR metabolomics of colorectal cancer. *Magn. Reson. Chem.* **2009**, *47*, S68–S73.

93. Leichtle, A.B.; Nuoffer, J.M.; Ceglarek, U.; Kase, J.; Conrad, T.; Witzigmann, H.; Thiery, J.; Fiedler, G.M. Serum amino acid profiles and their alterations in colorectal cancer. *Metabolomics* **2012**, *8*, 643–653.
94. Nishiumi, S.; Kobayashi, T.; Ikeda, A.; Yoshie, T.; Kibi, M.; Izumi, Y.; Okuno, T.; Hayashi, N.; Kawano, S.; Takenawa, T.; *et al.* A novel serum metabolomics-based diagnostic approach for colorectal cancer. *PLoS One* **2012**, *7*, e40459.
95. Farshidfar, F.; Weljie, A.M.; Kopciuk, K.; Buie, W.D.; Maclean, A.; Dixon, E.; Sutherland, F.R.; Molckovsky, A.; Vogel, H.J.; Bathe, O.F. Serum metabolomic profile as a means to distinguish stage of colorectal cancer. *Genome Med.* **2012**, *4*, doi:10.1186/gm341.
96. Tan, B.; Qiu, Y.; Zou, X.; Chen, T.; Xie, G.; Cheng, Y.; Dong, T.; Zhao, L.; Feng, B.; Hu, X.; *et al.* Metabonomics identifies serum metabolite markers of colorectal cancer. *J. Proteome Res.* **2013**, *12*, 3000–3009.
97. Bertini, I.; Cacciatore, S.; Jensen, B.V.; Schou, J.V.; Johansen, J.S.; Kruhoffer, M.; Luchinat, C.; Nielsen, D.L.; Turano, P. Metabolomic NMR fingerprinting to identify and predict survival of patients with metastatic colorectal cancer. *Cancer Res.* **2012**, *72*, 356–364.
98. Xia, J.; Psychogios, N.; Young, N.; Wishart, D.S. Metaboanalyst: A web server for metabolomic data analysis and interpretation. *Nucl. Acids Res.* **2009**, *37*, W652–W660.
99. Ingenuity Systems Pathway Analysis. Availbale online: http://www.ingenuity.com/ (aceessed on 23 December 2014).
100. Li, C.; Han, J.; Yao, Q.; Zou, C.; Xu, Y.; Zhang, C.; Shang, D.; Zhou, L.; Zou, C.; Sun, Z.; *et al.* Subpathway-GM: Identification of metabolic subpathways via joint power of interesting genes and metabolites and their topologies within pathways. *Nucl. Acids Res.* **2013**, *41*, e101.
101. Nibbe, R.K.; Koyutürk, M.; Chance, M.R. An integrative-omics approach to identify functional sub-networks in human colorectal cancer. *PLoS Comput. Biol.* **2010**, *6*, e1000639.
102. Kanani, H.; Dutta, B.; Klapa, M.I. Individual *vs.* Combinatorial effect of elevated CO_2 conditions and salinity stress on arabidopsis thaliana liquid cultures: Comparing the early molecular response using time-series transcriptomic and metabolomic analyses. *BMC Syst. Biol.* **2010**, *4*, doi:10.1186/1752-0509-4-177.
103. Grimplet, J.; Cramer, G.R.; Dickerson, J.A.; Mathiason, K.; van Hemert, J.; Fennell, A.Y. Vitisnet: "Omics" integration through grapevine molecular networks. *PLoS One* **2009**, *4*, e8365.
104. Oberbach, A.; Bluher, M.; Wirth, H.; Till, H.; Kovacs, P.; Kullnick, Y.; Schlichting, N.; Tomm, J.M.; Rolle-Kampczyk, U.; Murugaiyan, J.; *et al.* Combined proteomic and metabolomic profiling of serum reveals association of the complement system with obesity and identifies novel markers of body fat mass changes. *J. Proteome Res.* **2011**, *10*, 4769–4788.
105. Wong, C.K.; Vaske, C.J.; Ng, S.; Sanborn, J.Z.; Benz, S.C.; Haussler, D.; Stuart, J.M. The UCSC interaction browser: Multidimensional data views in pathway context. *Nucl. Acids Res.* **2013**, *41*, W218–W224.
106. Cbioportal for cancer genomics. Avaible online: http://www.cbioportal.org/public-portal/ (accessed on 14 January 2014).
107. Cerami, E.; Gao, J.; Dogrusoz, U.; Gross, B.E.; Sumer, S.O.; Aksoy, B.A.; Jacobsen, A.; Byrne, C.J.; Heuer, M.L.; Larsson, E.; *et al.* The cbio cancer genomics portal: An open platform for exploring multidimensional cancer genomics data. *Cancer Discov.* **2012**, *2*, 401–404.

108. Interaction browser. Avaible online: http://sysbio.soe.ucsc.edu/nets/ (accessed on 14 January 2014).
109. Qiu, Y.; Cai, G.; Zhou, B.D.O.M.; Li, D.; Zhao, A.; Xie, G.; Li, H.; Cai, S.; Xie, D.; Huang, C.; *et al.* A distinct metabolic signature of human colorectal cancer with prognostic potential. *Clin. Cancer Res.* **2014**, *20*, 2136–2146.
110. Manna, S.K.; Tanaka, N.; Krausz, K.W.; Haznadar, M.; Xue, X.; Matsubara, T.; Bowman, E.D.; Fearon, E.R.; Harris, C.C.; Shah, Y.M.; *et al.* Biomarkers of coordinate metabolic reprogramming in colorectal tumors in mice and humans. *Gastroenterology* **2014**, *146*, 1313–1324.
111. Douillard, J.Y.; Oliner, K.S.; Siena, S.; Tabernero, J.; Burkes, R.; Barugel, M.; Humblet, Y.; Bodoky, G.; Cunningham, D.; Jassem, J.; *et al.* Panitumumab-FOLFOX4 treatment and RAS mutations in colorectal cancer. *N. Engl. J. Med.* **2013**, *369*, 1023–1034.
112. Karapetis, C.S.; Khambata-Ford, S.; Jonker, D.J.; O'Callaghan, C.J.; Tu, D.; Tebbutt, N.C.; Simes, R.J.; Chalchal, H.; Shapiro, J.D.; Robitaille, S.; *et al.* K-ras mutations and benefit from cetuximab in advanced colorectal cancer. *N. Engl. J. Med.* **2008**, *359*, 1757–1765.
113. Leanza, L.; Zoratti, M.; Gulbins, E.; Szabo, I. Mitochondrial ion channels as oncological targets. *Oncogene* **2014**, doi:10.1038/onc.2013.578.
114. Evans, J.M.; Donnelly, L.A.; Emslie-Smith, A.M.; Alessi, D.R.; Morris, A.D. Metformin and reduced risk of cancer in diabetic patients. *Br. Med. J.* **2005**, *330*, 1304–1305.
115. Bowker, S.L.; Majumdar, S.R.; Veugelers, P.; Johnson, J.A. Increased cancer-related mortality for patients with type 2 diabetes who use sulfonylureas or insulin. *Diabetes Care* **2006**, *29*, 254–258.
116. Hirsch, H.A.; Iliopoulos, D.; Tsichlis, P.N.; Struhl, K. Metformin selectively targets cancer stem cells, and acts together with chemotherapy to block tumor growth and prolong remission. *Cancer Res.* **2009**, *69*, 7507–7511.
117. Jiralerspong, S.; Palla, S.L.; Giordano, S.H.; Meric-Bernstam, F.; Liedtke, C.; Barnett, C.M.; Hsu, L.; Hung, M.C.; Hortobagyi, G.N.; Gonzalez-Angulo, A.M. Metformin and pathologic complete responses to neoadjuvant chemotherapy in diabetic patients with breast cancer. *J. Clin. Oncol.* **2009**, *27*, 3297–3302.
118. Higurashi, T.; Takahashi, H.; Endo, H.; Hosono, K.; Yamada, E.; Ohkubo, H.; Sakai, E.; Uchiyama, T.; Hata, Y.; Fujisawa, N.; *et al.* Metformin efficacy and safety for colorectal polyps: A double-blind randomized controlled trial. *BMC Cancer* **2012**, *12*, doi:10.1186/1471-2407-12-118.
119. Jardim, D.L.; Wheler, J.J.; Hess, K.; Tsimberidou, A.M.; Zinner, R.; Janku, F.; Subbiah, V.; Naing, A.; Piha-Paul, S.A.; Westin, S.N.; *et al.* FBXW7 mutations in patients with advanced cancers: Clinical and molecular characteristics and outcomes with mtor inhibitors. *PLoS One* **2014**, *9*, e89388.
120. Francipane, M.G.; Lagasse, E. mTOR Pathway in colorectal cancer: An update. *Oncotarget* **2014**, *5*, 49–66.
121. Francipane, M.G.; Lagasse, E. Selective targeting of human colon cancer stem-like cells by the mTOR inhibitor Torin-1. *Oncotarget* **2013**, *4*, 1948–1962.
122. Ewing, G.P.; Goff, L.W. The insulin-like growth factor signaling pathway as a target for treatment of colorectal carcinoma. *Clin. Colorectal Cancer* **2010**, *9*, 219–223.
123. Golan, T.; Javle, M. Targeting the insulin growth factor pathway in gastrointestinal cancers. *Oncology (Williston Park)* **2011**, *25*, 518–526, 529.

124. Gaglio, D.; Soldati, C.; Vanoni, M.; Alberghina, L.; Chiaradonna, F. Glutamine deprivation induces abortive S-phase rescued by deoxyribonucleotides in K-ras transformed fibroblasts. *PLoS One* **2009**, *4*, e4715.

125. Clem, B.; Telang, S.; Clem, A.; Yalcin, A.; Meier, J.; Simmons, A.; Rasku, M.A.; Arumugam, S.; Dean, W.L.; Eaton, J.; *et al.* Small-molecule inhibition of 6-phosphofructo-2-kinase activity suppresses glycolytic flux and tumor growth. *Mol. Cancer Ther.* **2008**, *7*, 110–120.

9

Characterization of the Genomic Architecture and Mutational Spectrum of a Small Cell Prostate Carcinoma

Alan F. Scott [1,*], David W. Mohr [1], Hua Ling [1], Robert B. Scharpf [2], Peng Zhang [1] and Gregory S. Liptak [3,†]

[1] McKusick-Nathans Institute of Genetic Medicine, Johns Hopkins University School of Medicine, Baltimore, MD 21287, USA; E-Mails: dwmohr@jhmi.edu (D.W.M.); hling1@jhmi.edu (H.L.); pzhang13@jhmi.edu (P.Z.)

[2] Department of Oncology, Johns Hopkins University School of Medicine, Baltimore, MD 21287, USA; E-Mail: rscharpf@jhsph.edu

[3] Department of Pediatrics, SUNY Upstate Medical Center, Golisano Children's Hospital, Syracuse, NY 13210, USA; E-Mail: gliptak@rochester.rr.com

† This author has been deceased.

* Author to whom correspondence should be addressed; E-Mail: afscott@jhmi.edu;

Abstract: We present the use of a series of laboratory, analytical and interpretation methods to investigate personalized cancer care for a case of small cell prostate carcinoma (SCPC), a rare and aggressive tumor with poor prognosis, for which the underlying genomic architecture and mutational spectrum has not been well characterized. We performed both SNP genotyping and exome sequencing of a Virchow node metastasis from a patient with SCPC. A variety of methods were used to analyze and interpret the tumor genome for copy number variation, loss of heterozygosity (LOH), somatic mosaicism and mutations in genes from known cancer pathways. The combination of genotyping and exome sequencing approaches provided more information than either technique alone. The results showed widespread evidence of copy number changes involving most chromosomes including the possible loss of both alleles of CDKN1B (p27/Kip1). LOH was observed for the regions encompassing the tumor suppressors TP53, RB1, and CHD1. Predicted damaging somatic mutations were observed in the retained TP53 and RB1 alleles. Mutations in other genes that may be functionally relevant were noted, especially the recently reported high confidence cancer drivers FOXA1 and CCAR1. The disruption of multiple cancer drivers underscores why SCPC may be such a difficult cancer to manage.

Keywords: small cell prostate cancer (SCPC); exome sequencing; Illumina SNP array; Ingenuity; NGS (next generation sequencing)

---

1. Introduction

The promise of the Human Genome Project (HGP), for which we mark the tenth anniversary, was that individualized genomics would become a reality for medical diagnosis and care. However, only recently have methods for sequencing, data analysis and the interpretation of variation with respect to medically relevant sequence information become sufficiently robust to make this approach useful. In this paper we compared different methods to investigate the genomic architecture and mutational spectrum of a rare tumor, small cell prostate cancer (SCPC). Our goal was to identify which methods were most informative and what information might provide the best guidance to the patient and his physician. Secondarily, we hoped to provide further characterization for this tumor type that may be of use to the community.

SCPC is a high-grade malignant tumor with neuroendrocrine differentiation sometimes referred to as neuroendocrine prostate cancer (NEPC) [1]. SCPC is often discovered after the occurrence of metastases, has been reported to account for 0.5%–2% of all prostate carcinomas and has a median survival from diagnosis of approximately 12.5 months [2]. The largest SCPC series was published by Wang and Epstein [3] who histologically examined 95 cases of which 92% showed expression of the neuroendocrine marker CD56 (NCAM1) and of which approximately 80% failed to show elevated PSA levels. Aparicio *et al.* [1] noted that, although rare as a primary diagnosis, NEPC may be more common than appreciated and could account for as much as 25% of lethal prostate cancer.

A few studies have looked at the genomic events characterizing NEPC. Beltran *et al.* [4] measured gene expression using NGS RNA-sequencing and oligonucleotide arrays in NEPC tumors and observed a correlation between overexpression of MYCN and AURKA both of which were amplified at the gene level in 40% of NEPCs. The authors also noted evidence for a TMPRSS2-ERG gene fusion, a lack of the ERG protein marker, high expression of the neuroendocrine genes CGA and SYP, and low expression of the androgen-regulated genes KLK3 (PSA), TMPRSS2 and NXK3.1. Beltran *et al.* [4] further showed that NEPC cell lines were sensitive to the AURKA inhibitor danusertib which produced a suppression of neuroendocrine expression. However, phase II clinical trials of men with castration-resistant prostate cancer were disappointing [5]. Tzelepi *et al.* [6] produced SCPC xenografts and performed expression studies and genomic profiling using array-CGH (comparative genomic hybridization) which showed up-regulation of UBE2C and other mitotic genes along with the absence of expression of the androgen receptor (AR), RB1, and cyclin D1. A subset of tumors showed microdeletions of RB1. Grasso *et al.* [7] sequenced 50 lethal metastatic castration-resistant prostate cancers (CRPC) which include SCPC. The authors identified subsets of tumors with either disruptions in CHD1 (chromodomain helicase DNA-binding protein 1) or in ETS2 (usually from fusions of ETS2 with TMPRSS2). The authors also found mutations in multiple genes whose protein products physically interact with androgen receptors such as the ERG gene fusion, the chromatin modifying protein MLL2, and FOXA1 among others. Grasso *et al.* [7] further showed that mutated FOXA1 repressed androgen signaling and enhanced tumor growth. The importance of FOXA1 in tumor progression was also demonstrated by Imamura *et al.* [8] who were able to reduce proliferation in cell culture with an siRNA directed against FOXA1. Van Allen *et al.* [9] performed whole exome sequencing on a CRPC bone metastasis and identified a homozygous deletion in PTEN and a nonsense mutation in BRCA2, both of which suggested clinical treatment strategies.

The diversity of findings in the studies cited above likely results from both tumor heterogeneity and the different laboratory methods used. In this study we have used array based genotyping to examine the overall genomic architecture, exome sequencing to identify somatic mutations, various software tools to analyze the resulting data and both public and commercial interpretation tools to attempt to understand the findings.

2. Experimental

2.1. Sample

The patient was a consented 63-year old male of European ancestry who presented with hematuria and without elevated levels of prostate specific antigen (PSA). The cancer was detected after the development of metastases and the diagnosis of SCPC was made at the patient's primary care hospital and confirmed at the Johns Hopkins Hospital. High molecular weight DNA was isolated from two needle aspirates of a metastatic Virchow node and from saliva (Scope™ mouthwash, Procter & Gamble) using standard methods but with extended proteinase K digestion time for the biopsies. Cells from the aspirates were examined by a pathologist at the time of collection and the remaining tissue was transferred, unfixed, to the laboratory for DNA isolation. A total of 25.5 µg of DNA was obtained from the tumor and 60 µg from the mouthwash collection.

2.2. Genomic SNP Array and Analysis

The DNA from the metastasis was adjusted to 50 ng/µL and 5 µL (250 ng) were genotyped on an Illumina HumanOmni2.5S BeadChip™ array at the SNP Center of the Genetic Resources Core Facility [10] at the Johns Hopkins School of Medicine. Illumina GenomeStudio software (Illumina Inc., San Diego, CA, USA) was used to process the array data and calculate B-allele frequencies (BAF) and log R Ratios (LRR). The BAF and LRR values generated by GenomeStudio ver. 1.7.4 (Illumina, Inc.) are plotted in Figure 1 (panels A and B, respectively) for all the autosomes. The LRR values were segmented using the circular binary segmentation algorithm implemented in the R package DNAcopy (ver. 1.36.0) [11]. The black lines in the LRR plots are the average LRRs for those segments of the chromosome.

Figure 1. Allele frequencies (**A**) and log R ratios (**B**) estimated by GenomeStudio. Autosomal log R ratios were segmented by circular binary segmentation [11] as indicated in black; (**C**) Log R values from whole exome sequencing were obtained by the EXCAVATOR program [12] and aligned to panels A and B, providing a qualitatively similar profile of the copy number alterations. Black lines depict the segmentation of the log R values.

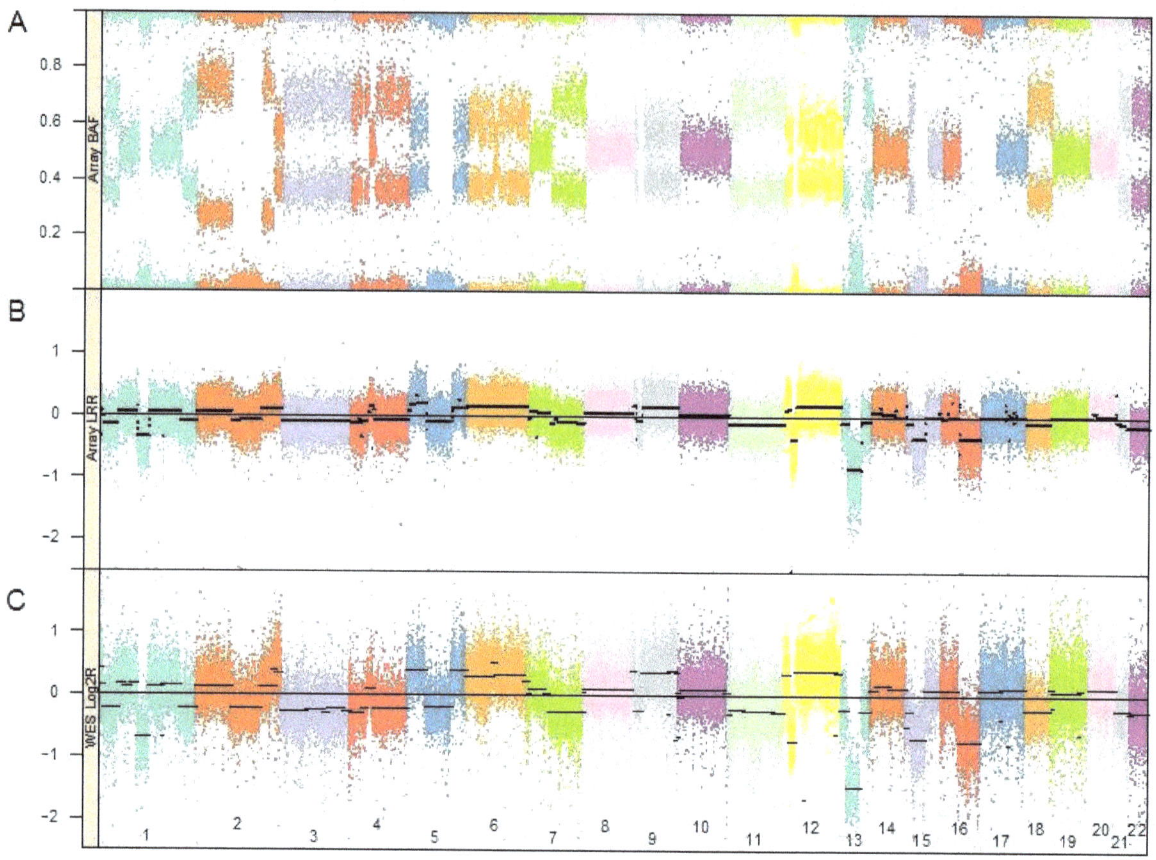

2.3. Exome Capture and Sequencing

Exome-sequencing was performed at the High-Throughput Sequencing facility of the GRCF. DNA (3 μg) from the tumor and saliva were sheared to a size of 150 to 200 bp using a Covaris E210 system (Covaris Inc., Woburn, MA, USA). End repair and addition of an overhanging "A" base was performed using a NEBNext^{TM} reagent kit (New England Biolabs, Ipswich, MA, USA). DNA fragments were ligated to library adapters (Illumina). The ligated fragments were then size selected through purification using SPRI beads and PCR amplified to prepare the libraries. An Agilent Bioanalyzer DNA1000 assay was used for quality control of the libraries to ensure adequate concentration and appropriate fragment size. Sequencing was performed on an Illumina HiSeq^{TM} 2000 following library capture with an Agilent SureSelect All Exon v3 kit. Sample indexing was applied to distinguish the source of the libraries. Sequence data was processed using CIDRSeqsuite v2.3.0 [13] as follows. Sequence reads were processed through Illumina software generating base calls and corresponding base-call quality scores. Resulting data was aligned to hg19 with the Burrows-Wheeler Alignment (BWA; [14]) tool resulting in a SAM/BAM file. Molecular and optical duplicate reads were flagged using software from the Picard program suite [15]. Post-processing of the aligned data included local realignment around SNPs and indels and base-call quality score recalibration using the Genome Analysis Tool Kit ver. 2 (GATK2; [16]). Single sample calling was done using GATK2 HaplotypeCaller with hard filtering and outputted in VCF 4.0 format. Analyses were performed in accordance with GATK Best Practices recommendations [17,18]. All positions reported are with respect to the hg19 reference sequence.

2.4. Sequence Interpretation

Differences of SNVs and indels between the tumor and normal exomes were computed using both open source and commercial software to identify somatic mutations. The open source programs included ANNOVAR [19], SG-adviser, a suite of web-based tool offered by The Scripps Translational Science Institute [20] Strelka [21] and Seurat [22] which were used with their default settings. In addition, we used the IntOGen-mutations platform [23,24] to identify genes mutated in the TCGA/ICCG cancer genome projects. The Condel online tool [25] was used to obtain scores for missense mutations [26] shown in the exome sequencing data tables. We also used the Ingenuity Variant Analysis web-based application to compare sequence between the tumor and matched normal exomes with differing filtering parameters (see Supplementary Methods: Analysis and Variant Filtering).

3. Results and Discussion

3.1. Genomic Landscape Detailed from Genotyping Data

The genotyping array showed a large degree of copy number variation of chromosomal segments and loss of heterozygosity with 17 of the 22 autosomes grossly affected (Figure 1). The genotypes from the Illumina 2.5 M BeadChip processed by circular binary segmentation indicated a modest range in copy number. Visual inspection of the BAF plots clearly identifies blocks of homozygosity on chromosomes 1, 2, 5, 9, 12, 13, 15, 16, and 17. At least four critical tumor suppressors are within these regions of LOH; TP53 on Chr 17, RB1 on Chr 13, CDKN1B on Chr 12 and CHD1 on chromosome 5 (Figure 2a). While the regions that include TP53 and CHD1 are essentially copy neutral, the LRR plot and segmentation showed a reduced copy number for the chromosome 13 block containing RB1 and a likely homozygous deletion of the chromosome 12 region containing CDKN1B (Figure 2b).

Grasso *et al.* [7] recently described the mutational landscape of castration-resistant prostate cancer (CRPC) based on exome sequencing of 50 lethal metastatic cases. An important finding of their study was that tumors involving CHD1 lacked ETS2 gene fusions and ETS2 mutations. CHD1 is an ATP-dependent chromatin-remodeling enzyme that recognizes histone H3 lysine 4 methylation and is associated with the promoters of active genes where it presumably acts in nucleosome disassembly [27]. In this tumor the regions around ETS2 and TMPRSS2 have normal copy number although SNP arrays are not capable of identifying contiguous chromosomal events and a translocation in non-coding DNA is certainly possible. In contrast, CHD1 clearly falls within a region of LOH and although the exon sequences for the gene are the same as reference we do not know if there might have been mutations in regulatory regions. Unfortunately, we were unable to obtain RNA from the limited biopsy specimen so we could not measure changes in CHD1 expression.

Figure 2. The location of the key genes described in the text. CHD1 (Chr 5: 98.190–98.265 Mb) was not mutated but occurs in a block of LOH; CDKN1B (Chr 12: 12.870–12.875 Mb) has low copy number, RB1 (Chr 13: 48.878 Mb–49.056 Mb) occurs in a block of LOH, has reduced copy number and the retained allele is predicted to be damaging; TP53 (Chr 17: 75.712 Mb–75.909 Mb) occurs in a copy neutral block of LOH and the retained allele is predicted to be damaged. Figure 2b, magnified view of the CDKN1B region showing a likely deletion of the gene supported by both the array (top and middle panels) and exome-sequencing platforms (bottom).

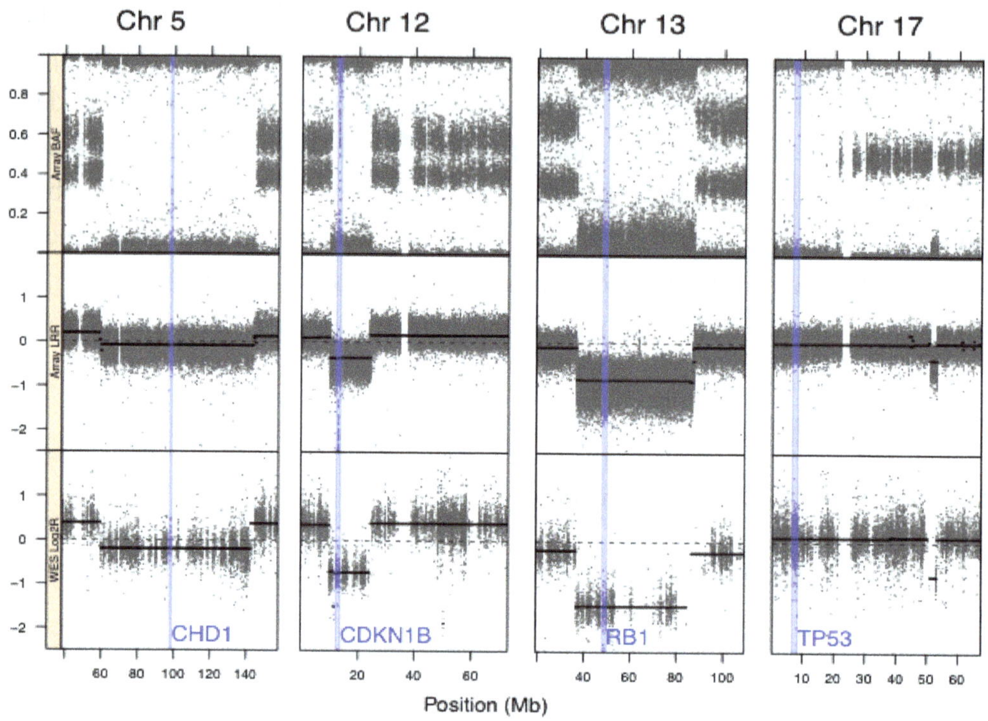

(a)

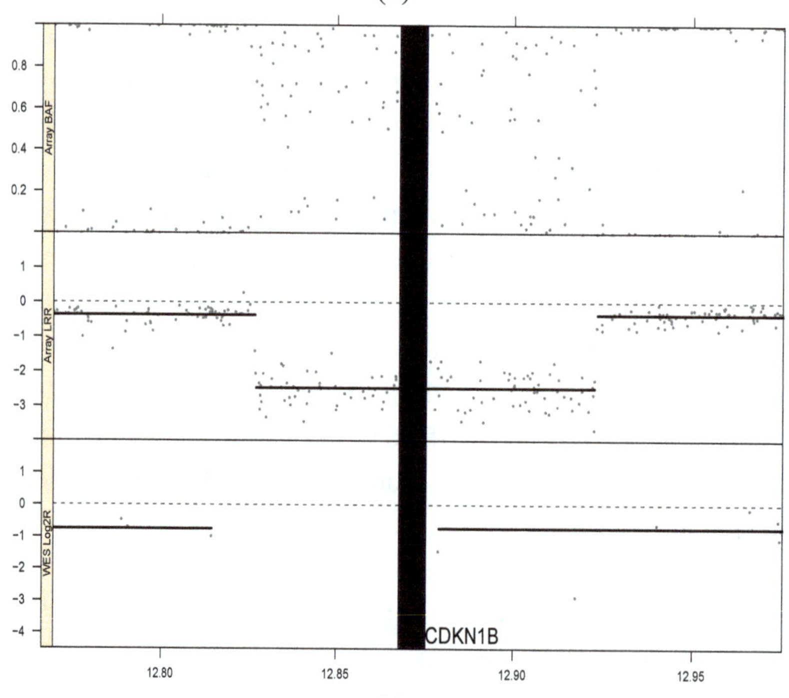

(b)

The genotyping data identified a region of LOH and markedly reduced copy number on chromosome 12p (~Chr 12: 10–24 Mb) that includes CDKN1B (Chr 12: 12.870–12.875 Mb) the gene which encodes the p27 cyclin-dependent kinase inhibitor also referred to as p27(KIP1). The exome reads were also significantly reduced and it is possible that the gene is completely absent in the tumor and that the observed reads represent those from contaminating or infiltrating normal cells. CDKN1B blocks cell division in G_0/G_1, regulates cell motility and apoptosis and is classified as a tumor suppressor [28]. Although CDKN1B was not mutated in this tumor, decreased copy number has been associated with tumor pathology in mice (e.g., [29]) and in lethal human epithelial cancers with a poor outcome [30].

Tumor Purity

Several regions of LOH identified by genotyping showed very few spurious mutant sequencing reads indicating that normal cells were not present in the tumor to any significant degree (e.g., TP53 Chr 17:5,578,394). Nevertheless, the mean read depth for the tumor library was 353X and 138X for the normal exome. Although the range in the frequency of mutant reads (Table 1) varies considerably, the fact that 97% of the reads for TP53 are mutant confirms that the tumor was unlikely to have been contaminated with a significant number of normal cells and that differences in allelic fraction at other positions most likely represent tumor heterogeneity. Because of this high level of tumor purity we felt that our exome sequencing provided a good representation of the genomic events without the need to sequence to extraordinary depth to distinguish tumor from infiltrating non-tumor cells.

3.2. Exome Variant Interpretation

We used open source tools to generate variant calls and a mixture of open-source and commercial programs to evaluate the significance of the somatic mutations. All variant filtering is a trade-off between sensitivity and specificity and the risk of missing variants of biological significance must be weighed against a larger number of false calls. The variant calling programs included Haplotype Caller [16,17], Strelka [21] and Seurat [22]. Haplotype Caller was run under GATK best practices with hard filtering [18], producing single sample calls for both tumor and normal. Strelka and Seurat are somatic variant callers that identify SNVs and indels present in a tumor but not the matched normal sample. Both were run using default settings. The default filter for Seurat removes sequences with a mapping quality score less than 10 while Strelka removes all read pairs with a mapping quality below 40. Seurat identified 3577 somatic SNVs and 2290 indels. In comparison, Strelka found 535 SNVs and 11 indels. Lists from both somatic callers were submitted to the Integrative Onco Genomics single tumor analysis web tool [23] which searches somatic mutations, genes and pathways identified, at the time of the analysis, from 4623 tumor/normal exomes by the International Cancer Genome Consortium (ICCG; [31]) and The Cancer Genome Anatomy (TCGA; [32]) initiatives. Because the Seurat results appeared to have high sensitivity but low specificity we decided to focus on the Strelka list. We manually inspected the Strelka /IntOGen dataset by examining each of the positions using the Integrative Genomic Browser (IGV, ver. 2.3.23; [33]).

Table 1. Principal findings: Protein coding genes with somatic mutations. The copy number and location within regions of LOH are noted. The SNV in the tumor *vs.* normal cells is indicated in the allele column and the percent of variant reads or mosaicism is given. Condel scores (D = deleterious, N = neutral) and protein changes are shown for each predicted isoform. Premature stop mutations are indicated and assumed to be damaging. Genes confirmed by the Ingenuity analysis are indicated along with the Ingenuity "assessment".

Chr	Position	Gene	Driver	CN	EXC	LOH	Ref	SNV	% Var	Condel Score	Protein change	Ingenuity Assesment	Gene Description	Comments
17	7,578,394	TP53	HCD	2.02	2.12	YES	C	T	0.97	D(0.97–1.0)	H179R, H86R, H47R	Pathogenic	Tumor protein p53	Damaging in all alternate translation products
13	48,941,657	RB1	HCD	1.12	0.72	YES	G	T	0.75	STOP GAIN	E323*	Pathogenic	Retinoblastoma 1	Premature termination codon is inferred damaging
14	38,061,334	FOXA1	HCD	2.11	2.25	NO	G	T	0.44	D(1.0)	R219S, R186S	Likely Pathogenic	Forkhead box A1	Damaging in two alternate translation products
10	70,508,917	CCAR1	HCD	2.05	2.13	NO	G	A	0.24	N(0.02), D(0.81), D(0.83), N(0.02)	R269H, R258H, R284H, R89H	No Assessment	Cell division cycle and apoptosis regulator 1	Probably damaging in two alternate translation products
12	12.870 Mb–12.875 Mb	CDKN1B	HCD	0.36	0.13	YES		None		Same as hg19 reference	DELETION	Not flagged	Cyclin-dependent kinase inhibitor 1B (p27/KIP1)	No mutations in coding regions

Ingenuity Variant Analysis (QIAGEN, Redwood City, CA, USA) was also used to filter and interpret somatic variants under different filtering criteria and single-sample variant call files were uploaded, parsed, and comparatively queried initially for rare (<3% allele frequency in public genome/exome datasets) missense, nonsense, coding indel, or clinically classified (pathogenic/likely pathogenic) variants, confidently called (PHRED-scaled variant call quality >20 in either sample) in genes directly or indirectly (within 2 upstream interaction hops) implicated in "prostate cancer" or "small cell adenocarcinoma" (interactive supplement at https://variants.ingenuity.com/Scott-etal-2014). Other filtering parameters (e.g., 1 upstream interaction hop, frequency in 1000 genomes or Complete Genomics data of less than 0.001%, and broader disease terms including "small cell adenocarcinoma", "castration-refractory prostate cancer", and "metastases") were also evaluated (interactive supplement at https://variants.ingenuity.com/Scott2014ver2).

We grouped SNVs into four categories: (1) The principal findings (Table 1) that we speculate have a strong likelihood of causing or contributing to SCPC; (2) Potentially implicated genes (Table 2) for which there is some evidence of an involvement in cancer but are less certain; (3) Genes with probable passenger mutations (Table S1) whose involvement in cancer is less obvious or lacking and (4) Possibly inherited risk factors (Table S2) for cancer susceptibility. The distinction between each of the first three categories is somewhat arbitrary.

Table 1 shows the top six genes based on their designation as high-confidence or candidate drivers (HCD, CD) of cancer by Tamborero *et al.* [34], their consensus deleteriousness (Condel) scores [26] or their reduced copy number. Premature nonsense mutations were presumed to be deleterious. The fraction of mutant reads at each position was also calculated from the BWA alignment. As discussed above, the predicted copy numbers in Table 1 and whether the gene fell into a region of LOH was based on both the genotyping array data as well as normalized exome capture read depth.

The most obvious findings from the genotyping and sequencing data are that the classic tumor suppressors TP53 and RB1 both occur in blocks of LOH and the retained alleles were mutated. In the case of TP53 the His to Arg missense substitution is damaging by both SIFT and Condel. The RB1 mutation was a premature stop codon at amino acid 323. As with other treatment resistant cancers, mutations in TP53 and RB1 have been reported in lethal prostate cancers [6,7]. Mutations in MLL2 have been reported in about 9% of CRPC while mutations in FOXA1 occurred in about 3.4% of tumors [7]. In this case of SCPC we did not find mutations in MLL2 but did detect a mutation in FOXA1. FOXA1 is a nuclear protein that promotes tumor progression through its interaction with the androgen receptor, which in turn, induces several prostate-specific genes. FOXA1 levels are positively correlated with PSA, Gleason scores and AR expression [8]. The damaging FOXA1 mutation reported here occurred in about half of the sequence reads and is presumed to result in lower activity which may, in part, explain the fact that the patient's PSA levels were not elevated. Grasso *et al.* [7] also showed that FOXA1 mutations repressed androgen signaling and increased tumor growth.

Table 2. Potentially implicated genes. Selection criteria included whether the gene function is cancer-related, a member of a gene family with established cancer drivers, or assessed as damaged by Condel or the Ingenuity Variant Caller.

Chr	Position	Gene	CN	EXC	LOH	Ref	SNV	% Var	Condel Score	Protein change	Ingenuity Assesment	Gene Description	Comments
2	106,498,240	NCK2	1.89	1.70	YES	C	G	1.00	D(0.91)	P228R	Uncertain	NCK adaptor protein 2	Promotes melanoma cell proliferation, migration and invasion
2	107,041,278	RGPD3	1.89	1.70	YES	C	A	0.90	SIFT = Damaging	E1049*	Likely Pathogenic	RANBP2-like and GRIP domain containing 3	Reported expression in testis and HeLa cells
1	109,742,795–109,742,798	KIAA1324, EIG21	1.57	1.24	YES	G	4 bp del	0.85	Frameshift	G829fs*10	Uncertain	KIAA1324; Estrogen induced gene 121	High expression is associated with shorter survival in ovarian cancer
2	102,407,183	MAP4K4	1.89	1.70	YES	G	T	0.49	D(0.88) or N(0.02)	G42V, G4V	Likely Pathogenic	Mitogen-activated protein kinase kinase kinase kinase 4	Often overexpressed in cancer and has roles in various cancer processes
19	50,247,621	TSKS	2.00	2.14	NO	C	T	0.45	N(0.05)	E410K	Likely Pathogenic	Testis-specific kinase substrate	Low expression in some embryonal carcinoma lines
15	88,678,358	NTRK3	2.02	2.14	NO	C	T	0.41	D(0.72–0.87)	G295D, G393D	Likely Pathogenic	Neurotrophic tyrosine kinase, receptor, type 3	Potential tumor suppressor, often fused with ETV6 in thyroid cancer
16	7,759,062	RBFOX1	2.10	2.14	NO	G	A	0.40	D(0.91–1.0)	G307R, G334R, G355R, G339R, G377R	Uncertain	RNA binding protein, fox-1 homolog (*C. elegans*) 1	Related gene RBFOX2 is a Candidate Driver (CD)
19	38,865,389	PSMD8	2.00	2.10	NO	C	T	0.38	N(0.00)	R50C	Uncertain	Proteasome (prosome, macropain) 26S subunit, non-ATPase, 8	Upregulated in a choriocarcinoma cell line
12	31,254,871	DDX11	2.25	2.61	NO	C	G	0.34	N(0.37)	H693Q, H317Q	Likely Pathogenic	DEAD/H box helicase 11	Associated with small-cell carcinoma

Table 2. *Cont.*

Chr	Position	Gene	CN	EXC	LOH	Ref	SNV	% Var	Condel Score	Protein change	Ingenuity Assesment	Gene Description	Comments
1	36,290,920	AGO4	1.83	1.71	NO	G	A	0.24	N(0.00)	M105V	Uncertain	Argonaute RISC catalytic component 4	Down-regulated in hepatocellular cancer
12	31,250,875	DDX11	2.25	2.61	NO	G	C	0.14	D(0.49)	A607P	Pathogenic	DEAD/H box helicase 11	Expressed at high levels in melanoma
19	4,689,651	DPP9	2.00	2.09	NO	G	T	0.12	D(0.50)	S560R	Uncertain	Dipeptidyl-peptidase 9	Expressed in breast and ovarian cancers
10	81,921,760	ANXA11	2.05	2.13	NO	G	A	0.04	D(0.85–0.95)	R338C, R371C, R4C	Not flagged	Annexin A11	May enhance metastasis and invasion; related gene ANXA6 is a Candidate Driver (CD)
9	100,843,284	TRIM14	2.23	2.59	NO	C	T	0.03	D(0.82–0.94)	R264W	Not flagged	Tripartite motif containing 14	Related gene TRIM7 is a High Confidence Driver (HCD)

Barbieri *et al.* [35] examined 112 prostate cancer tumor-normal pairs by exome sequencing and found recurrent somatic mutations in the genes FOXA1 and MED12 (~5% of tumors each) and in SPOP (~13% of tumors) in individuals with metastatic disease. The authors observed three different FOXA1 missense mutations in the forkhead (FH) domain, the DNA-binding domain of the protein [35,36]. FOXA1 binds to the androgen receptor and regulates the transcription of prostatic genes and is required for development of the prostate. The damaging mutation reported here also occurred in the FH domain. We did not observe mutations in SPOP and it does not occur in the region of LOH we observed on chromosome 17. Likewise, MED12 on the X chromosome did not appear to have somatic mutations when compared to the normal DNA sample.

Recurrent deletions of 5q21 have been reported in prostate cancer [35,37], and correlated with loss of the tumor suppressor CHD1. Further, Burkhardt *et al.* [37] showed a strong correlation between the loss of CHD1 and the biochemical failure to detect prostate-specific antigen. Similarly, we observed a region of LOH on chromosome 5 (~60–145 Mb, 5q12.1–31.3) which includes the CHD1 gene (Chr 5: 98,188,908–98,264,238) and, as noted, PSA levels were also not elevated in this cancer. While the remaining allele of CHD1 appears to have a normal sequence the genotyping array shows a reduced LRR. We were unable to perform studies to determine if RNA or protein levels were concomitantly reduced. The LOH region in our patient also includes the PIK3R1 gene, mutations in which are associated with various tumors (e.g., [9,38]). We observed no somatic mutations in the retained PIK3R1 allele.

The list of potentially implicated genes with somatic mutations is shown in Table 2 and ordered by the percent of mosaicism of the variant. These were selected based on the Ingenuity assessment, being a member of a gene family in which a related gene is a known or candidate cancer driver or from published literature implicating them in some aspect of cancer. Among these are NCK2 whose potentially damaging mutation occurs in 100% of reads. NCK2 is reported to promote melanoma cell proliferation, migration and invasion [39]. RGPD3 is expressed in the testis and HeLa cells [40], KIAA1324 or estrogen-expressed gene 21 is associated with ovarian cancer survival [41] and overexpression of EIG121 was observed to cause "profound suppression" of cell growth [42]. Presumably, reduced expression would have the opposite effect. MAP4K4 is a serine/threonine kinase that is overexpressed in many cancers where it is implicated in migration and invasion [43] and RBFOX1 is related to the candidate driver RBFOX2 [44]. Zhou *et al.* [45] reported a mutation in RBFOX1 in a colorectal adenoma. Decreased expression of testis-specific kinase substrate, TSKS has been observed in cancerous testicular tissue and in very low levels in various embryonal carcinomas [46]. NTRK3 is a potential tumor suppressor [47] often fused with ETV6 in thyroid cancer [48]. PSMD8 is up-regulated in a choriocarcinoma cell line [49]. AGO4 or EIF2C4 is down-regulated in hepatocellular cancer [50]. DDX11 is required for sister chromatid cohesion and is expressed at high levels in primary and metastatic melanomas [51] and DDP9 is expressed in breast and ovarian cancer [52]. ANXA11 plays an important role in cell division and disruption of the gene "may lead to or enhance the metastasis, invasion and drug resistance of cancers" [53]. A literature survey for TRIM14 did not identify a link to cancer but the protein shares homology to the reported high confidence driver TRIM7 [34].

We noted LOH and slightly reduced copy number for the region on chromosome 2 (181.5–181.8 Mb) containing the long non-coding RNAs SChLAP1 (LINC00913) [54] and for PCGEM1 (193.6 Mb) [55] both of which have been reported to be overexpressed in aggressive prostate cancer. We did not find evidence that CDKN2A was deleted or that CCNE1, E2F3, UBE2C, or MYCC were amplified as seen in other cancers (e.g., [56]). Our patient did not receive castration therapy and the androgen receptor gene (AR) was not deleted. We did not find evidence for a fusion between TRPSSC2 and ERG based on exon sequences although, as noted above, we cannot rule out a translocation outside of coding regions. Cyclin D1 (CCND1), a gene often altered in cancer and a modifier of androgen receptor function [57], may have reduced copy numbers based on the array data but had no obvious sequence differences from the normal sample. We also found no evidence for copy number or somatic mutations in AURKA, KLK3, CGA, SYP, NXK3.1, NCAM1, CD56, ETS or UBE2C. In fact, the total estimated mutational burden is low (<1/50,000 bp).

The patient's normal genome was also studied for inherited SNPs that might confer an increased risk for cancer (Table S2). A heterozygous SNP in FOXC1 that creates a P321Q variant that is predicted to be deleterious by SIFT and Condel (score = 0.92) was observed. Overexpression of FOXC1 has been correlated with poor outcome (e.g., [58,59]) and as a promoter of invasion in breast cancer [60]. The patient was also heterozygous for a known rare variant in DND1 (rs72800920) that is predicted to be damaging by both SIFT and CONDEL (score = 0.96). In mice, a premature stop mutation in Dnd1 has been shown to markedly increase the risk of testicular germ line tumors [61]. We do not know if rs72800920 is a cancer risk factor or simply a private rare variant. Other possible risk factors for which the patient was heterozygous were ALK, NCK2, DDX11 and CBWD3. Somatic mutations in ALK (anaplastic lymphoma receptor tyrosine kinase) have been seen in neuoblastomas [62]. NCK adaptor protein 2, NCK2, has been reported to promote melanoma cell proliferation [39]. DEAD/DEAH box helicase 11, DDX11, is required for sister chromatid cohesion and has been reported to be essential for the survival of advanced melanomas [51]. It is curious that the patient was a carrier for a likely pathogenic inherited variant and his tumor showed two somatic mutations in DDX11. Each of the germline alleles in these genes remained heterozygous in the tumor (*i.e.*, did not show evidence for selection) and we have no formal evidence that they conferred risk for disease or its progression.

4. Conclusions

The main goal of this study was to assess how the new genomic technologies, analysis methods and interpretation tools might be used to provide clinical utility. Secondarily, we hoped to better characterize a SCPC metastasis in a single case using these approaches. The combination of genotyping arrays, to provide a broad overview of the genomic landscape, and exome sequencing, to identify specific mutations, was more useful than either method alone. The genotyping array highlighted key regions of the genome that showed abnormal copy number or loss of heterozygosity. In general, these changes in genomic architecture are clues to underlying genes that may be implicated in cancer. LOH is commonly associated with the loss of tumor suppressors and by identifying those regions first on an array we were able to focus attention on the somatic sequence variants found there.

Because of limited sample we were unable to perform karyotyping or expression studies (either arrays or RNAseq). However, such limitations are likely to be expected in routine clinical testing so maximizing information from samples is critical. Going forward it would be preferable to do dual RNA and DNA isolations from fresh needle biopsies and perform RNA sequencing to measure relative expression levels, identify the main splice variants and any fusion transcripts. Given the good correlation between exome read depth and copy number from the array shown here it may be unnecessary to perform high-density genotyping in the future. However, we would likely replace exome-capture with PCR-free whole genome sequencing in order to eliminate biases related to the capture reagents and be able to potentially identify chromosomal translocation events and other genomic rearrangements. Currently, we feel it is important to manually review the sequence data at key positions and perform Sanger sequencing as confirmation for actionable mutations. However, improvements in laboratory methods and analysis may soon make this unnecessary (e.g., [63]).

We found that both open source and commercial tools were invaluable for interpreting somatic variants although it is essential that the analysis pipeline that produces the variants for interpretation be as rigorous as possible. Interpretation software essentially performs two tasks: it matches lists of variants within a study to those reported in various databases or in the literature in a way that is meaningful for the disease or mode of inheritance and it uses one or more algorithms to predict the effects of mutations. The first function will only be as good as the databases referenced and for commercial databases the details are usually not available. In general, there was excellent concordance between the open source tools and the Ingenuity Variant Analysis although the latter identified several somatic mutations in genes that were not flagged by the IntOGen analysis. This is not surprising given that IntOGen is based on data from large cancer sequencing projects while Ingenuity also includes literature-based gene information and predictive algorithms that infer change in protein function.

A current limitation to variant interpretation is, as seen in Table 1, that many genes produce multiple alternative transcripts and may have deleterious mutations predicted in some isoforms but not others. In the absence of RNASeq data we do not know which isoforms may predominate in a given cancer type. Information about the splice variants and fusion transcripts will have to be included in a comprehensive analysis. Further, because the interpretation tools used in this analysis were based on VCFs they did not take copy number or LOH into account. As shown from the Excavator analysis this is something that could certainly be added. A clear advantage of the Ingenuity Variant Analysis tool was the ability of a user to easily link to the biomedical literature and pathway information that included potential drugs for targets it identified. Providing such information will be a valuable adjunct to physicians acting on exome and genome test results.

As noted above, as with other types of genetic testing, NGS approaches need to be standardized, accurate and have practical utility. Perhaps more than other genetic tests, whole genome or whole exome sequencing blurs the borders between clinical testing and research. This was also true of other methods when they first appeared (e.g., FISH, comparative hybridization arrays, *etc.*) and only with the accumulation of large datasets and more standardized methods in the laboratory and during analysis will the utility of genome sequencing become routine. As more correlations are made between patterns of the genomic landscape and mutational profiles we should be better able to tailor treatments or predict the course of disease. Already, sequencing data are being used to design patient-specific tests to follow response to treatment [64] and gene or mutation-specific treatments have been and are being developed.

SCPC is a lethal cancer with a poor prognosis. Ciriello *et al.* [56], in summarizing the ICGC and TCGA oncogenic signatures from over 3000 tumors, concluded that cancers generally fall into one of two classifications; "M" class cancers with, often large numbers of somatic mutations, and the "C" class with chromosomal abnormalities and fewer variants but which often involve somatic mutations in TP53, the likely cause of the genomic instability [65]. This tumor clearly falls into the C class. It is remarkable that while the overall somatic mutation rate was relatively low given that so many cancer driver genes were mutated. Perhaps the rarity of SCPC reflects the need to accumulate many separate deleterious mutations. Unfortunately, nothing in our sequencing or interpretation analysis offered a useful treatment strategy but, hopefully, the approaches and results reported here will be of use to others studying this and other aggressive cancers. By redefining cancers based on their genomic, expression and mutational architectures we may be able to markedly improve cancer diagnosis and therapy.

Acknowledgments

The authors thank Rachel Pattiglio and Laura Kasch of the Johns Hopkins Genetic Resources Core Facility for isolation of DNA from the biopsy and mouthwash samples and Nathan Pearson and Shela Shah from Ingenuity for assistance with the Ingenuity Variant Caller and recommended filtering criteria.

This work is dedicated to the memory of Gregory S. Liptak, physician, academic and friend.

Author Contributions

Designed study: Alan F. Scott and Gregory S. Liptak; Provided clinical information: Gregory S. Liptak; Wrote manuscript: Alan F. Scott, Robert B. Scharpf, David W. Mohr, Hua Ling, and Peng Zhang; Performed laboratory work: David W. Mohr; Implemented and ran software: David W. Mohr, Robert B. Scharpf, Hua Ling, and Peng Zhang.

References

1. Aparicio, A.; Logothetis, C.J.; Maity, S.N. Understanding the lethal variant of prostate cancer: Power of examining extremes. *Cancer Discov.* **2011**, *1*, 466–468.
2. Spiess, P.E.; Pettaway, C.A.; Vakar-Lopez, F.; Kassouf, W.; Wang, X.; Busby, J.E.; Do, K.A.; Davuluri, R.; Tannir, N.M. Treatment outcomes of small cell carcinoma of the prostate: A single-center study. *Cancer* **2007**, *110*, 1729–1737.
3. Wang, W.; Epstein, J.I. Small cell carcinoma of the prostate. A morphologic and immunohistochemical study of 95 cases. *Am. J. Surg. Pathol.* **2008**, *32*, 65–71.
4. Beltran, H.; Rickman, D.S.; Park, K.; Chae, S.S.; Sboner, A.; MacDonald, T.Y.; Wang, Y.; Sheikh, K.L.; Terry, S.; Tagawa, S.T.; *et al.* Molecular characterization of neuroendocrine prostate cancer and identification of new drug targets. *Cancer Discov.* **2011**, *1*, 487–495.

5. Meulenbeld, H.J.; Bleuse, J.P.; Vinci, E.M.; Raymond, E.; Vitali, G.; Santoro, A.; Dogliotti, L.; Berardi, R.; Cappuzzo, F.; Tagawa, S.T.; *et al.* Randomized phase II study of danusertib in patients with metastatic castration-resistant prostate cancer after docetaxel failure. *BJU Int.* **2013**, *111*, 44–52.
6. Tzelepi, V.; Zhang, J.; Lu, J.F.; Kleb, B.; Wu, G.; Wan, X.; Hoang, A.; Efstathiou, E.; Sircar, K.; Navone, N.M.; *et al.* Modeling a lethal prostate cancer variant with small-cell carcinoma features. *Clin. Cancer Res.* **2012**, *18*, 666–677.
7. Grasso, C.S.; Wu, Y.M.; Robinson, D.R.; Cao, X.; Dhanasekaran, S.M.; Khan, A.P.; Quist, M.J.; Jing, X.; Lonigro, R.J.; Brenner, J.C.; *et al.* The mutational landscape of lethal castration-resistant prostate cancer. *Nature* **2012**, *487*, 239–243.
8. Imamura, Y.; Sakamoto, S.; Endo, T.; Utsumi, T.; Fuse, M.; Suyama, T.; Kawamura, K.; Imamoto, T.; Yano, K.; Uzawa, K.; *et al.* FOXA1 promotes tumor progression in prostate cancer via the insulin-like growth factor binding protein 3 pathway. *PLoS One* **2012**, *7*, e42456.
9. Van Allen, E.M.; Foye, A.; Wagle, N.; Kim, W.; Carter, S.L.; McKenna, A.; Simko, J.P.; Garraway, L.A.; Febbo, P.G. Successful whole-exome sequencing from a prostate cancer bone metastasis biopsy. *Prostate Cancer Prostatic Dis.* **2014**, *17*, 23–27.
10. Genetic Resources Core Facility. Available online: http://grcf.jhmi.edu/ (accessed on 2 January 2014).
11. Olshen, A.B.; Venkatraman, E.S.; Lucito, R.; Wigler, M. Circular binary segmentation for the analysis of array-based DNA copy number data. *Biostatistics* **2004**, *5*, 557–572.
12. Magi, A.; Tattini, L.; Cifola, I.; D'Aurizio, R.; Benelli, M.; Mangano, E.; Battaglia, C.; Bonora, E.; Kurg, A.; Seri, M.; *et al.* EXCAVATOR: Detecting copy number variants from whole-exome sequencing data. *Genome Biol.* **2013**, doi:10.1186/gb-2013-14-10-r120.
13. Barnhart, M.G.S.; Hetrick, K.; Goldstein, J.; Marosy, D.; Mohr, D.; Craig, B.; Watkins, L., Jr.; Doheny, K. CIDRSeqSuite 2.0: An automated analysis pipeline for next-generation sequencing. In Proceedings of the presented at the 61st annual meeting of the American society for human genetics, Montreal, QC, Canada, 12 October 2011.
14. Li, H.; Durbin, R. Fast and accurate long-read alignment with Burrows-Wheeler transform. *Bioinformatics* **2010**, *26*, 589–595.
15. Picard. Available online: http://picard.sourceforge.net/ (accessed on 2 January 2014).
16. McKenna, A.; Hanna, M.; Banks, E.; Sivachenko, A.; Cibulskis, K.; Kernytsky, A.; Garimella, K.; Altshuler, D.; Gabriel, S.; Daly, M.; *et al.* The Genome Analysis Toolkit: A MapReduce framework for analyzing next-generation DNA sequencing data. *Genome Res.* **2010**, *20*, 1297–1303.
17. DePristo, M.A.; Banks, E.; Poplin, R.; Garimella, K.V.; Maguire, J.R.; Hartl, C.; Philippakis, A.A.; del Angel, G.; Rivas, M.A.; Hanna, M.; *et al.* A framework for variation discovery and genotyping using next-generation DNA sequencing data. *Nat. Genet.* **2011**, *43*, 491–498.
18. Van der Auwera, G.A.; Carneiro, M.O.; Hartl, C.; Poplin, R.; del Angel, G.; Levy-Moonshine, A.; Jordan, T.; Shakir, K.; Roazen, D.; Thibault, J.; *et al.* From FastQ data to high-confidence variant calls: The genome analysis toolkit best practices pipeline. *Curr. Protoc. Bioinform.* **2013**, *11*, 1–33.
19. Wang, K.; Li, M.; Hakonarson, H. ANNOVAR: Functional annotation of genetic variants from high-throughput sequencing data. *Nucleic Acids Res.* **2010**, *38*, e164.
20. Scripps Genome Adviser. Available online: http://genomics.scripps.edu/ADVISER/ (accessed on 2 January 2014).

21. Saunders, C.T.; Wong, W.S.; Swamy, S.; Becq, J.; Murray, L.J.; Cheetham, R.K. Strelka: Accurate somatic small-variant calling from sequenced tumor-normal sample pairs. *Bioinformatics* **2012**, *28*, 1811–1817.
22. Christoforides, A.; Carpten, J.D.; Weiss, G.J.; Demeure, M.J.; von Hoff, D.D.; Craig, D.W. Identification of somatic mutations in cancer through Bayesian-based analysis of sequenced genome pairs. *BMC Genomics* **2013**, doi:10.1186/1471-2164-14-302.
23. IntOgen: Interactive Onco Genomics Mutations Server. Available online: http://www.intogen.org/mutations/ (accessed 4 January 2014).
24. Gonzalez-Perez, A.; Perez-Llamas, C.; Deu-Pons, J.; Tamborero, D.; Schroeder, M.P.; Jene-Sanz, A.; Santos, A.; Lopez-Bigas, N. IntOGen-mutations identifies cancer drivers across tumor types. *Nat. Methods* **2013**, *10*, 1081–1082.
25. Condel: CONsensus DELeteriousness Score of Missense SNVs Server. Available online: http://bg.upf.edu/condel/analysis/ (accessed 4 January 2014).
26. Gonzalez-Perez, A.; Lopez-Bigas, N. Improving the assessment of the outcome of nonsynonymous SNVs with a consensus deleteriousness score, Condel. *Am. J. Hum. Genet.* **2011**, *88*, 440–449.
27. Petty, E.; Pillus, L. Balancing chromatin remodeling and histone modifications in transcription. *Trends Genet.* **2013**, *29*, 621–629.
28. Cuesta, R.; Martinez-Sanchez, A.; Gebauer, F. miR-181a regulates cap-dependent translation of p27(kip1) mRNA in myeloid cells. *Mol. Cell. Biol.* **2009**, *29*, 2841–2851.
29. Fero, M.L.; Randel, E.; Gurley, K.E.; Roberts, J.M.; Kemp, C.J. The murine gene p27Kip1 is haplo-insufficient for tumor suppression. *Nature* **1998**, *396*, 177–180.
30. Chu, I.M.; Hengst, L.; Slingerland, J.M. The Cdk inhibitor p27 in human cancer: Prognostic potential and relevance to anticancer therapy. *Nat. Rev. Cancer* **2008**, *8*, 253–267.
31. Hudson, T.J.; Anderson, W.; Artez, A.; Barker, A.D.; Bell, C.; Bernabe, R.R.; Bhan, M.K.; Calvo, F.; Eerola, I.; Gerhard, D.S.; *et al.* International network of cancer genome projects. *Nature* **2010**, *464*, 993–998.
32. Weinstein, J.N.; Collisson, E.A.; Mills, G.B.; Shaw, K.R.; Ozenberger, B.A.; Ellrott, K.; Shmulevich, I.; Sander, C.; Stuart, J.M. The Cancer Genome Atlas Pan-Cancer analysis project. *Nat. Genet.* **2013**, *45*, 1113–1120.
33. Robinson, J.T.; Thorvaldsdottir, H.; Winckler, W.; Guttman, M.; Lander, E.S.; Getz, G.; Mesirov, J.P. Integrative genomics viewer. *Nat. Biotechnol.* **2011**, *29*, 24–26.
34. Tamborero, D.; Gonzalez-Perez, A.; Perez-Llamas, C.; Deu-Pons, J.; Kandoth, C.; Reimand, J.; Lawrence, M.S.; Getz, G.; Bader, G.D.; Ding, L.; *et al.* Comprehensive identification of mutational cancer driver genes across 12 tumor types. *Sci. Rep.* **2013**, doi:10.1038/srep02650.
35. Barbieri, C.E.; Baca, S.C.; Lawrence, M.S.; Demichelis, F.; Blattner, M.; Theurillat, J.P.; White, T.A.; Stojanov, P.; van Allen, E.; Stransky, N.; *et al.* Exome sequencing identifies recurrent SPOP, FOXA1 and MED12 mutations in prostate cancer. *Nat. Genet.* **2012**, *44*, 685–689.
36. Clark, K.L.; Halay, E.D.; Lai, E.; Burley, S.K. Co-crystal structure of the HNF-3/fork head DNA-recognition motif resembles histone H5. *Nature* **1993**, *364*, 412–420.
37. Burkhardt, L.; Fuchs, S.; Krohn, A.; Masser, S.; Mader, M.; Kluth, M.; Bachmann, F.; Huland, H.; Steuber, T.; Graefen, M.; *et al.* CHD1 is a 5q21 tumor suppressor required for ERG rearrangement in prostate cancer. *Cancer Res.* **2013**, *73*, 2795–2805.

38. Quayle, S.N.; Lee, J.Y.; Cheung, L.W.; Ding, L.; Wiedemeyer, R.; Dewan, R.W.; Huang-Hobbs, E.; Zhuang, L.; Wilson, R.K.; Ligon, K.L.; *et al.* Somatic mutations of PIK3R1 promote gliomagenesis. *PLoS One* **2012**, *7*, e49466.
39. Labelle-Cote, M.; Dusseault, J.; Ismail, S.; Picard-Cloutier, A.; Siegel, P.M.; Larose, L. Nck2 promotes human melanoma cell proliferation, migration and invasion *in vitro* and primary melanoma-derived tumor growth *in vivo*. *BMC Cancer* **2011**, doi:10.1186/1471-2407-11-443.
40. Ciccarelli, F.D.; von Mering, C.; Suyama, M.; Harrington, E.D.; Izaurralde, E.; Bork, P. Complex genomic rearrangements lead to novel primate gene function. *Genome Res.* **2005**, *15*, 343–351.
41. Schlumbrecht, M.P.; Xie, S.S.; Shipley, G.L.; Urbauer, D.L.; Broaddus, R.R. Molecular clustering based on ERalpha and EIG121 predicts survival in high-grade serous carcinoma of the ovary/peritoneum. *Mod. Pathol.* **2011**, *24*, 453–462.
42. Deng, L.; Feng, J.; Broaddus, R.R. The novel estrogen-induced gene EIG121 regulates autophagy and promotes cell survival under stress. *Cell Death Dis.* **2010**, doi:10.1038/cddis.2010.9.
43. Qiu, M.H.; Qian, Y.M.; Zhao, X.L.; Wang, S.M.; Feng, X.J.; Chen, X.F.; Zhang, S.H. Expression and prognostic significance of MAP4K4 in lung adenocarcinoma. *Pathol. Res. Pract.* **2012**, *208*, 541–548.
44. Venables, J.P.; Brosseau, J.P.; Gadea, G.; Klinck, R.; Prinos, P.; Beaulieu, J.F.; Lapointe, E.; Durand, M.; Thibault, P.; Tremblay, K.; *et al.* RBFOX2 is an important regulator of mesenchymal tissue-specific splicing in both normal and cancer tissues. *Mol. Cell. Biol.* **2013**, *33*, 396–405.
45. Zhou, D.; Yang, L.; Zheng, L.; Ge, W.; Li, D.; Zhang, Y.; Hu, X.; Gao, Z.; Xu, J.; Huang, Y.; *et al.* Exome capture sequencing of adenoma reveals genetic alterations in multiple cellular pathways at the early stage of colorectal tumorigenesis. *PLoS One* **2013**, *8*, e53310.
46. Scorilas, A.; Yousef, G.M.; Jung, K.; Rajpert-De Meyts, E.; Carsten, S.; Diamandis, E.P. Identification and characterization of a novel human testis-specific kinase substrate gene which is downregulated in testicular tumors. *Biochem. Biophys. Res. Commun.* **2001**, *285*, 400–408.
47. Luo, Y.; Kaz, A.M.; Kanngurn, S.; Welsch, P.; Morris, S.M.; Wang, J.; Lutterbaugh, J.D.; Markowitz, S.D.; Grady, W.M. NTRK3 is a potential tumor suppressor gene commonly inactivated by epigenetic mechanisms in colorectal cancer. *PLoS Genet.* **2013**, *9*, e1003552.
48. Kralik, J.M.; Kranewitter, W.; Boesmueller, H.; Marschon, R.; Tschurtschenthaler, G.; Rumpold, H.; Wiesinger, K.; Erdel, M.; Petzer, A.L.; Webersinke, G. Characterization of a newly identified ETV6-NTRK3 fusion transcript in acute myeloid leukemia. *Diagn. Pathol.* **2011**, doi:10.1186/1746-1596-6-19.
49. Kobayashi, Y.; Banno, K.; Shimizu, T.; Ueki, A.; Tsuji, K.; Masuda, K.; Kisu, I.; Nomura, H.; Tominaga, E.; Nagano, O.; *et al.* Gene expression profile of a newly established choriocarcinoma cell line, iC3-1, compared to existing choriocarcinoma cell lines and normal placenta. *Placenta* **2013**, *34*, 110–118.
50. Kitagawa, N.; Ojima, H.; Shirakihara, T.; Shimizu, H.; Kokubu, A.; Urushidate, T.; Totoki, Y.; Kosuge, T.; Miyagawa, S.; Shibata, T. Downregulation of the microRNA biogenesis components and its association with poor prognosis in hepatocellular carcinoma. *Cancer Sci.* **2013**, *104*, 543–551.
51. Bhattacharya, C.; Wang, X.; Becker, D. The DEAD/DEAH box helicase, DDX11, is essential for the survival of advanced melanomas. *Mol. Cancer* **2012**, doi:10.1186/1476-4598-11-82.

52. Wilson, C.H.; Abbott, C.A. Expression profiling of dipeptidyl peptidase 8 and 9 in breast and ovarian carcinoma cell lines. *Int. J. Oncol.* **2012**, *41*, 919–932.
53. Wang, J.; Guo, C.; Liu, S.; Qi, H.; Yin, Y.; Liang, R.; Sun, M.Z.; Greenaway, F.T. Annexin A11 in disease. *Clin. Chim. Acta* **2014**, *431*, 164–168.
54. Prensner, J.R.; Iyer, M.K.; Sahu, A.; Asangani, I.A.; Cao, Q.; Patel, L.; Vergara, I.A.; Davicioni, E.; Erho, N.; Ghadessi, M.; *et al.* The long noncoding RNA SChLAP1 promotes aggressive prostate cancer and antagonizes the SWI/SNF complex. *Nat. Genet.* **2013**, *45*, 1392–1398.
55. Yang, L.; Lin, C.; Jin, C.; Yang, J.C.; Tanasa, B.; Li, W.; Merkurjev, D.; Ohgi, K.A.; Meng, D.; Zhang, J.; *et al.* lncRNA-dependent mechanisms of androgen-receptor-regulated gene activation programs. *Nature* **2013**, *500*, 598–602.
56. Ciriello, G.; Miller, M.L.; Aksoy, B.A.; Senbabaoglu, Y.; Schultz, N.; Sander, C. Emerging landscape of oncogenic signatures across human cancers. *Nat. Genet.* **2013**, *45*, 1127–1133.
57. Comstock, C.E.; Augello, M.A.; Schiewer, M.J.; Karch, J.; Burd, C.J.; Ertel, A.; Knudsen, E.S.; Jessen, W.J.; Aronow, B.J.; Knudsen, K.E. Cyclin D1 is a selective modifier of androgen-dependent signaling and androgen receptor function. *J. Biol. Chem.* **2011**, *286*, 8117–8127.
58. Xu, Y.; Shao, Q.S.; Yao, H.B.; Jin, Y.; Ma, Y.Y.; Jia, L.H. Up-expression of FOXC1 correlates with poor prognosis in gastric cancer patients. *Histopathology* **2013**, doi:10.1111/his.12347.
59. Wei, L.X.; Zhou, R.S.; Xu, H.F.; Wang, J.Y.; Yuan, M.H. High expression of FOXC1 is associated with poor clinical outcome in non-small cell lung cancer patients. *Tumour Biol.* **2013**, *34*, 941–946.
60. Sizemore, S.T.; Keri, R.A. The forkhead box transcription factor FOXC1 promotes breast cancer invasion by inducing matrix metalloprotease 7 (MMP7) expression. *J. Biol. Chem.* **2012**, *287*, 24631–24640.
61. Youngren, K.K.; Coveney, D.; Peng, X.; Bhattacharya, C.; Schmidt, L.S.; Nickerson, M.L.; Lamb, B.T.; Deng, J.M.; Behringer, R.R.; Capel, B.; *et al.* The Ter mutation in the dead end gene causes germ cell loss and testicular germ cell tumors. *Nature* **2005**, *435*, 360–364.
62. Chen, Y.; Takita, J.; Choi, Y.L.; Kato, M.; Ohira, M.; Sanada, M.; Wang, L.; Soda, M.; Kikuchi, A.; Igarashi, T.; *et al.* Oncogenic mutations of ALK kinase in neuroblastoma. *Nature* **2008**, *455*, 971–974.
63. Schmitt, M.W.; Kennedy, S.R.; Salk, J.J.; Fox, E.J.; Hiatt, J.B.; Loeb, L.A. Detection of ultra-rare mutations by next-generation sequencing. *Proc. Natl. Acad. Sci. USA* **2012**, *109*, 14508–14513.
64. Leary, R.J.; Sausen, M.; Kinde, I.; Papadopoulos, N.; Carpten, J.D.; Craig, D.; O'Shaughnessy, J.; Kinzler, K.W.; Parmigiani, G.; Vogelstein, B.; *et al.* Detection of chromosomal alterations in the circulation of cancer patients with whole-genome sequencing. *Sci Transl. Med.* **2012**, doi:10.1126/scitranslmed.3004742.
65. Rausch, T.; Jones, D.T.; Zapatka, M.; Stutz, A.M.; Zichner, T.; Weischenfeldt, J.; Jager, N.; Remke, M.; Shih, D.; Northcott, P.A.; *et al.* Genome sequencing of pediatric medulloblastoma links catastrophic DNA rearrangements with TP53 mutations. *Cell* **2012**, *148*, 59–71.

10

Pharmacogenomics: Current State-of-the-Art

Daniel F. Carr, Ana Alfirevic and Munir Pirmohamed *

Wolfson Centre for Personalised Medicine, Department of Molecular and Clinical Pharmacology, University of Liverpool, Block A Waterhouse Buildings, 1-5 Brownlow Street, Liverpool, L69 3GL, UK; E-Mails: d.carr@liv.ac.uk (D.F.C.); Ana.Alfirevic@liv.ac.uk (A.A.)

* Author to whom correspondence should be addressed; E-Mail: munirp@liv.ac.uk;

Abstract: The completion of the human genome project 10 years ago was met with great optimism for improving drug therapy through personalized medicine approaches, with the anticipation that an era of genotype-guided patient prescribing was imminent. To some extent this has come to pass and a number of key pharmacogenomics markers of inter-individual drug response, for both safety and efficacy, have been identified and subsequently been adopted in clinical practice as pre-treatment genetic tests. However, the universal application of genetics in treatment guidance is still a long way off. This review will highlight important pharmacogenomic discoveries which have been facilitated by the human genome project and other milestone projects such as the International HapMap and 1000 genomes, and by the continued development of genotyping and sequencing technologies, including rapid point of care pre-treatment genetic testing. However, there are still many challenges to implementation for the many other reported biomarkers which continue to languish within the discovery phase. As technology advances over the next 10 years, and the costs fall, the field will see larger genetic data sets, including affordable whole genome sequences, which will, it is hoped, improve patient outcomes through better diagnostic, prognostic and predictive biomarkers.

Keywords: pharmacogenetics; personalized medicine; pre-treatment genetic testing

---

1. Introduction

The field of pharmacogenomics can trace its roots back to significantly earlier than the first draft publication human genome sequence in 2001 [1] and the subsequent completion in 2003. Indeed, the term "Pharmacogenetics" was first coined by Friedrich Vogel in 1959 [2], just 6 years after Watson and Crick's discovery of the structure of DNA [3].

Though significant progress has been made in the field since 2003, it could be argued that pharmacogenomics has failed to live up to expectations. A vast number of discoveries relating to genomic variability and drug response have been made in the last 10 years. The challenge remains to translate these findings into clinical practice for the benefit of the patient.

2. The International HapMap Project

The completion of the first phase of the International Hapmap project [4], a catalogue of common genetic variations within individuals of diverse ethnicities, in 2003, provided a rich data resource which enabled researchers to investigate the association of variants across the human genome with a wide range of clinical phenotypes. Indeed, the data derived from this allowed the creation of SNP arrays whereby researchers could analyze patient genotype for 100,000 s to millions of SNPs at a time. Thus, for the first time, unbiased genetic analysis of clinical phenotype/genotype associations was possible using platforms created for the Genome Wide Association Study (GWAS). More recent advances in genomics have seen the development of next generation sequencing methodologies which do not require *a priori* knowledge of genetic variation.

Mining of the National Human Genome Institute (NHGRI) genome wide association study (GWAS) catalogue [5] (accessed on 22nd April 2014) showed that a total of 1885 peer-reviewed articles have been published reporting GWAS findings. However, only 93 (4.9%) are related to inter-individual variability of drug response phenotypes (either safety or efficacy).

Despite the small numbers of studies to date, which have investigated common genetic variant associations with pharmacological phenotypes, a number of important genotype-phenotype associations have been identified in pharmacogenomics using GWAS approaches (Table 1). The interesting phenomenon, when compared with GWAS of complex diseases, is that the effect size for pharmacogenomics phenotypes on the whole seems to be larger than that observed for complex diseases, which allows for (a) smaller sample sizes to be studied, which is more time- and cost-efficient, and (b) some variants to be considered for clinical implementation in terms of use prior to drug prescription, which contrasts with complex diseases, where the relative risks identified are usually below 1.5 and have not been used clinically. Whilst common genetic variation has been shown, in a number of examples to be important factors determining inter-individual variability in drug response, the role of rare, or private, variants is unclear in pharmacogenomics phenotypes, and is an important area for further research.

Table 1. Genetic biomarkers of (**A**) adverse drug reactions and (**B**) inter-individual variability of drug efficacy identified, or confirmed, from genome wide association studies identifying.

A	Year	Drug	Indication	Phenotype	Population	Associated Loci	SNP/Allele	Ref.
	2008	Simvastatin	Hypercholesterolaemia	Myopathy	Caucasian	SLCO1B1	rs4149056 (c.521T>C/*5)	[6]
	2008	Bisphosphonate	Multiple Myeloma	Osteonecrosis of the jaw	Spanish	CYP2C8	rs1934951	[7]
	2008	Ximelagatran	Anticoagulant	Hepatotoxicity	Caucasian	HLA-DRB1	*07 and *02	[8]
	2009	Flucloxacillin	Macrolide antibiotic	Hepatotoxicity	Caucasian	HLA-B	B*57:01	[9]
	2011	Carbamazepine	Epilepsy	Skin Rash/Hypersensitivity	Japanese Causasian	HLA-A	A*31:01	[10,11]
	2013	Allopurinol	Gout	SJS-TEN	Japanese	HLA-B	B*58:01	[12]
B	**Year**	**Drug**	**Indication**	**Phenotype**	**Population**	**Gene Loci**	**Allele**	**Ref.**
	2009	Clopidogrel	Antiplatelet		Amish	CYP2C19	*2	[13]
	2009	Pegylated Interferon	Hepatitis C		Caucasian	*IFNL3* (IL-28B)	rs12979860	[14]
	2009	Warfarin	Anticoagulant		Caucasian	CYP2C9 and VKORC1	*2,*3/c.-1639G>A	[15]

3. The 1000 Genomes Project

The first findings from the 1000 genomes project were reported in 2010 [16]. This has provided researchers with a population scale map of rare variants to complement and enrich existing knowledge of common variants gained from the HapMap project. With the per-base cost of sequencing using "next generation sequencing (NGS)" platforms continuing to fall, it is likely that studies will further investigate the role of rare genetic variants in defining variation in drug response phenotypes.

A recent study reported the mapping of rare and common variants within 12 Cytochrome P450 genes, thought to be responsible for metabolizing 75% of prescribed drugs [17]. Using whole exome sequence data for 2203 African Americans and 4300 Caucasians, researchers were able to identify novel, potentially deleterious alleles in major drug metabolizing enzymes in 7.6%–11.7% of individuals. The power of NGS technologies to identify rare variants could allow for greater understanding of their contribution to inter-individual variability in drug responses where common variants explain a limited degree of variability. One such example where rare variants may enhance our understanding of variability is the case of warfarin dose prediction. Incorporating CYP2C9*2, *3 and VKORC1 polymorphisms along with clinical confounding factors into a dosing algorithm allows clinicians to predict ~60% [18] of dose variability. However, it is entirely plausible that incorporating the contribution of rare functional genetic variants into such algorithms may in the future allow for even greater accuracy in warfarin dose prediction. Indeed a number of small scale studies and case reports have already identified rare missense variants in the VKORC1 gene in warfarin resistant patients [19–22]. Another example is drug-induced torsades de pointes where at least 10% of cases of may be due to rare mutations in the congenital long QT syndrome genes [23]. At least 23% of Caucasian subjects with drug-induced torsades de pointes carry a variant within 22 congenital arrhythmia genes (which include the 13 congenital long QT syndrome genes), compared with a background rate of 1.7% in 60 control subjects from the 1000 Genomes CEU data [24]. As greater numbers of NGS-based analyses in pharmacogenetic studies are undertaken, so the contribution of rare variants in other drug safety and efficacy phenotypes will be better understood.

4. Non-Coding RNAs

Large-scale, coordinated, international, research efforts, such as the ENCODE project [25] have expelled the myth that large regions of our human genome are "junk" DNA. Indeed, the presence of non-coding RNAs has altered the scientific community's perception of the central dogma of molecular biology. To date, only a very small number of studies have investigated the application of small non-coding RNA molecules as biomarkers of variable drug response.

However, studies have shown a potential utility for specific microRNAs as markers of both drug-induced liver injury (miR-122 and -192) [26] and severe skin reactions (miR-18a-5p) [27]. Though such biomarkers are still at the discovery stage, further clinical validation may see these and other non-coding RNAs enter clinical practice as early-stage pharmacogenomics predictors of adverse drug reactions.

5. Clinical Utilization of Pharmacogenomics

The biggest challenge that has faced the field of pharmacogenomics in the 10 years since the completion of the human genome project is clearly the application of genetic markers of variable drug response to decision-making in relation to prescribing. Indeed, as recently as 2011 it has been estimated that, of the >150,000 papers reporting claimed biomarkers, less than 100 has made it into clinical utility [28]. This of course refers to all biomarkers, not just pharmacogenomic biomarkers; whether pharmacogenomics biomarkers have been more successful is unclear, and would require formal analysis.

There are currently 121 Food and Drug Administation (FDA) drug labels referring to pharmacogenetic biomarkers of drug safety or efficacy [29]. Only a very small proportion of these drug labels mandate clinicians to test for pharmacogenetic markers (e.g., abacavir and *HLA-B*57:01*; and carbamazepine and HLA-*B*15:02* in Southern Asians). However, a large number of testing guideline position papers have been published by the Clinical Pharmacogenetics Implementation Consortium [30]. The aim of this collaborative effort is to help clinicians understand how genetic tests may be used to guide treatment decisions.

One reason why many pharmacogenetic biomarkers have failed to move from discovery to clinical implementation is that many genotype/phenotype associations fail to be independently replicated. It has been recognized that variability in phenotype definition could contribute to this, particularly in relation to adverse drug reactions. To this end a number of phenotype standardization efforts have been undertaken in recent years. These include drug-induced skin injury [31], liver injury [32] and Torsade de Pointes (long-QT syndrome) [33]. It is hoped that, with studies applying consistent phenotype definitions, future pharmacogenetic studies may identify replicable pharmacogenetic biomarkers that, with sufficient weight of evidence, could find their way into clinical utility. Another reason for lack of replication is the inability to find replication sample sets, particularly where the phenotype is a rare adverse event. In such cases, functional genomic analyses may reduce false positives, and provide more confidence for implementation because of insights into the mechanism of effect of that biomarker.

6. The Future...

6.1. Point of Care Genetic Testing

As our understanding of the human genome has grown, so the technology with which we can analyze it has developed in terms of speed and fidelity. The polymerase chain reaction, first described by Kary Mullis in 1983 [34], allowed sensitive analysis of DNA and ultimately yielded a number of pharmacogenetic biomarkers in the pre-genome era based on prior knowledge of gene function (candidate gene studies). As with our understanding of the genome, PCR-based technology has advanced significantly and now allows for rapid and accurate genotype detection. In recent years a number of studies have investigated the potential utility of rapid point-of-care (POC) genetic testing.

Two key randomized control trials have recently utilized pre-treatment genetic testing, with POC devices, to guide the use of drugs used in cardiovascular medicine, the anticoagulant coumarin derivatives (e.g., warfarin) [35] and the antiplatelet drug clopidogrel [36]. The EU-PACT study [35] randomized patients to either genotype-guided or standard dosing. The genotype guided group, utilized molecular beacon technology to genotype for the *CYP2C9*2* and *CYP2C9*3* variants and a promoter polymorphism of *VKORC1* prior to the patient receiving warfarin. The results generated were then incorporated into a warfarin dose calculator, incorporating a number of key non-genetic factors affecting dose requirement. The warfarin trial found that patients randomized to pharmacogenetic-guided dosing spent a mean percentage time in the therapeutic range of 67.4% compared with 60.3% on the standard dosing protocol. In this example, the healthcare professional was presented with an individualized, genotype-guided dosing regimen for the patient based on not only clinical variables but also genetic factors. The technology used in the EU-PACT trial had a lead time of approximately 2 h to obtain the required genotypes.

In the case of the RAPID-GENE trial [36], patients were typed for the *CYP2C19*2* allele, which has been associated with a lack of efficacy of clopidogrel following percutaneous coronary intervention (PCI). Patients were randomized to standard (75 mg/day clopidogrel) or genotype-guided arms. In the genotype guided arm, patients carrying the *CYP2C19*2* allele were prescribed 10 mg prasugrel while wild-type patients were given the standard therapy of 75 mg/day clopidogrel. The trial demonstrated that genotype-guided dosing significantly reduced adverse events (platelet reactivity) compared to standard treatment. The genotyping technology utilized for RAPID-GENE, the Spartan RX CYP2C19, is able to produce genotypes in less than 60 min. In August 2013, the Spartan RX CYP2C19 device became the first on-demand rapid genetic testing device for prescribing guidance to gain Food and Drug Administration (FDA) approval. However, it is important to note that it is not approved as a POC device and must be operated within a clinical laboratory environment.

POC Genotype technologies continue to advance and molecular biology methodologies, such as SmartAmp 2 [37] are reducing the time to obtain a genotype further. It is already possible to determine genotypes for pharmacogenomics-related variants such as those associated with warfarin dose requirement (*VKORC1* and *CYP2C9*) in 30–40 min [37,38]. With addition of microfluidics technology to devices [39,40], the size of the equipment for genotyping will decrease and subsequently, the portability will also increase. It is the ease of use and interpretation of results which are the key attributes for POC genetic test which will facilitate their adoption by health-care professionals. In future years, it is likely that many more devices and tests will be applied to clinical care and obtain regulatory approval.

6.2. Companion Diagnostics

For a number of years now the notion of "one size fits all" in both drug prescribing and drug development has seemed an outdated concept. Indeed, the pharmaceutical industry has re-focused its efforts away from the previous block buster drug model to develop more "niche-busting" products which are licensed alongside a companion diagnostic assay (Table 2). This allows for drugs which may be largely considered ineffective in the wider population to be targeted to a subset of patients likely to respond well to the treatment. To date, the vast majority of these have been developed in the oncology field but it is likely that many more examples of population stratification using genomics methodologies for targeted treatment will emerge for other indications. An added advantage of undertaking targeted therapies in cancer using companion diagnostics has been the rapid approvals obtained from the FDA, despite pivotal efficacy trials testing smaller numbers than is usual in non-stratified trials [41].

While the majority of companion diagnostics products to date have focused on variation in pharmacodynamic factors for stratification, there is also a growing interest in individualizing drug doses based on pharmacokinetic variability. For example, dose escalation of tivantinib, a non-small cell lung cancer therapy, is based on stratification for the *CYP2C19* genotype [42]. This is consistent with European Medicines Agency guidance on pharmacogenetic effects on drug pharmacokinetics [43]. Thus, in the future, it is likely that regulatory approval for a new drug could be dependent on dose stratification based on the underlying metabolizer phenotypes.

6.3. Pre-Emptive Genotyping

The possibility of pre-emptying genotyping is being explored in the U.S., for example, through the Vanderbilt Personalized Medicines Program, where patients are genotyped on a 184-variant platform [44]. The theory behind this approach is that if a genotype is available to the physician in the electronic medical record, at the point of prescribing, they will be able to make a rational decision as to the choice and/or dose of the drug. It also highlights an important issue in relation to the evidence for clinical implementation: we cannot possibly undertake randomised controlled trials or prospective studies for every genomic variant that is identified, and other methodologies for evaluating the clinical utility of a genomic biomarker will have to be utilized. It is also important to note that such information is already used by physicians with respect to non-pharmacogenomic tests, for example renal function tests allow clinicians to reduce the dose of a drug that is renally excreted [45]. Whether and how such pre-prescription genotyping will be used by prescribers, and whether it leads to improvement in clinical outcomes, will require careful evaluation over the next few years. Indeed it can be seen as a prelude to a time when whole genome sequencing is so cheap that is undertaken routinely with the data being available within electronic medical records.

6.4. Personal Genomes and Clinical Applications

The per-base cost of sequencing has fallen exponentially since 2003 and has for many years exceeded the trajectory of Moores Law (the number of transistors on integrated circuits approximately doubles every 2 years) [46]. With this is mind it is feasible to imagine in the not too distant future that whole genome sequencing will be a cost effective option for healthcare providers. However, with the >3 billion base pairs confirmed in 2003 by the human genome project and the complexity of interpreting the role of genetic variation in inter-individual drug response, it is likely the challenge for the next 10 years will be in producing the tools with which to interpret this data and provide meaningful outputs that can be utilized by healthcare professionals.

Table 2. *In vitro* companion diagnostic kits currently licensed by the Food and Drug Administration (FDA).

Biomarker	Prevalance	Indication	Drug	Assay Kit	Technology	Manufacturer
HER2 gene amplification	22.2% [47]	Breast cancer	trastuzumab (Herceptin)	Inform	FISH	Ventana Medical Systems
				PathVysion #	FISH	Abbott Molecular Inc.
				SPOT-Light HER2	CISH	Life Technologies
				InSite HER2	IHC	Biogenex Laboratories, Inc.
				HercepTest	IHC	Dako Denmark
				HER2 PharmDx	CISH	Dako Denmark
				HER2 PharmDx *	FISH	Dako Denmark
				PATHWAY Her2	IHC	Ventana Medical Systems
				Bond Oracle	IHC	Leica Biosystems
EGFR protein expression	60%–80% [48]	Colorectal cancer	cetuximab (Erbitux) panitumumab (Vectibix)	EGFR pharmDx	IHC	Dako North America
c-Kit protein expression	100% [49]	Gastrointestinal stromal tumours	imatinib (Gleevac)	c-kit pharmDx	IHC	Dako North America
ALK gene rearrangement	7.5% [50]	Non-small cell lung cancer	crizotinib (Xalkori)	VYSIS	FISH	Abbott Molecular Inc.
BRAF p.V600E mutation	75.4% [51]	Melanoma	vemurafenib (Zelboraf)	Cobas 4800	Real-time PCR	Roche Molecular Systems, Inc
KRAS mutation	30%–60% [52]	Colorectal cancer	cetuximb (Erbitux)	therascreen	Real-time PCR	Qiagen

* Indicated for assessment of breast cancer patients considered for pertuzumab (Perjeta) or # cyclophosphamide, doxorubicin, 5-FU treatment.

6.5. Ethical Considerations

A key ethical issue with regard to the implementation of personalised medicine relates to the fact that it may lead to health inequalities, within and between countries. Given the costs related to developing, manufacturing and obtaining approval for genetic tests, in addition to defining the role of ethnic genetic differences, it is easy to see how resource poor countries and communities could miss out on the benefits of personalised medicine advances in the future. This may be offset by the advances in genomics technologies and the vast reduction in costs of implementing them that has taken place over the last 10 years. Another issue which is becoming increasingly important with the use of next generation sequencing technologies is whether patients should be informed about incidental findings. An analysis of 1000 participants' exomes showed that the frequency of actionable (*i.e.*, pathogenic or likely pathogenic single nucleotide variants) incidental findings was 3.4% in European Caucasians and 1.2% in Africans [53]. Some guidelines have been produced [54], but there is still controversy [54,55], and the debate will no doubt become more intense as more people have their genomes sequenced, and more variants are classified as being actionable.

6.6. Educating Stakeholders

In order for pharmacogenomics, as one of the technologies that is important for personalised medicine, to realize wider uptake into healthcare provision, there needs to be greater awareness and education. This applies not only to healthcare professional but to patients who are the ultimate stakeholder in pharmacogenomics, and stand to gain the most if we can improve predictability of how patients will respond to drugs.

Clinicians need to be made aware of the availability of genetic tests relating to treatment decisions, and how to interpret them. Lack of familiarity with genetic tests may be one reason for the poor uptake into clinical practice [45]. The lack of training is amongst the most common reasons cited as a barrier to pharmacogenomic implementation [56].

6.7. Regulatory Environment

There are of course many other issues which need to be tackled in order to facilitate the implementation of pharmacogenomic testing into clinical practice—many of these also apply to other biomarker strategies that are important for personalised medicine [41]. Amongst this is the regulatory environment, which has a big influence on the diagnostics industry. With the development of companion diagnostics, it will be important for there to be streamlined procedures which allow for simultaneous approval of the drug and diagnostic. There are differences between the FDA and European Medicines Agency for the development of companion diagnostics, with the latter being less stringent, but likely to adopt similar procedures in a drive to ensure there is global harmonization of the regulatory procedures needed for approval [57]. For the diagnostics industry, the requirement for a clinical utility study may become prohibitive in terms of cost, unless there are clear pathways for protection of intellectual property and re-imbursement.

7. Conclusions

Since the completion of the human genome, there has been steady, albeit slow, progress in the identification and implementation of biomarkers into clinical practice. This progress is likely to continue, and hopefully accelerate as our ability to interrogate the human genome becomes more cost- and time-efficient, and we start embracing, and intelligently interpreting, different sources of data to define the clinical validity and utility of biomarkers. Outcomes research will thus become particularly important, and will depend on having access to curated electronic healthcare databases where patients can be followed longitudinally from the time of having a biomarker assessed to the time a drug is prescribed, and forward into the future to define the clinical outcome of the patient (in comparison to relevant *a priori* defined controls).

Acknowledgments

MP is a NIHR Senior Investigator. We thank the UK Department of Health (NHS Chair of Pharmacogenetics), Wolfson Foundation, EU-FP7 and the MRC (MRC Centre for Drug Safety Science) for grant funding.

Author Contributions

All authors contributed equally.

References

1. Lander, E.S.; Linton, L.M.; Birren, B.; Nusbaum, C.; Zody, M.C.; Baldwin, J.; Devon, K.; Dewar, K.; Doyle, M.; FitzHugh, W.; *et al.* Initial sequencing and analysis of the human genome. *Nature* **2001**, *409*, 860–921.
2. Vogel, F. Moderne probleme der humangenetik. *Ergeb. Inn. Med. Kinderheilkd* **1959**, *12*, 52–125.
3. Watson, J.D.; Crick, F.H. Molecular structure of nucleic acids; a structure for deoxyribose nucleic acid. *Nature* **1953**, *171*, 737–738.
4. International HapMap Consortium. The international hapmap project. *Nature* **2003**, *426*, 789–796.
5. Hindorff, L.A.; Sethupathy, P.; Junkins, H.A.; Ramos, E.M.; Mehta, J.P.; Collins, F.S.; Manolio, T.A. Potential etiologic and functional implications of genome-wide association loci for human diseases and traits. *Proc. Natl. Acad. Sci. USA* **2009**, *106*, 9362–9367.
6. Link, E.; Parish, S.; Armitage, J.; Bowman, L.; Heath, S.; Matsuda, F.; Gut, I.; Lathrop, M.; Collins, R. Slco1b1 variants and statin-induced myopathy—A genomewide study. *New Engl. J. Med.* **2008**, *359*, 789–799.
7. Sarasquete, M.E.; Garcia-Sanz, R.; Marin, L.; Alcoceba, M.; Chillon, M.C.; Balanzategui, A.; Santamaria, C.; Rosinol, L.; de la Rubia, J.; Hernandez, M.T.; *et al.* Bisphosphonate-related osteonecrosis of the jaw is associated with polymorphisms of the cytochrome p450 cyp2c8 in multiple myeloma: A genome-wide single nucleotide polymorphism analysis. *Blood* **2008**, *112*, 2709–2712.
8. Kindmark, A.; Jawaid, A.; Harbron, C.G.; Barratt, B.J.; Bengtsson, O.F.; Andersson, T.B.;

Carlsson, S.; Cederbrant, K.E.; Gibson, N.J.; Armstrong, M.; *et al.* Genome-wide pharmacogenetic investigation of a hepatic adverse event without clinical signs of immunopathology suggests an underlying immune pathogenesis. *Pharmacogenomics J.* **2008**, *8*, 186–195.

9. Daly, A.K.; Donaldson, P.T.; Bhatnagar, P.; Shen, Y.; Pe'er, I.; Floratos, A.; Daly, M.J.; Goldstein, D.B.; John, S.; Nelson, M.R.; *et al.* Hla-b*5701 genotype is a major determinant of drug-induced liver injury due to flucloxacillin. *Nat. Genet.* **2009**, *41*, 816–819.
10. McCormack, M.; Alfirevic, A.; Bourgeois, S.; Farrell, J.J.; Kasperaviciute, D.; Carrington, M.; Sills, G.J.; Marson, T.; Jia, X.; de Bakker, P.I.; *et al.* Hla-a*3101 and carbamazepine-induced hypersensitivity reactions in europeans. *New Engl. J. Med.* **2011**, *364*, 1134–1143.
11. Ozeki, T.; Mushiroda, T.; Yowang, A.; Takahashi, A.; Kubo, M.; Shirakata, Y.; Ikezawa, Z.; Iijima, M.; Shiohara, T.; Hashimoto, K.; *et al.* Genome-wide association study identifies hla-a*3101 allele as a genetic risk factor for carbamazepine-induced cutaneous adverse drug reactions in japanese population. *Hum. Mol. Gen.* **2011**, *20*, 1034–1041.
12. Tohkin, M.; Kaniwa, N.; Saito, Y.; Sugiyama, E.; Kurose, K.; Nishikawa, J.; Hasegawa, R.; Aihara, M.; Matsunaga, K.; Abe, M.; *et al.* A whole-genome association study of major determinants for allopurinol-related stevens-johnson syndrome and toxic epidermal necrolysis in japanese patients. *Pharmacogenomics J.* **2013**, *13*, 60–69.
13. Shuldiner, A.R.; O'Connell, J.R.; Bliden, K.P.; Gandhi, A.; Ryan, K.; Horenstein, R.B.; Damcott, C.M.; Pakyz, R.; Tantry, U.S.; Gibson, Q.; *et al.* Association of cytochrome p450 2c19 genotype with the antiplatelet effect and clinical efficacy of clopidogrel therapy. *JAMA* **2009**, *302*, 849–857.
14. Ge, D.; Fellay, J.; Thompson, A.J.; Simon, J.S.; Shianna, K.V.; Urban, T.J.; Heinzen, E.L.; Qiu, P.; Bertelsen, A.H.; Muir, A.J.; *et al.* Genetic variation in il28b predicts hepatitis c treatment-induced viral clearance. *Nature* **2009**, *461*, 399–401.
15. Takeuchi, F.; McGinnis, R.; Bourgeois, S.; Barnes, C.; Eriksson, N.; Soranzo, N.; Whittaker, P.; Ranganath, V.; Kumanduri, V.; McLaren, W.; *et al.* A genome-wide association study confirms vkorc1, cyp2c9, and cyp4f2 as principal genetic determinants of warfarin dose. *PLoS Genet.* **2009**, *5*, e1000433.
16. The 1000 Genomes Project Consortium. A map of human genome variation from population-scale sequencing. *Nature* **2010**, *467*, 1061–1073.
17. Gordon, A.S.; Tabor, H.K.; Johnson, A.D.; Snively, B.M.; Assimes, T.L.; Auer, P.L.; Ioannidis, J.P.; Peters, U.; Robinson, J.G.; Sucheston, L.E.; *et al.* Quantifying rare, deleterious variation in 12 human cytochrome p450 drug-metabolism genes in a large-scale exome dataset. *Hum. Mol. Gen.* **2014**. 23, 1957–1963.
18. Wadelius, M.; Chen, L.Y.; Lindh, J.D.; Eriksson, N.; Ghori, M.J.; Bumpstead, S.; Holm, L.; McGinnis, R.; Rane, A.; Deloukas, P. The largest prospective warfarin-treated cohort supports genetic forecasting. *Blood* **2009**, *113*, 784–792.
19. Ainle, F.N.; Mumford, A.; Tallon, E.; McCarthy, D.; Murphy, K. A vitamin k epoxide reductase complex subunit 1 mutation in an irish patient with warfarin resistance. *Irish J. Med. Sci.* **2008**, *177*, 159–161.

20. Harrington, D.J.; Gorska, R.; Wheeler, R.; Davidson, S.; Murden, S.; Morse, C.; Shearer, M.J.; Mumford, A.D. Pharmacodynamic resistance to warfarin is associated with nucleotide substitutions in vkorc1. *J. Thrombosis Haemost.* **2008**, *6*, 1663–1670.
21. Harrington, D.J.; Underwood, S.; Morse, C.; Shearer, M.J.; Tuddenham, E.G.; Mumford, A.D. Pharmacodynamic resistance to warfarin associated with a val66met substitution in vitamin k epoxide reductase complex subunit 1. *Thromb. Haemost.* **2005**, *93*, 23–26.
22. Rost, S.; Fregin, A.; Ivaskevicius, V.; Conzelmann, E.; Hortnagel, K.; Pelz, H.J.; Lappegard, K.; Seifried, E.; Scharrer, I.; Tuddenham, E.G.; *et al.* Mutations in vkorc1 cause warfarin resistance and multiple coagulation factor deficiency type 2. *Nature* **2004**, *427*, 537–541.
23. Behr, E.R.; Roden, D. Drug-induced arrhythmia: Pharmacogenomic prescribing? *Eur. Heart J.* **2013**, *34*, 89–95.
24. Ramirez, A.H.; Shaffer, C.M.; Delaney, J.T.; Sexton, D.P.; Levy, S.E.; Rieder, M.J.; Nickerson, D.A.; George, A.L., Jr.; Roden, D.M. Novel rare variants in congenital cardiac arrhythmia genes are frequent in drug-induced torsades de pointes. *Pharmacogenomics J.* **2013**, *13*, 325–329.
25. Consortium, E.P.; Bernstein, B.E.; Birney, E.; Dunham, I.; Green, E.D.; Gunter, C.; Snyder, M. An integrated encyclopedia of DNA elements in the human genome. *Nature* **2012**, *489*, 57–74.
26. Starkey Lewis, P.J.; Dear, J.; Platt, V.; Simpson, K.J.; Craig, D.G.; Antoine, D.J.; French, N.S.; Dhaun, N.; Webb, D.J.; Costello, E.M.; *et al.* Circulating micrornas as potential markers of human drug-induced liver injury. *Hepatology* **2011**, *54*, 1767–1776.
27. Ichihara, A.; Wang, Z.; Jinnin, M.; Izuno, Y.; Shimozono, N.; Yamane, K.; Fujisawa, A.; Moriya, C.; Fukushima, S.; Inoue, Y.; *et al.* Upregulation of mir-18a-5p contributes to epidermal necrolysis in severe drug eruptions. *J. Allergy Clin. Immunol.* **2013**, *133*, 1065–1074.
28. Poste, G. Bring on the biomarkers. *Nature* **2011**, *469*, 156–157.
29. US Food and Drug Administration. Table of Pharmacogenomic Biomarkers in Drug Labeling. Available online: http://www.fda.gov/drugs/scienceresearch/researchareas/pharmacogenetics/ucm083378.htm/ (6 November 2013).
30. Relling, M.V.; Klein, T.E. Cpic: Clinical pharmacogenetics implementation consortium of the pharmacogenomics research network. *Clin. Pharmacol. Ther.* **2011**, *89*, 464–467.
31. Pirmohamed, M.; Friedmann, P.S.; Molokhia, M.; Loke, Y.K.; Smith, C.; Phillips, E.; La Grenade, L.; Carleton, B.; Papaluca-Amati, M.; Demoly, P.; *et al.* Phenotype standardization for immune-mediated drug-induced skin injury. *Clin. Pharmacol. Ther.* **2011**, *89*, 896–901.
32. Pirmohamed, M.; Aithal, G.P.; Behr, E.; Daly, A.; Roden, D. The phenotype standardization project: Improving pharmacogenetic studies of serious adverse drug reactions. *Clin. Pharmacol. Ther.* **2011**, *89*, 784–785.
33. Behr, E.R.; January, C.; Schulze-Bahr, E.; Grace, A.A.; Kaab, S.; Fiszman, M.; Gathers, S.; Buckman, S.; Youssef, A.; Pirmohamed, M.; *et al.* The international serious adverse events consortium (isaec) phenotype standardization project for drug-induced torsades de pointes. *Eur. Heart J.* **2012**. *34*, 1958–1963
34. Saiki, R.K.; Gelfand, D.H.; Stoffel, S.; Scharf, S.J.; Higuchi, R.; Horn, G.T.; Mullis, K.B.; Erlich, H.A. Primer-directed enzymatic amplification of DNA with a thermostable DNA

polymerase. *Science* **1988**, *239*, 487–491.
35. Pirmohamed, M.; Burnside, G.; Eriksson, N.; Jorgensen, A.L.; Toh, C.H.; Nicholson, T.; Kesteven, P.; Christersson, C.; Wahlstrom, B.; Stafberg, C.; *et al.* A randomized trial of genotype-guided dosing of warfarin. *New Engl. J. Med.* **2013**, *369*, 2294–2303.
36. Roberts, J.D.; Wells, G.A.; Le May, M.R.; Labinaz, M.; Glover, C.; Froeschl, M.; Dick, A.; Marquis, J.F.; O'Brien, E.; Goncalves, S.; *et al.* Point-of-care genetic testing for personalisation of antiplatelet treatment (rapid gene): A prospective, randomised, proof-of-concept trial. *Lancet* **2012**, *379*, 1705–1711.
37. Lezhava, A.; Ishidao, T.; Ishizu, Y.; Naito, K.; Hanami, T.; Katayama, A.; Kogo, Y.; Soma, T.; Ikeda, S.; Murakami, K.; *et al.* Exciton primer-mediated snp detection in smartamp2 reactions. *Hum. Mut.* **2010**, *31*, 208–217.
38. Aomori, T.; Yamamoto, K.; Oguchi-Katayama, A.; Kawai, Y.; Ishidao, T.; Mitani, Y.; Kogo, Y.; Lezhava, A.; Fujita, Y.; Obayashi, K.; *et al.* Rapid single-nucleotide polymorphism detection of cytochrome p450 (cyp2c9) and vitamin k epoxide reductase (vkorc1) genes for the warfarin dose adjustment by the smart-amplification process version 2. *Clin. Chem.* **2009**, *55*, 804–812.
39. Burn, J. Company profile: Quantumdx group limited. *Pharmacogenomics* **2013**, *14*, 1011–1015.
40. Stedtfeld, R.D.; Tourlousse, D.M.; Seyrig, G.; Stedtfeld, T.M.; Kronlein, M.; Price, S.; Ahmad, F.; Gulari, E.; Tiedje, J.M.; Hashsham, S.A. Gene-z: A device for point of care genetic testing using a smartphone. *Lab. Chip* **2012**, *12*, 1454–1462.
41. *Realising the potential of stratified medicine*; Academy of Medical Sciences: London, UK, 2013.
42. Yamamoto, N.; Murakami, H.; Hayashi, H.; Fujisaka, Y.; Hirashima, T.; Takeda, K.; Satouchi, M.; Miyoshi, K.; Akinaga, S.; Takahashi, T.; *et al.* Cyp2c19 genotype-based phase i studies of a c-met inhibitor tivantinib in combination with erlotinib, in advanced/metastatic non-small cell lung cancer. *Br. J. Cancer* **2013**, *109*, 2803–2809.
43. European Medicines Agency. Guideline on the Use of Pharmacogenetic Methodologies in the Pharmacokinetic Evaluation of Medicinal Products. Available online: http://www.ema.europa.eu/docs/en_GB/document_library/Scientific_guideline/2012/02/WC500121954.pdf (accessed on 30 October 2012).
44. Pulley, J.M.; Denny, J.C.; Peterson, J.F.; Bernard, G.R.; Vnencak-Jones, C.L.; Ramirez, A.H.; Delaney, J.T.; Bowton, E.; Brothers, K.; Johnson, K.; *et al.* Operational implementation of prospective genotyping for personalized medicine: The design of the vanderbilt predict project. *Clin. Pharmacol. Ther.* **2012**, *92*, 87–95.
45. Pirmohamed, M. Pharmacogenetics: Past, present and future. *Drug Discov. Today* **2011**, *16*, 852–861.
46. Moore, G.E. Cramming more components onto integrated circuits. *Electronics* **1965**, *38*, 114–117.
47. Ross, J.S.; Slodkowska, E.A.; Symmans, W.F.; Pusztai, L.; Ravdin, P.M.; Hortobagyi, G.N. The her-2 receptor and breast cancer: Ten years of targeted anti-her-2 therapy and personalized medicine. *Oncologist* **2009**, *14*, 320–368.
48. Goldstein, N.S.; Armin, M. Epidermal growth factor receptor immunohistochemical reactivity in patients with american joint committee on cancer stage iv colon adenocarcinoma: Implications for a standardized scoring system. *Cancer* **2001**, *92*, 1331–1346.

49. Went, P.T.; Dirnhofer, S.; Bundi, M.; Mirlacher, M.; Schraml, P.; Mangialaio, S.; Dimitrijevic, S.; Kononen, J.; Lugli, A.; Simon, R.; *et al.* Prevalence of kit expression in human tumors. *J. Clin. Oncol.* **2004**, *22*, 4514–4522.
50. Martelli, M.P.; Sozzi, G.; Hernandez, L.; Pettirossi, V.; Navarro, A.; Conte, D.; Gasparini, P.; Perrone, F.; Modena, P.; Pastorino, U.; *et al.* Eml4-alk rearrangement in non-small cell lung cancer and non-tumor lung tissues. *Am. J. Pathol.* **2009**, *174*, 661–670.
51. Greaves, W.O.; Verma, S.; Patel, K.P.; Davies, M.A.; Barkoh, B.A.; Galbincea, J.M.; Yao, H.; Lazar, A.; Aldape, K.D.; Medeiros, L.J., *et al.* Frequency and spectrum of braf mutations in a retrospective, single-institution study of 1112 cases of melanoma. *J. Mol. Diagn.* **2012**, *15*, 220–226.
52. Brink, M.; de Goeij, A.F.; Weijenberg, M.P.; Roemen, G.M.; Lentjes, M.H.; Pachen, M.M.; Smits, K.M.; de Bruine, A.P.; Goldbohm, R.A.; van den Brandt, P.A. K-ras oncogene mutations in sporadic colorectal cancer in the netherlands cohort study. *Carcinogenesis* **2003**, *24*, 703–710.
53. Dorschner, M.O.; Amendola, L.M.; Turner, E.H.; Robertson, P.D.; Shirts, B.H.; Gallego, C.J.; Bennett, R.L.; Jones, K.L.; Tokita, M.J.; Bennett, J.T., *et al.* Actionable, pathogenic incidental findings in 1,000 participants' exomes. *Am. J. Hum. Genet.* **2013**, *93*, 631–640.
54. Green, R.C.; Berg, J.S.; Grody, W.W.; Kalia, S.S.; Korf, B.R.; Martin, C.L.; McGuire, A.L.; Nussbaum, R.L.; O'Daniel, J.M.; Ormond, K.E., *et al.* Acmg recommendations for reporting of incidental findings in clinical exome and genome sequencing. *Genet. Med.* **2013**, *15*, 565–574.
55. Kocarnik, J.M.; Fullerton, S.M. Returning pleiotropic results from genetic testing to patients and research participants. *JAMA* **2014**, *311*, 795–796.
56. Patel, H.N.; Ursan, I.D.; Zueger, P.M.; Cavallari, L.H.; Pickard, A.S. Stakeholder views on pharmacogenomic testing. *Pharmacotherapy* **2014**, *34*, 151–165.
57. Senderowicz, A.M.; Pfaff, O. Similarities and differences in the oncology drug approval process between fda and european union with emphasis on in vitro companion diagnostics. *Clin. Cancer Res.* **2014**, *20*, 1445–1452.

11

Imprinted Genes and the Environment: Links to the Toxic Metals Arsenic, Cadmium and Lead

Lisa Smeester [1], Andrew E. Yosim [1], Monica D. Nye [2,3], Cathrine Hoyo [4], Susan K. Murphy [3] and Rebecca C. Fry [1,2,5,*]

1 Department of Environmental Sciences and Engineering, Gillings School of Global Public Health, The University of North Carolina, 135 Dauer Drive, CB 7431, UNC, Chapel Hill, NC 27599, USA; E-Mails: smesta@unc.edu (L.S.); yosim@unc.edu (A.E.Y.)

2 Lineberger Comprehensive Cancer Center, The University of North Carolina, 450 West Street, CB 7295, UNC, Chapel Hill, NC 27599, USA; E-Mail: mnye@email.unc.edu

3 Department of Obstetrics and Gynecology, Duke University Medical Center, B226 LSRC, Box 91012, Research Drive, Durham, NC 27708, USA; E-Mail: susan.murphy@duke.edu

4 Department of Biological Sciences, Center for Human Health and Environment, Campus Box 7633, NC State University, Raleigh, NC 27695, USA; E-Mail: choyo@ncstate.edu

5 Curriculum in Toxicology, The University of North Carolina, 135 Dauer Drive, CB 7431, UNC, Chapel Hill, NC 27599, USA

* Author to whom correspondence should be addressed; E-Mail: rfry@unc.edu;

Abstract: Imprinted genes defy rules of Mendelian genetics with their expression tied to the parent from whom each allele was inherited. They are known to play a role in various diseases/disorders including fetal growth disruption, lower birth weight, obesity, and cancer. There is increasing interest in understanding their influence on environmentally-induced disease. The environment can be thought of broadly as including chemicals present in air, water and soil, as well as food. According to the Agency for Toxic Substances and Disease Registry (ATSDR), some of the highest ranking environmental chemicals of concern include metals/metalloids such as arsenic, cadmium, and lead. The complex relationships between toxic metal exposure, imprinted gene regulation/expression and health outcomes are understudied. Herein we examine trends in imprinted gene biology, including an assessment of the imprinted genes and their known functional roles in the cell, particularly as they relate to toxic metals exposure and disease. The data highlight that many of the imprinted genes have known associations to developmental diseases and are enriched for their role in the TP53 and AhR pathways. Assessment of the promoter regions of the imprinted genes resulted in the identification of an enrichment of binding sites for two transcription factor families, namely the zinc finger family II and PLAG transcription factors. Taken together these data contribute insight into the complex relationships between toxic metals in the environment and imprinted gene biology.

Keywords: imprinted genes; epigenetics; DNA methylation; TP53; aryl hydrocarbon receptor; systems biology; environmental health; toxic metals; epigenomics; environment

---

1. Introduction

There is heightened interest in understanding the role of epigenetic mechanisms in cell signaling regulation and disease. This is particularly the case when attempting to discern the etiology of disease where a cause is hitherto unknown. Recent studies suggest that disease can be influenced by the environment via epigenetic mechanisms [1–3], a seemingly Lamarkian notion discordant with the tenets set forth by Mendel's work. Yet the advent of modern epigenetics as a distinct field of study is much more storied than simply pitting Lamark's theory of "soft inheritance" against Mendel's firm genetic basis of heredity. While controversial to this day, some of the earliest experiments that suggested non-Mendelian inheritance, and possibly the first indication of parent-of-origin phenomena, came from Kammerer's midwife toad experiments [4] which pointed to the ability of a modified environment to modulate heritable shifts in mating.

However, it was not until the early 1940s, the same time Huxley's Modern Synthesis sought an interdisciplinary approach to inheritance and evolution [5], that the term epigenetics was first introduced by developmental biologist Conrad Waddington [6]. Describing mechanisms that determine cell fate and differentiation during development, Waddington's epigenetics arose as a conceptual means to describe how a complex network of genes and gene-environment interactions brought about phenotype in an evolutionary context [7].

It was more recently though, with Holliday linking the term epigenetics with the ability of DNA methylation to modulate gene activity, that the definition of epigenetics began to shift to being inclusive of any mechanism with the ability to modulate gene activity without a change in DNA sequence [8]. These epigenetic modifications to an individual's genome include, but are not limited to, three commonly studied mechanisms: DNA methylation, histone modification, and non-coding RNA expression [9].

While such marks are generally stable and heritable in nature, there is the potential for epigenetic modifications to be reversible, as seen with their innate reversibility during critical stages of fetal development, as epigenetic tags are added and removed [10–12]. Such changes to the epigenetic landscape are vital in the process of normal development [13], yet imprinted genes are protected from this process [14].

Changes to the epigenome can also be induced by the environment resulting in abnormal physiologic changes [3,15]. Stable epigenetic modifications are implicated in adult onset disease such as cancer, neurodevelopmental/neurodegenerative disorders, and autoimmune disorders [16] among others, and also have the potential to be trans-generational in nature [17,18]. However, it should be noted that the stability of such epigenetic alterations and their link to later life health outcomes is still under active debate [19,20].

Current research has shown that epigenetic modifications can be induced by exposure to environmental contaminants, such as toxic metals [21–23]. Such modifications have been linked to later life health outcomes including cancers, heart disease, kidney disease, and various neurological conditions [20,24,25]. While cadmium and lead are metals, arsenic is a metalloid, with shared properties of both metals and non-metals. For the purposes of this article, all three elements will hereby be referred to as "metals." While these metals are among the most studied, many other metals have demonstrated toxicity and are associated with epigenetic alterations including nickel and chromium [21]. In addition, new research is currently investigating various under-studied metals for toxicity and associated epigenetic alterations including tungsten and cobalt [26,27].

While research continues into the stability of epigenetic alterations associated with environmental contaminants and their associations with negative health endpoints, many have theorized that such reversibility may provide the opportunity for therapeutic targets for disease prevention following environmental exposure [24,28].

1.1. What is Genomic Imprinting?

Evidence that parental genomes are not equivalent was first described in mouse models by both the Surani and McGrath groups in the early 1980s [29,30]. The researchers attempted to generate viable embryos using only maternal or paternal chromosomes. They found that normal development required genetic material from both parents; maternal and paternal genomes were not interchangeable, indicating the first experiments to demonstrate mammalian imprinting. While the exact mechanism of these imprinting phenomena was unknown at that time, they hypothesized a process that operated pre-fertilization, yet impacted post-fertilization expression. To date, researchers have used a variety of strategies including genome-wide studies, gene-specific experiments, and transcriptome analysis to determine which human or mammalian genes are imprinted [31–33]. Imprinted genes are vulnerable to genetic and epigenetic perturbation and have been tied to adverse health outcomes. As imprinted genes are monoallelically expressed with one of the copies of the gene silenced in a parent-of-origin dependent manner, only one copy is functional. As a result, mutations or epigenetic alterations on one allele that would normally have minimal impact for a biallelically expressed gene may lead to detrimental consequences for an imprinted gene.

As it is a critical part of the epigenome, the inheritance and manifestation of traits associated with imprinted genes is regulated through epigenetic marks. The term "imprintome" was first coined to describe a set of "*cis*-acting imprint regulatory elements" [34]. This term refers to the mechanisms needed for modifying expression including DNA methylation and histone modification, which are two such mechanisms that are well established as being required for the appropriate maintenance of imprinted gene expression [35]. The imprintome is vulnerable to the environment and potentially modified by a host of environmental chemicals and contaminants [36].

Many imprinted genes are grouped in clusters and possess imprinting control regions (ICRs) or a central control region [37]. These ICRs, as well as other regulatory regions associated with imprinted genes, are referred to as differentially methylated regions (DMRs) and display ~50% methylation, where one of the parental alleles is methylated and the other unmethylated in a manner based on parent of origin. These DMRs represent discrete DNA elements that carry a heritable epigenetic mark that distinguishes the parental alleles.

1.2. Evolution of Imprinted Genes: A Fight between the Parental Chromosomes

Perhaps the most widely accepted hypothesis for evolutionary underpinnings of the origins of genomic imprinting is the parental conflict theory or the "battle of the sexes" [38,39]. Central to this hypothesis is the struggle to control maternal resources during fetal development, with paternal genes favoring increased use of maternal resources in order to promote the fittest possible offspring, and to divert resources from offspring of other males. Contextually, this hypothesis posits that imprinting arose during early mammalian evolution, where females were able to simultaneously gestate offspring from multiple males. The basis then for the desired growth and resource extraction of the offspring carrying the male's DNA is an attempt to out-compete offspring from other males. In contrast, maternal genes will suppress fetal growth to ensure equal reproductive success among all her offspring. Supporting this argument, many paternally expressed genes are growth promoting and metabolism-related, whereas maternally expressed genes tend to be growth limiting. The placenta, which can serve to regulate nutrients and growth for the developing fetus, is thought to play a pivotal role in this maternal-paternal fight over resources and control of fetal growth [40]. Within the placenta, a large number of the known imprinted genes are expressed, and genomic imprinting has been confirmed in all placental mammals studied thus far [41].

1.3. Imprinted Genes and Their Relationship to Human Health

Imprinted genes have been associated with various human adverse outcomes including diabetes, cancer, developmental disorders, behavioral disorders, and reproductive diseases [42,43]. Prader-Willi and Angelman syndromes were the first disorders to suggest an imprinted mechanism. Though each syndrome manifests differently, it was shown they both arise from deletions in the same region of chromosome fifteen [44]; Prader-Willi results from the loss of a cluster of paternally expressed genes, whereas Angelman syndrome results from the loss of maternal expression within the 15q11-q13 region.

Imprinted genes have also been associated with altered cellular growth resulting in cancer [45,46]. Imprinted genes are particularly vulnerable because they are functionally haploid, thus any epigenetic or genetic perturbations may have a greater impact. For example, epigenetically-induced silencing of the active allele of an imprinted tumor suppressor gene could result in complete loss of expression which in turn would influence cell growth or proliferation [47]. Conversely, epigenetic alterations can also result in activation of the otherwise silent copy of an imprinted growth-promoting gene, contributing to loss of growth regulation. Both of these types of alterations are referred to as "loss of imprinting" (LOI). In fact, LOI has been found across a broad spectrum of tumors and is one of the most common alterations in cancer [46]. As examples, in cancer, the active copy of tumor suppressor cyclin-dependent kinase inhibitor 1C (*CDKN1C*) is frequently aberrantly silenced and the silent copy of the growth promoting insulin-like growth factor II (*IGF2*) gene is often inappropriately activated [48].

1.4. Links between Imprinted Genes and Toxic Environmental Metals

Environmental contaminants are currently estimated to be responsible for almost five million deaths and over eighty million Disability-Adjusted Life Years (DALYs) globally [49], and are thought to be involved in 13% to 37% of the global disease burden [50]. Toxic metals represent some of the highest priority contaminants as determined by the Agency for Toxic Substances and Disease Registry (ATSDR) [51].

A number of studies have observed links between exposure to toxic metals and epigenetic events tied to imprinted genes. For example, quantitative analysis was conducted on multiple imprinted gene DMRs in peripheral blood from individuals followed as part of the Cincinnati Lead Study [52]. The researchers identified early childhood lead exposure was associated with hypomethylation of the gene pleiomorphic adenoma gene-like 1 (*PLAGL1*) [53]. Prenatal lead exposure has also been linked to decreases in global DNA methylation in cord blood [54]. In addition, lead has been shown to disrupt global DNA methylation patterns in embryonic stem cells [55].

Similarly, exposure to arsenic and cadmium has been observed to alter the methylation of both experimentally validated and predicted imprinted genes. Specifically, in adults exposed to inorganic arsenic, the known imprinted gene anoctamin 1, calcium activated chloride channel (*ANO1*) and predicted imprinted gene forkhead box F1 (*FOXF1*) shows increased promoter methylation in leukocytes [56]. Conversely, in a separate study, the imprinted gene insulin (*INS*) exhibits decreased promoter methylation as a result of arsenic exposure [57]. In a cohort of mother-newborn pairs, the putative imprinted gene zic family member 1 (*ZIC1*), as well as *ANO1*, are differentially methylated in leukocytes, where cadmium exposure was associated with hypermethylation of *ZIC1* in mothers and *ANO1* in newborns [58].

Tobacco is a common source of cadmium exposure [59,60]. Tobacco smoke exposure in utero has been associated with altered DNA methylation [61,62]. These studies showed that among smoking mothers, ten imprinted genes including aryl-hydrocarbon receptor repressor (AHRR), growth factor independent 1 transcription repressor (GFI1), and cytochrome P450, family 1, subfamily A, polypeptide 1 (CYP1A1) were hypomethylated in newborn cord blood [61]. Additionally, the imprinted gene *IGF2* was found to be hypermethylated in cord blood of newborns born to smoking mothers [62]. Infant gender was correlated with differential methylation, as males exhibited smoking related methylation changes at *IGF2* while female newborns did not [62]. While the mechanisms of such sex-associated methylation patterning are unknown, other studies have shown similar sex-differentiated changes in methylation [63,64]. Further research is needed to understand gender specific epigenetic alterations, including imprinting, and may help to inform both sex-linked susceptibility to disease, as well as potentially being predictive of severity and/or prognosis. Additionally, research into the biological basis underlying the roll an individual's sex plays in disease development may afford the ability to develop targeted therapies based on differential responses.

1.5. Study Aim

In the present study, we set out to analyze imprinted genes for their involvement in shared biological pathways and to determine their known interactions with arsenic, cadmium, and lead. This analysis included an assessment of: (i) known relationships between the proteins encoded by the imprinted genes; (ii) common functionality of the proteins encoded by the imprinted genes in the cell; and (iii) common transcription factor-based regulatory regions present in the promoter regions of the imprinted genes.

2. Methods

Imprinted Gene List and Network, Pathway, and Functional Enrichment Analysis

We analyzed imprinted genes derived from a publically available database from GeneImprint [65] that were filtered for experimentally validated and computationally predicted imprinted genes. The genes within the GeneImprint database are classified as predicted based upon chromosomal location [66]. From those, individuals have confirmed some to be imprinted based on actual experiments with cDNA showing parent-of-origin monoallelic expression from humans or other methodologies. Specifically, of the 197 genes, 90 were experimentally confirmed. These two imprinted gene sets were further analyzed for known metals relationships using the comparative toxicogenomics database (CTD) resulting in two additional gene lists with n = 43, and n = 14, respectively. Toxic metals were prioritized based on their 2011 ATSDR rankings [51]. Metal and imprinted gene relationships were determined using the CTD [67]. The CTD is a public resource that synthesizes current scientific literature on interactions between chemicals, genes, proteins, and the diseases associated with each. However, it should be noted that while the CTD is a useful tool that may be utilized to query a centralized public repository for known gene-metal interactions, the database may be limited in use for emerging findings, as there may be a delay between recent publications and inclusion in the database.

In order to identify biological pathways enriched within the imprinted gene sets, the four gene lists were analyzed for enrichment using Ingenuity Pathway Analysis (IPA) [68]. Ingenuity allows for the mapping of genes, proteins, and their corresponding regulatory networks as a useful tool for the identification of molecular pathways in disease. As a secondary method, for verification, The Database for Annotation, Visualization and Integrated Discovery (DAVIDv6.7) was also used to analyze the four gene lists.

As a method to identify transcription factor binding site enrichment within the imprintome, the imprinted gene set(s) were analyzed using the Genomatix Matinspector module (Genomatix Software Inc., Ann Arbor, MI, USA) [69]. The analysis was used to examine the four gene sets for common regulatory sequences and/or known regulation by common transcription factors. Where multiple promoter regions were possible for a given gene, a single promoter region was selected to maximize the number of experimentally verified 5' complete transcripts. The promoter regions were analyzed with the additional search criteria of 1000 base pairs upstream, and 50 base pairs downstream relative to the transcription start site. The genes were analyzed with a minimum core and matrix similarity of 1.00, the highest level of sensitivity possible. The p-value generated is the probability to obtain an equal or greater number of sequences with a match in a randomly drawn sample of the same size as the input sequence set. The lower this probability the higher is the importance of the observed common transcription factor.

3. Results and Discussion

3.1. Enriched Biological Trends within the Imprinted Gene Set

We analyzed imprinted genes derived from a publically available database that were filtered for experimentally validated and computationally predicted imprinted genes. Of the 197 genes, 90 were experimentally confirmed. These gene lists can be found in Table S1. These two imprinted gene sets were further analyzed for metals relationships using the CTD. Specifically, these associations include cellular perturbations such as altered mRNA and/or protein expression as well as epigenetic modifications (e.g., DNA methylation). The analysis identified known metals interactions for $n = 43$ predicted and experimentally validated imprinted genes and $n = 14$ experimentally validated imprinted genes with known metals interactions (Table S2).

The resulting four imprinted gene lists (both computationally predicted and validated) and the metals-associated genes were analyzed using IPA for known molecular interactions and for enrichment of diseases and canonical pathways. Many of the genes are involved in gene expression, embryonic development, organismal development, gastrointestinal disease, endocrine system disorders, hereditary disorders, cell morphology, cellular development, cell growth and proliferation, cell death and survival (Table 1). The results of IPA analysis were supported by analysis performed using DAVID which found similar enrichment of several of the functional pathways such as embryonic development and gene regulation (Table 1).

In addition, enrichment analyses for canonical pathways were performed using IPA. The imprinted gene set is enriched for two canonical pathways, namely the TP53 and aryl-hydrocarbon receptor (AhR) signalling pathways (Table 1). A total of five genes were associated with TP53 including cyclin-dependent kinase 4 (*CDK4*), *PLAGL1*, retinoblastoma 1 (*RB1*), tumor protein p73 (*TP73*), and Wilms tumor 1 (*WT1*) (enrichment p value = 0.00107). A total of five genes were also associated with AhR including aldehyde dehydrogenase 1 family, member L1 (*ALDH1L1*), *CDK4*, cytochrome P450, family 1, subfamily B, polypeptide 1 (*CYP1B1*), *RB1*, and *TP73* (enrichment p value < 0.01). The following three genes were common to both pathways: *CDK4*, *RB1*, and *TP73*. This shared gene set perhaps is not surprising, as imprinted genes play a large role in development, and these pathways are fundamental in regulating development, signaling, and cellular responses to stress. However, it should be noted that the observation of the links between the imprinted gene set and these two critical biological pathways has not been previously reported.

Enrichment analyses for canonical pathways were also performed for the experimentally validated imprinted gene set ($n = 90$), where the TP53 signalling pathway was enriched ($p \leq 0.001$) represented by the following four genes *PLAGL1*, *RB1*, *TP73*, and *WT1*. Imprinted genes involved in the AhR signalling pathway were also present including *RB1* and *TP73*, and the pathway was marginally significant ($p = 0.09$). Further enrichment analysis was performed on the two imprinted gene sets filtered for known association with our prioritized metals; those genes that are predicted or experimentally validated ($n = 43$), and just those experimentally validated ($n = 14$). The TP53 signaling pathway was also enriched in both the predicted/validated and experimentally validated gene sets ($p < 5.7 \times 10^{-5}$, and $p < 0.0023$, respectively). Enrichment of the AhR signaling pathway was significant ($p < 0.00023$) represented by the genes *CDK4*, *CYP1B1*, *RB1*, and *TP73* in the computationally predicted/experimentally validated gene set, but not in the set of only experimentally validated metal-associated imprinted genes.

Table 1. Summary of enriched biological processes/functions of the imprinted gene set.

Networks	*p*-value
Embryonic Development, Organismal Development, Gene Expression *	1×10^{-39}
Embryonic Organ Development **	1.7×10^{-12}
Gene Expression, Developmental Disorder, Endocrine System Disorders *	1×10^{-39}
Diseases and Disorders	**Average *p*-value**
Developmental Disorder *	0.001
Endocrine System Disorders *	0.001
Organismal Injury and Abnormalities *	0.002
Gastrointestinal Disease *	0.002
Hereditary Disorder *	0.006
Prader-Willi Syndrome **	0.02
Beckwith-Wiedemann Syndrome **	0.0007
Molecular and Cellular Functions	**Average *p*-value**
Gene Expression *	<0.001
Transcription Regulation **	8.2×10^{-8}
Cell Morphology *	0.002
Cell Morphogenesis Involved in Differentiation	0.005
Cellular Development *	0.003
Development-associated Proteins **	1.8×10^{-10}
Cell Death and Survival *	0.003
Cell Signaling *	0.003
Canonical Pathways	***p*-value**
TP53 Signaling *	0.001
Aryl Hydrocarbon Receptor Signaling *	0.006

(*) Ingenuity Pathway Analysis (IPA) results; (**) Database for Annotation, Visualization and Integrated Discovery (DAVID) results.

3.1.1. TP53 Signaling-Associated Imprinted Genes

The TP53 tumor suppressor protein is known as the guardian of the genome. It is a key transcriptional regulator that responds to a variety of cellular stresses including damage induced by various environmental contaminants. It serves to control key cellular processes such as DNA repair, cell-cycle progression, angiogenesis, and apoptosis pathways critical for influencing apoptosis or cell-cycle arrest [70]. In addition to its critical roles in DNA repair pathways, TP53 can act as a transcriptional regulator. Mutations within the *TP53* gene are responsible for Li-Fraumeni syndrome [71] and loss in functionality of the tumor suppressor is thought to be a contributing factor in the majority of cancer cases [72,73].

Of the five imprinted genes found to be associated with TP53, TP73 has been shown to increase activation of phosphorylated TP53 in mouse embryo fibroblasts [74] and PLAGL1 was shown to act as a transcriptional co-activator and enhance the activity of TP53 in both human carcinoma $P53^{+/-}$ and HeLa cells [75]. In *in vitro* models of rat kidney and human osteoblast-like cells, binding of WT1 to

TP53 has been shown to stabilize TP53, as well as inhibit TP53-mediated apoptosis [76]. Further, TP53 has been shown to modulate CDK4/RB1 through CDKN1A in human colon cancer cells [77]. TP53 can increase the expression of CDKN1A, which in turn decreases phosphorylation of RB1 [78]. Additionally, CDKN1A may also decrease phosphorylation of RB1 via CDK2-Cyclin D1 complex [79].

3.1.2. AhR Signaling-Associated Imprinted Genes

The aryl-hydrocarbon receptor is a transcription factor involved in cell cycle regulation, and an initiator of biological responses to xenobiotics. AhR has been shown to regulate enzymes such as cytochrome P450 and other xenobiotic metabolizing enzyme genes including putative imprinted gene *CYP1B1* [80]. In addition to xenobiotic metabolism, the AhR pathway plays a key role in organismal development processes [81]. In vertebrates, AhR is important for cellular proliferation and differentiation as well as many developmental pathways [82]. In addition to its role in mediating xenobiotics, the AhR pathway contributes to gene regulation and carcinogenesis [83].

Of the five genes found to be associated with AhR, CDK4 is involved in regulation of RB1 through phosphorylation [84]. It has been shown that RB1 increases activation of the AhR-Arnt complex in hepatoma cells [85] and this heterodimer complex is known to regulate CYP1B1 [86] and ALDH1 [87]. AhR has also been shown to decrease activation of phosphorylated E2F1 in tandem with RB1 [88,89]. This is worth noting, as expression of *TP73*, as well as many cyclins, is regulated by E2F1 [90,91].

3.1.3. Relationships between Imprinted Genes and Toxic Metals

Using data collected from the CTD, the seven unique genes within the TP53 and AhR pathways were detailed according to their known relationships to metals (Figure 1). The current body of literature pertaining to these relationships includes associations of the metals with altered gene regulation. Of the three genes (*CDK4*, *RB1*, and *TP73*) common to both pathways, upregulation of *CDK4* is observed upon exposure to both arsenic in a glioma cell line [92] and lead measured in peripheral blood [93]. Similarly, exposure to inorganic arsenic is associated with increased TP73 protein expression and activation of the downstream TP73-TP53 pathway in leukemia cells [94]. Conversely, there is decreased expression of *RB1* from arsenic [95] and cadmium exposure [96].

Of the genes specific to the TP53 pathway, exposure to inorganic arsenic is associated with decreased expression of the *WT1* gene in leukemia cell lines [97]. To date, the only known interaction between *PLAGL1* and environmental metals is hypomethylation associated with lead exposure [53]. Importantly, changes to signaling within the TP53 pathway by environmental contaminants such as inorganic arsenic and cadmium can impact the balance between apoptosis and proliferation in epithelial cancer cell lines [98,99].

Gene-metal interactions that disrupt the AhR pathway can disturb xenobiotic metabolism and increase an individual's susceptibility to a range of negative health outcomes. Studies have shown decreased expression of *CYP1B1* in response to metals such as arsenic [100] and cadmium in acute promyelocytic leukemia cell lines [101]. Interactions between *ALDH1L1* and environmental metals have not been studied. It should be mentioned that the analyses for gene-metal interactions were performed using the CTD database, which may not be fully comprehensive due to querying limitations and time delays between published findings and inclusion in the database. Tools such as CTD or IPA's canonical pathway analysis may be prone to inherent selection bias. Resources such as these that index interactions from published research may disproportionately display enrichment for a particular outcome or pathway as a consequence of overrepresentation of that particular interaction within the literature. Nevertheless, the alteration of these putative imprinted genes within the AhR pathway by inorganic arsenic, cadmium, and possibly other toxic metals may then serve as a potential mechanism of later life health outcomes including cancers and susceptibility to exogenous chemicals.

Figure 1. Top canonical pathways and their relationships to toxic metals. The aryl-hydrocarbon receptor (AhR) (p = 0.001) and TP53 (p = 0.007) networks display known interactions between pathway genes and priority metals (arsenic, cadmium, and lead). Imprinted status is also noted; abbreviations are shown in the figure legend.

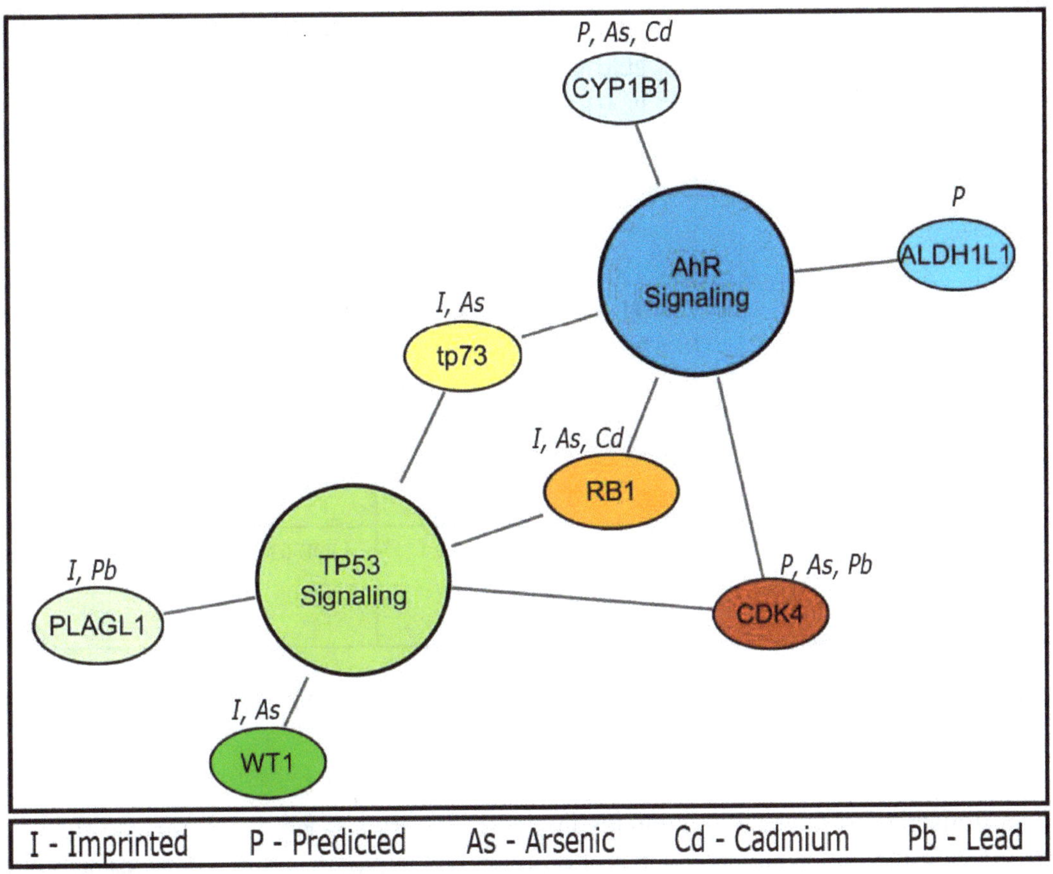

3.2. Sequence Specific Patterns of TF Elements in the Imprinted Genes

An analytical method was used to explore the four separate gene sets including experimentally validated and predicted imprinted genes (n = 197), experimentally validated imprinted genes (n = 90), experimentally validated and predicted imprinted genes associated with metals enrichment (n = 43), and experimentally validated imprinted genes associated with metals enrichment (n = 14) for common regulatory sequences and/or known regulation by common transcription factors. These results demonstrated that two transcription factor families were identified across all four gene lists, namely the family containing TF2B (Transcription Factor II B) and the PLAG family (Table 2).

Table 2. Enriched transcription factors amongst the imprinted gene sets.

Gene set	Transcription factor families	Transcription factors	Genes	*p*-value	Representative consensus sequence
Experimentally Validated & Predicted IGs (*n* = 197)	TF2B	GTF2B	93/171 (54%)	3.79E-45	ccgCGCC [1]
	PLAG	PLAG1 PLAGL1 PLAGL2	133/171 (78%)	2.10E-08	gaGGGGgcgggggggggggggg [2]
Experimentally Validated IGs (*n* = 90)	TF2B	GTF2B	38/71 (54%)	4.72E-19	ccgCGCC
	PLAG	PLAG1 PLAGL1 PLAGL2	55/71 (77%)	3.48E-04	gaGGGGgcgggggggggggggg
Experimentally Validated & Predicted IGs Metals-associated (*n* = 43)	TF2B	GTF2B	21/43 (49%)	1.09E-10	ccgCGCC
	PLAG	PLAG1 PLAGL1 PLAGL2	35/43 (81%)	1.31E-04	gaGGGGgcgggggggggggggg
Experimentally Validated IGs Metals-associated (*n* = 14)	TF2B	GTF2B	7/14 (50%)	2.62E-04	ccgCGCC
	PLAG	PLAG1 PLAGL1 PLAGL2	13/14 (93%)	4.90E-03	gaGGGGgcgggggggggggggg

[1] Consensus sequence based on most conserved nucleotide at each position for 210 sequences; [2] Consensus sequence based on most conserved nucleotide at each position for 337 sequences.

TF2B is a transcription factor that mediates interactions between RNA polymerase II and promoter regions [102]. The other significantly enriched transcription factor family was the PLAG family. The PLAG family contains the transcription factors pleiomorphic adenoma gene 1 (PLAG1), pleiomorphic adenoma gene-like 2 (PLAGL2), and Pleiomorphic Adenoma Gene-Like 1 (PLAGL1, also known as ZAC1) encoded by the imprinted gene *PLAGL1*.

Interestingly, PLAGL1 is a zinc finger protein transcription factor that has been implicated as a regulatory hub in an "imprinted gene network" (IGN) controlling embryonic growth and cell proliferation, and proposed as a regulator of expression of other imprinted genes including *IGF2*, *H19*, and *CDKN1C* [103,104]. The data here expand upon this notion and in fact support that many, specifically 133 of the 171 analyzed imprinted genes (*i.e.*, 77%–78%) have binding sites for PLAG transcription factors, thus greatly expanding the current list of potential imprinted gene targets of PLAG.

Likewise, when only the metal-associated imprinted genes ($n = 14$) were analyzed, the most significantly enriched transcription factor families were once again the families of *TF2B* ($p < 2.62 \times 10^{-4}$) and *PLAG* ($p < 4.9 \times 10^{-3}$). TF2B motifs were enriched in 7 of 14 (50%) metal associated gene sequences.

Transcription factors are known regulators of gene expression but they also serve an important role in regulating access to the DNA within genes. This access has consequences for both transcription and DNA methylation patterning. The binding of transcription factors to specific sites within the promoter region may protect CpG islands from methylation [105]. Recent work has found that areas with high transcription factor binding tend to have lower methylation [106].

Expanding upon previous work linking transcription factor occupancy with footprints left by DNase 1 [107,108], the Stamatoyannopoulos group studied numerous cell and tissue lines and found such footprint occupancy represented a viable quantitative measure of transcription factor occupancy and such occupancy afforded protection from DNA methylation [109].

In this context, our laboratory has recently hypothesized that in the case of cadmium and likely other environmental contaminants as well, distinct methylation patterns may represent "environmental footprints" or indicators of transcription factor occupancy during times of DNA methylation [58]. Based on the results from the present study, it can be hypothesized that specific transcription factor families may impact occupancy related to imprinted gene regulatory regions and subsequently their potential for DNA methylation. The identification of these regulators of the methylation "footprints" may serve as important biomarkers of environmental exposure, and may help to support a mechanistic link between toxic metals exposure, imprinted gene alterations, and later life health outcomes.

4. Conclusions and Future Research Directions

Using a survey of current experimentally validated and predicted imprinted genes, we set out to examine shared functionality and pathway enrichment within the imprinted gene set. Notably, we found: (i) common functionality among proteins encoded by the imprinted gene set; (ii) several of the putative and known imprinted genes play a role in the TP53 and AhR pathways and many of these imprinted genes have been shown to interact with toxic environmental metals specifically in their ability to modify, and in certain cases disrupt, apoptosis, cell cycle arrest, ligand metabolism, DNA repair, and may promote the development of certain cancers; and (iii) common transcription factor-based regulatory regions for TF2B and PLAG present in the promoter regions of the experimentally validated and predicted imprinted genes.

As a result of their impact of access to DNA, transcription factors may contribute to specific DNA methylation patterning upon exposure to metals. Our lab has recently hypothesized that distinct methylation patterns due to exposure to environmental contaminants may represent "environmental footprints" of transcription factor occupancy during DNA methylating events. Based on our results, it can be hypothesized that specific transcription factor families such as TF2B and PLAG may impact the occupancy and subsequent methylation of the imprinted gene set. Additional research is needed to understand the mechanisms linking metals exposure, transcription factor occupancy and the imprintome. It is also worth noting that we prioritized the current research on cadmium, lead, and arsenic. While these metals represent some of the most studied, the lack of inclusion the emerging roll additional metals play in terms of impact on the epigenome is not a reflection of their toxicity, or relationship to genetic imprinting. Furthermore, metals are only one class of the broad range of environmental contaminants that have been shown, or may be associated with genetic imprinting and other epigenetic alterations.

Our novel finding that *CDK4*, *RB1*, and *TP73* are associated with two biologically critical pathways, TP53 and AhR may have implications for understanding the biological mechanisms of how these imprinted genes may ultimately be associated with negative health outcomes. Importantly, as more databases are populated with published work on imprinted genes and their interactions with toxic metals, other statistically significant pathways may be associated with the imprintome, or subsets of imprinted genes, and may provide further mechanistic links between genetic imprinting and disease.

Furthermore, our current findings have implications for understanding perturbed biological pathways that may provide insight into early life exposures and later life health consequences. In addition, it is worth noting, that while the idea is still under debate, further studies are needed to determine the extent of trans-generational inheritability of specific patterns of methylation and their contributions to disease. Combined with the dysregulation of imprinted genes by toxic-metals, it may be possible to link ancestral exposure to environmental contaminants with current patterns of disease.

Based on our findings, further research is recommended to investigate the biological consequences of the imprinted gene set and its relationship to transcription factor occupancy. Specifically, transcription factor knockdown experiments or controlled toxicological experiments may help to elucidate the relationship between imprinted genes, transcription factor occupancy, environmental exposures, and associated health consequences.

Acknowledgments

This research was supported in part by the National Institute of Environmental Health Sciences (NIEHS) (ES019315, ES010126, ES005948, ES016772, ES022831), the National Cancer Institute (CA057726), the National Institute of Diabetes and Digestive and Kidney Diseases (DK085173) and by the USEPA (RD-83543701). The content is solely the responsibility of the authors and does not necessarily represent the official views of the NIH or the USEPA.

Author Contributions

Rebecca C. Fry conceived of and oversaw the study; Lisa Smeester and Andrew E. Yosim performed analyses; all authors contributed to the writing of the manuscript.

References

1. Hou, L.; Zhang, X.; Wang, D.; Baccarelli, A. Environmental chemical exposures and human epigenetics. *Int. J. Epidemiol.* **2012**, *41*, 79–105.
2. Gluckman, P.D.; Hanson, M.A.; Cooper, C.; Thornburg, K.L. Effect of in utero and early-life conditions on adult health and disease. *N. Engl. J. Med.* **2008**, *359*, 61–73.

3. Jirtle, R.L.; Skinner, M.K. Environmental epigenomics and disease susceptibility. *Nat. Rev. Genet.* **2007**, *8*, 253–262.
4. Vargas, A.O. Did paul kammerer discover epigenetic inheritance? A modern look at the controversial midwife toad experiments. *J. Exp. Zool. B Mol. Dev. Evol.* **2009**, *312*, 667–678.
5. Lamm, E. Review of: Julian Huxley, evolution: The modern synthesis—The definitive edition, with a new forward by Massimo Pigliucci and Gerd B. Müller. MIT Press, 2010. *Integr. Psych. Behav.* **2011**, *45*, 154–159.
6. Jablonka, E.; Lamb, M.J. The changing concept of epigenetics. *Ann. N. Y. Acad. Sci.* **2002**, *981*, 82–96.
7. Haig, D. The (dual) origin of epigenetics. *Cold Spring Harb. Symp. Quant. Biol.* **2004**, *69*, 67–70.
8. Holliday, R. The inheritance of epigenetic defects. *Science* **1987**, *238*, 163–170.
9. Jaenisch, R.; Bird, A. Epigenetic regulation of gene expression: How the genome integrates intrinsic and environmental signals. *Nat. Genet.* **2003**, *33*, 245–254.
10. Li, E. Chromatin modification and epigenetic reprogramming in mammalian development. *Nat. Rev. Genet.* **2002**, *3*, 662–673.
11. Dean, W.; Santos, F.; Reik, W. Epigenetic reprogramming in early mammalian development and following somatic nuclear transfer. *Semin. Cell Dev. Biol.* **2003**, *14*, 93–100.
12. Kiefer, J.C. Epigenetics in development. *Dev. Dyn.* **2007**, *236*, 1144–1156.
13. Reik, W.; Dean, W.; Walter, J. Epigenetic reprogramming in mammalian development. *Science* **2001**, *293*, 1089–1093.
14. Delaval, K.; Feil, R. Epigenetic regulation of mammalian genomic imprinting. *Curr. Opin. Genet. Dev.* **2004**, *14*, 188–195.
15. Feinberg, A.P. Phenotypic plasticity and the epigenetics of human disease. *Nature* **2007**, *447*, 433–440.
16. Portela, A.; Esteller, M. Epigenetic modifications and human disease. *Nat. Biotechnol.* **2010**, *28*, 1057–1068.
17. Jablonka, E.; Raz, G. Transgenerational epigenetic inheritance: Prevalence, mechanisms, and implications for the study of heredity and evolution. *Q. Rev. Biol.* **2009**, *84*, 131–176.
18. Skinner, M.K.; Manikkam, M.; Guerrero-Bosagna, C. Epigenetic transgenerational actions of environmental factors in disease etiology. *Trends Endocrinol. Metab.* **2010**, *21*, 214–222.
19. Feinberg, A.P.; Tycko, B. The history of cancer epigenetics. *Nat. Rev. Cancer* **2004**, *4*, 143–153.
20. Perera, F.; Herbstman, J. Prenatal environmental exposures, epigenetics, and disease. *Reprod. Toxicol.* **2011**, *31*, 363–373.
21. Salnikow, K.; Zhitkovich, A. Genetic and epigenetic mechanisms in metal carcinogenesis and cocarcinogenesis: Nickel, arsenic, and chromium. *Chem. Res. Toxicol.* **2008**, *21*, 28–44.
22. Baccarelli, A.; Bollati, V. Epigenetics and environmental chemicals. *Curr. Opin. Pediatr.* **2009**, *21*, 243–251.
23. Cheng, T.F.; Choudhuri, S.; Muldoon-Jacobs, K. Epigenetic targets of some toxicologically relevant metals: A review of the literature. *J. Appl. Toxicol.* **2012**, *32*, 643–653.
24. Egger, G.; Liang, G.; Aparicio, A.; Jones, P.A. Epigenetics in human disease and prospects for epigenetic therapy. *Nature* **2004**, *429*, 457–463.

25. Aguilera, O.; Fernandez, A.F.; Munoz, A.; Fraga, M.F. Epigenetics and environment: A complex relationship. *J. Appl. Physiol.* **2010**, *109*, 243–251.
26. Verma, R.; Xu, X.; Jaiswal, M.K.; Olsen, C.; Mears, D.; Caretti, G.; Galdzicki, Z. *In vitro* profiling of epigenetic modifications underlying heavy metal toxicity of tungsten-alloy and its components. *Toxicol. Appl. Pharmacol.* **2011**, *253*, 178–187.
27. Li, Q.; Ke, Q.; Costa, M. Alterations of histone modifications by cobalt compounds. *Carcinogenesis* **2009**, *30*, 1243–1251.
28. Kelly, T.K.; de Carvalho, D.D.; Jones, P.A. Epigenetic modifications as therapeutic targets. *Nat. Biotechnol.* **2010**, *28*, 1069–1078.
29. Moore, G.E.; Oakey, R. The role of imprinted genes in humans. *Genome Biol.* **2011**, *12*, 106.
30. Munshi, A.; Duvvuri, S. Genomic imprinting—The story of the other half and the conflicts of silencing. *J. Genet. Genomics* **2007**, *34*, 93–103.
31. Babak, T.; Deveale, B.; Armour, C.; Raymond, C.; Cleary, M.A.; van der Kooy, D.; Johnson, J.M.; Lim, L.P. Global survey of genomic imprinting by transcriptome sequencing. *Curr. Biol.* **2008**, *18*, 1735–1741.
32. Wang, X.; Sun, Q.; McGrath, S.D.; Mardis, E.R.; Soloway, P.D.; Clark, A.G. Transcriptome-wide identification of novel imprinted genes in neonatal mouse brain. *PLoS One* **2008**, *3*, e3839.
33. Bartolomei, M.S.; Ferguson-Smith, A.C. Mammalian genomic imprinting. *Cold Spring Harb. Perspect Biol.* **2011**, *3*, a002592.
34. Jirtle, R.L. Epigenome: The program for human health and disease. *Epigenomics* **2009**, *1*, 13–16.
35. Skaar, D.A.; Li, Y.; Bernal, A.J.; Hoyo, C.; Murphy, S.K.; Jirtle, R.L. The human imprintome: Regulatory mechanisms, methods of ascertainment, and roles in disease susceptibility. *ILAR J.* **2012**, *53*, 341–358.
36. Murphy, S.; Hoyo, C. Sculpting our future: Environmental nudging of the imprintome. In *Environmental Epigenomics in Health and Disease*; Jirtle, R.L., Tyson, F.L., Eds.; Springer: Berlin/Heidelberg, Germany, 2013; pp. 51–73.
37. Barlow, D.P. Genomic imprinting: A mammalian epigenetic discovery model. *Annu. Rev. Genet.* **2011**, *45*, 379–403.
38. Moore, T.; Haig, D. Genomic imprinting in mammalian development: A parental tug-of-war. *Trends Genet.* **1991**, *7*, 45–49.
39. Kinoshita, T.; Ikeda, Y.; Ishikawa, R. Genomic imprinting: A balance between antagonistic roles of parental chromosomes. *Semin. Cell Dev. Biol.* **2008**, *19*, 574–579.
40. Frost, J.M.; Moore, G.E. The importance of imprinting in the human placenta. *PLoS Genet.* **2010**, *6*, e1001015.
41. Renfree, M.B.; Suzuki, S.; Kaneko-Ishino, T. The origin and evolution of genomic imprinting and viviparity in mammals. *Philos. Trans. R. Soc. Lond. B Biol. Sci.* **2013**, doi: 10.1098/rstb.2012.0151.
42. Úbeda, F.; Wilkins, J. Imprinted genes and human disease: An evolutionary perspective. In *Genomic Imprinting*; Wilkins, J., Ed.; Springer: New York, NY, USA, 2008; Volume 626, pp. 101–115.
43. Wilkins, J.F.; Ubeda, F. Diseases associated with genomic imprinting. *Prog. Mol. Biol. Transl. Sci.* **2011**, *101*, 401–445.

44. Knoll, J.H.M.; Nicholls, R.D.; Magenis, R.E.; Graham, J.M.; Lalande, M.; Latt, S.A.; Opitz, J.M.; Reynolds, J.F. Angelman and prader-willi syndromes share a common chromosome 15 deletion but differ in parental origin of the deletion. *Am. J. Med. Genet.* **1989**, *32*, 285–290.
45. Monk, D. Deciphering the cancer imprintome. *Brief. Funct. Genomics* **2010**, *9*, 329–339.
46. Jelinic, P.; Shaw, P. Loss of imprinting and cancer. *J. Pathol.* **2007**, *211*, 261–268.
47. Jones, P.A.; Baylin, S.B. The fundamental role of epigenetic events in cancer. *Nat. Rev. Genet.* **2002**, *3*, 415–428.
48. Leick, M.B.; Shoff, C.J.; Wang, E.C.; Congress, J.L.; Gallicano, G.I. Loss of imprinting of IGF2 and the epigenetic progenitor model of cancer. *Am. J. Stem Cells* **2012**, *1*, 59–74.
49. Pruss-Ustun, A.; Vickers, C.; Haefliger, P.; Bertollini, R. Knowns and unknowns on burden of disease due to chemicals: A systematic review. *Environ. Health* **2011**, *10*, 9.
50. Pruss-Ustun, A.; Bonjour, S.; Corvalan, C. The impact of the environment on health by country: A meta-synthesis. *Environ. Health* **2008**, *7*, 7.
51. Agency for toxic substances and disease registry (ATSDR). Available online: http://www.atsdr.cdc.gov/ (accessed on 28 Feburary 2014).
52. Murphy, S.K.; Hoyo, C. *Sculpting Our Future: Environmental Nudging of the Imprintome*; Jirtle, R.L., Tyson, F.L., Eds.; Springer: New York, NY, USA, 2013; pp. 51–73.
53. Hoyo, C. Department of Biological Sciences, Center for Human Health and Environment, Campus Box 7633, NC State University, Raleigh, NC 27695, USA. Unpublished work, 2014.
54. Pilsner, J.R.; Hu, H.; Ettinger, A.; Sanchez, B.N.; Wright, R.O.; Cantonwine, D.; Lazarus, A.; Lamadrid-Figueroa, H.; Mercado-Garcia, A.; Tellez-Rojo, M.M.; *et al.* Influence of prenatal lead exposure on genomic methylation of cord blood DNA. *Environ. Health Perspect.* **2009**, *117*, 1466–1471.
55. Senut, M.-C.; Sen, A.; Cingolani, P.; Shaik, A.; Land, S.J.; Ruden, D.M. Lead exposure disrupts global DNA methylation in human embryonic stem cells and alters their neuronal differentiation. *Toxicol. Sci.* **2014**, 142–161.
56. Smeester, L.; Rager, J.E.; Bailey, K.A.; Guan, X.; Smith, N.; Garcia-Vargas, G.; del Razo, L.M.; Drobna, Z.; Kelkar, H.; Styblo, M.; *et al.* Epigenetic changes in individuals with arsenicosis. *Chem. Res. Toxicol.* **2011**, *24*, 165–167.
57. Bailey, K.A.; Wu, M.C.; Ward, W.O.; Smeester, L.; Rager, J.E.; García-Vargas, G.; del Razo, L.M.; Drobná, Z.; Stýblo, M.; Fry, R.C. Arsenic and the epigenome: Interindividual differences in arsenic metabolism related to distinct patterns of DNA methylation. *J. Biochem. Mol. Toxicol.* **2013**, *27*, 106–115.
58. Sanders, A.P.; Smeester, L.; Rojas, D.; Debussycher, T.; Wu, M.C.; Wright, F.A.; Zhou, Y.H.; Laine, J.E.; Rager, J.E.; Swamy, G.K.; *et al.* Cadmium exposure and the epigenome: Exposure-associated patterns of DNA methylation in leukocytes from mother-baby pairs. *Epigenetics* **2013**, doi:10.4161/epi.26798.
59. Agency for Toxic Substances and Disease Registry (ATSDR). *Toxicological Profile for Cadmium*; ATSDR: Atlanta, GA, USA, 2012.
60. International Agency for Research on Cancer (IARC). Cadmium and cadmium compounds. IARC: Lyon, France, 2012.

61. Joubert, B.R.; Haberg, S.E.; Nilsen, R.M.; Wang, X.; Vollset, S.E.; Murphy, S.K.; Huang, Z.; Hoyo, C.; Midttun, O.; Cupul-Uicab, L.A.; *et al.* 450k epigenome-wide scan identifies differential DNA methylation in newborns related to maternal smoking during pregnancy. *Environ. Health Perspect.* **2012**, *120*, 1425–1431.
62. Murphy, S.K.; Adigun, A.; Huang, Z.; Overcash, F.; Wang, F.; Jirtle, R.L.; Schildkraut, J.M.; Murtha, A.P.; Iversen, E.S.; Hoyo, C. Gender-specific methylation differences in relation to prenatal exposure to cigarette smoke. *Gene* **2012**, *494*, 36–43.
63. Tobi, E.W.; Lumey, L.H.; Talens, R.P.; Kremer, D.; Putter, H.; Stein, A.D.; Slagboom, P.E.; Heijmans, B.T. DNA methylation differences after exposure to prenatal famine are common and timing- and sex-specific. *Hum. Mol. Genet.* **2009**, *18*, 4046–4053.
64. Kundakovic, M.; Gudsnuk, K.; Franks, B.; Madrid, J.; Miller, R.L.; Perera, F.P.; Champagne, F.A. Sex-specific epigenetic disruption and behavioral changes following low-dose in utero bisphenol a exposure. *Proc. Natl. Acad. Sci. USA* **2013**, *110*, 9956–9961.
65. Genomic Imprinting Database. Available online: http://www.geneimprint.org/ (accessed on 19 Feburary 2014).
66. Luedi, P.P.; Dietrich, F.S.; Weidman, J.R.; Bosko, J.M.; Jirtle, R.L.; Hartemink, A.J. Computational and experimental identification of novel human imprinted genes. *Genome Res.* **2007**, *17*, 1723–1730.
67. Comparative Toxicogenomics Database. Available online: http://ctdbase.org/ (accessed on 19 March 2014).
68. Ingenuity Pathway Analysis (IPA). Available online: http://www.ingenuity.com/ (Accessed on 18 March 2014).
69. Genomatix Software Suite v3.1. Available online: http://www.genomatix.de/ (Accessed on 18 March 2014).
70. Levine, A.J. P53, the cellular gatekeeper for growth and division. *Cell* **1997**, *88*, 323–331.
71. Birch, J.M. Li-fraumeni syndrome. *Eur. J. Cancer* **1994**, *30A*, 1935–1941.
72. Soussi, T.; Legros, Y.; Lubin, R.; Ory, K.; Schlichtholz, B. Multifactorial analysis of p53 alteration in human cancer: A review. *Int. J. Cancer* **1994**, *57*, 1–9.
73. Royds, J.A.; Iacopetta, B. P53 and disease: When the guardian angel fails. *Cell Death Differ.* **2006**, *13*, 1017–1026.
74. Flores, E.R.; Tsai, K.Y.; Crowley, D.; Sengupta, S.; Yang, A.; McKeon, F.; Jacks, T. P63 and p73 are required for p53-dependent apoptosis in response to DNA damage. *Nature* **2002**, *416*, 560–564.
75. Huang, S.M.; Schonthal, A.H.; Stallcup, M.R. Enhancement of p53-dependent gene activation by the transcriptional coactivator ZAC1. *Oncogene* **2001**, *20*, 2134–2143.
76. Maheswaran, S.; Englert, C.; Bennett, P.; Heinrich, G.; Haber, D.A. The WT1 gene product stabilizes p53 and inhibits p53-mediated apoptosis. *Genes Dev.* **1995**, *9*, 2143–2156.
77. Draus, J.M.; Elliott, M.J.; Atienza, C., Jr.; Stilwell, A.; Wong, S.L.; Dong, Y.; Yang, H.; McMasters, K.M. P53 gene transfer does not enhance E2F-1-mediated apoptosis in human colon cancer cells. *Exp. Mol. Med.* **2001**, *33*, 209–219.
78. Shav-Tal, Y.; Zipori, D. The role of activin a in regulation of hemopoiesis. *Stem Cells* **2002**, *20*, 493–500.

79. Adams, P.D.; Sellers, W.R.; Sharma, S.K.; Wu, A.D.; Nalin, C.M.; Kaelin, W.G., Jr. Identification of a cyclin-cdk2 recognition motif present in substrates and p21-like cyclin-dependent kinase inhibitors. *Mol. Cell Biol.* **1996**, *16*, 6623–6633.
80. Nebert, D.W.; Roe, A.L.; Dieter, M.Z.; Solis, W.A.; Yang, Y.; Dalton, T.P. Role of the aromatic hydrocarbon receptor and [*Ah*] gene battery in the oxidative stress response, cell cycle control, and apoptosis. *Biochem. Pharmacol.* **2000**, *59*, 65–85.
81. Hankinson, O. The aryl hydrocarbon receptor complex. *Annu. Rev. Pharmacol. Toxicol.* **1995**, *35*, 307–340.
82. Tijet, N.; Boutros, P.C.; Moffat, I.D.; Okey, A.B.; Tuomisto, J.; Pohjanvirta, R. Aryl hydrocarbon receptor regulates distinct dioxin-dependent and dioxin-independent gene batteries. *Mol. Pharmacol.* **2006**, *69*, 140–153.
83. Barouki, R.; Coumoul, X.; Fernandez-Salguero, P.M. The Aryl hydrocarbon receptor, more than a xenobiotic-interacting protein. *FEBS Lett.* **2007**, *581*, 3608–3615.
84. Gao, N.; Flynn, D.C.; Zhang, Z.; Zhong, X.S.; Walker, V.; Liu, K.J.; Shi, X.; Jiang, B.H. G1 cell cycle progression and the expression of G1 cyclins are regulated by Pi3k/Akt/Mtor/P70s6k1 signaling in human ovarian cancer cells. *Am. J. Physiol. Cell Physiol.* **2004**, *287*, C281–C291.
85. Huang, G.; Elferink, C.J. Multiple mechanisms are involved in ah receptor-mediated cell cycle arrest. *Mol. Pharmacol.* **2005**, *67*, 88–96.
86. Vrzal, R.; Ulrichova, J.; Dvorak, Z. Aromatic hydrocarbon receptor status in the metabolism of xenobiotics under normal and pathophysiological conditions. *Biomed. Pap. Med. Fac. Univ. Palacky Olomouc Czech Repub.* **2004**, *148*, 3–10.
87. Lindros, K.O.; Oinonen, T.; Kettunen, E.; Sippel, H.; Muro-Lupori, C.; Koivusalo, M. Aryl hydrocarbon receptor-associated genes in rat liver: Regional coinduction of aldehyde dehydrogenase 3 and glutathione transferase YA. *Biochem. Pharmacol.* **1998**, *55*, 413–421.
88. Marlowe, J.L.; Puga, A. Aryl hydrocarbon receptor, cell cycle regulation, toxicity, and tumorigenesis. *J. Cell Biochem.* **2005**, *96*, 1174–1184.
89. Puga, A.; Barnes, S.J.; Dalton, T.P.; Chang, C.; Knudsen, E.S.; Maier, M.A. Aromatic hydrocarbon receptor interaction with the retinoblastoma protein potentiates repression of E2f-dependent transcription and cell cycle arrest. *J. Biol. Chem.* **2000**, *275*, 2943–2950.
90. Urist, M.; Tanaka, T.; Poyurovsky, M.V.; Prives, C. P73 induction after DNA damage is regulated by checkpoint kinases Chk1 and Chk2. *Genes Dev.* **2004**, *18*, 3041–3054.
91. Murray, A.W. Recycling the cell cycle: Cyclins revisited. *Cell* **2004**, *116*, 221–234.
92. Zhao, S.; Zhang, J.; Zhang, X.; Dong, X.; Sun, X. Arsenic trioxide induces different gene expression profiles of genes related to growth and apoptosis in glioma cells dependent on the p53 status. *Mol. Biol. Rep.* **2008**, *35*, 421–429.
93. Tian, Y.; Green, P.; Stamova, B.; Hertz-Picciotto, I.; Pessah, I.; Hansen, R.; Yang, X.; Gregg, J.; Ashwood, P.; Jickling, G.; *et al.* Correlations of gene expression with blood lead levels in children with autism compared to typically developing controls. *Neurotox Res.* **2011**, *19*, 1–13.
94. Lunghi, P.; Costanzo, A.; Levrero, M.; Bonati, A. Treatment with arsenic trioxide (Ato) and Mek1 inhibitor activates the p73-p53aip1 apoptotic pathway in leukemia cells. *Blood* **2004**, *104*, 519–525.

95. Wu, X.; Shi, J.; Wu, Y.; Tao, Y.; Hou, J.; Meng, X.; Hu, X.; Han, Y.; Jiang, W.; Tang, S.; *et al.* Arsenic trioxide-mediated growth inhibition of myeloma cells is associated with an extrinsic or intrinsic signaling pathway through activation of trail or trail receptor 2. *Cancer Biol. Ther.* **2010**, *10*, 1201–1214.
96. Choi, Y.-J.; Yin, H.-Q.; Suh, H.-R.; Lee, Y.-J.; Park, S.-R.; Lee, B.-H. Involvement of e2f1 transcriptional activity in cadmium-induced cell-cycle arrest at G1 in human lung fibroblasts. *Environ. Mol. Mutagen.* **2011**, *52*, 145–152.
97. Glienke, W.C.; Chow, K.U.; Bauer, N.; Bergmann, L. Down-regulation of Wt1 expression in leukemia cell lines as part of apoptotic effect in arsenic treatment using two compounds. *Leuk. Lymphoma* **2006**, *47*, 1629–1638.
98. Sandoval, M.; Morales, M.; Tapia, R.; del Carmen Alarcón, L.; Sordo, M.; Ostrosky-Wegman, P.; Ortega, A.; López-Bayghen, E. P53 response to arsenic exposure in epithelial cells: Protein kinase B/Akt involvement. *Toxicol. Sci.* **2007**, *99*, 126–140.
99. Meplan, C.; Mann, K.; Hainaut, P. Cadmium induces conformational modifications of wild-type p53 and suppresses p53 response to DNA damage in cultured cells. *J. Biol. Chem.* **1999**, *274*, 31663–31670.
100. Zheng, P.-Z.; Wang, K.-K.; Zhang, Q.-Y.; Huang, Q.-H.; Du, Y.-Z.; Zhang, Q.-H.; Xiao, D.-K.; Shen, S.-H.; Imbeaud, S.; Eveno, E.; *et al.* Systems analysis of transcriptome and proteome in retinoic acid/arsenic trioxide-induced cell differentiation/apoptosis of promyelocytic leukemia. *Proc. Natl. Acad. Sci. USA* **2005**, *102*, 7653–7658.
101. Andrew, A.S.; Warren, A.J.; Barchowsky, A.; Temple, K.A.; Klei, L.; Soucy, N.V.; O'Hara, K.A.; Hamilton, J.W. Genomic and proteomic profiling of responses to toxic metals in human lung cells. *Environ. Health Perspect.* **2003**, *111*, 825–835.
102. Roeder, R.G. The role of general initiation factors in transcription by RNA polymerase II. *Trends Biochem. Sci.* **1996**, *21*, 327–335.
103. Lui, J.C.; Finkielstain, G.P.; Barnes, K.M.; Baron, J. An imprinted gene network that controls mammalian somatic growth is down-regulated during postnatal growth deceleration in multiple organs. *Am. J. Physiol. Regul. Integr. Comp. Physiol.* **2008**, *295*, R189–R196.
104. Varrault, A.; Gueydan, C.; Delalbre, A.; Bellmann, A.; Houssami, S.; Aknin, C.; Severac, D.; Chotard, L.; Kahli, M.; le Digarcher, A.; *et al.* Zac1 regulates an imprinted gene network critically involved in the control of embryonic growth. *Dev. Cell* **2006**, *11*, 711–722.
105. Brandeis, M.; Frank, D.; Keshet, I.; Siegfried, Z.; Mendelsohn, M.; Nemes, A.; Temper, V.; Razin, A.; Cedar, H. SP1 elements protect a CPG island from *de novo* methylation. *Nature* **1994**, *371*, 435–438.
106. Feldmann, A.; Ivanek, R.; Murr, R.; Gaidatzis, D.; Burger, L.; Schubeler, D. Transcription factor occupancy can mediate active turnover of DNA methylation at regulatory regions. *PLoS Genet.* **2013**, *9*, e1003994.
107. Galas, D.J.; Schmitz, A. Dnase footprinting: A simple method for the detection of protein-DNA binding specificity. *Nucleic Acids Res.* **1978**, *5*, 3157–3170.
108. Hesselberth, J.R.; Chen, X.; Zhang, Z.; Sabo, P.J.; Sandstrom, R.; Reynolds, A.P.; Thurman, R.E.; Neph, S.; Kuehn, M.S.; Noble, W.S.; *et al.* Global mapping of protein-DNA interactions *in vivo* by digital genomic footprinting. *Nat. Methods* **2009**, *6*, 283–289.

109. Neph, S.; Vierstra, J.; Stergachis, A.B.; Reynolds, A.P.; Haugen, E.; Vernot, B.; Thurman, R.E.; John, S.; Sandstrom, R.; Johnson, A.K.; *et al.* An expansive human regulatory lexicon encoded in transcription factor footprints. *Nature* **2012**, *489*, 83–90.

PERMISSIONS

All chapters in this book were first published in MDPI; hereby published with permission under the Creative Commons Attribution License or equivalent. Every chapter published in this book has been scrutinized by our experts. Their significance has been extensively debated. The topics covered herein carry significant findings which will fuel the growth of the discipline. They may even be implemented as practical applications or may be referred to as a beginning point for another development.

The contributors of this book come from diverse backgrounds, making this book a truly international effort. This book will bring forth new frontiers with its revolutionizing research information and detailed analysis of the nascent developments around the world.

We would like to thank all the contributing authors for lending their expertise to make the book truly unique. They have played a crucial role in the development of this book. Without their invaluable contributions this book wouldn't have been possible. They have made vital efforts to compile up to date information on the varied aspects of this subject to make this book a valuable addition to the collection of many professionals and students.

This book was conceptualized with the vision of imparting up-to-date information and advanced data in this field. To ensure the same, a matchless editorial board was set up. Every individual on the board went through rigorous rounds of assessment to prove their worth. After which they invested a large part of their time researching and compiling the most relevant data for our readers.

The editorial board has been involved in producing this book since its inception. They have spent rigorous hours researching and exploring the diverse topics which have resulted in the successful publishing of this book. They have passed on their knowledge of decades through this book. To expedite this challenging task, the publisher supported the team at every step. A small team of assistant editors was also appointed to further simplify the editing procedure and attain best results for the readers.

Apart from the editorial board, the designing team has also invested a significant amount of their time in understanding the subject and creating the most relevant covers. They scrutinized every image to scout for the most suitable representation of the subject and create an appropriate cover for the book.

The publishing team has been an ardent support to the editorial, designing and production team. Their endless efforts to recruit the best for this project, has resulted in the accomplishment of this book. They are a veteran in the field of academics and their pool of knowledge is as vast as their experience in printing. Their expertise and guidance has proved useful at every step. Their uncompromising quality standards have made this book an exceptional effort. Their encouragement from time to time has been an inspiration for everyone.

The publisher and the editorial board hope that this book will prove to be a valuable piece of knowledge for researchers, students, practitioners and scholars across the globe.

LIST OF CONTRIBUTORS

Toshihiko Eki
Molecular Genetics Laboratory, Division of Bioscience and Biotechnology, Department of Environmental and Life Sciences, Toyohashi University of Technology, 1-1 Hibarigaoka, Tempaku-cho, Toyohashi, Aichi 441-8580, Japan
Cellular Physiology Laboratory, RIKEN, 2-1 Hirosawa, Wako, Saitama 351-0198, Japan

Yasufumi Murakami
Cellular Physiology Laboratory, RIKEN, 2-1 Hirosawa, Wako, Saitama 351-0198, Japan
Department of Biological Science and Technology, Faculty of Industrial Science and Technology, Tokyo University of Science, 2641 Yamazaki, Noda, Chiba 278-8510, Japan

Fumio Hanaoka
Cellular Physiology Laboratory, RIKEN, 2-1 Hirosawa, Wako, Saitama 351-0198, Japan
Faculty of Science, Gakushuin University, 1-5-1 Mejiro, Toshima-ku, Tokyo 171-8588, Japan

Thomas Mikeska
Genetic Technologies Ltd., Fitzroy, Victoria 3065, Australia

Jeffrey M. Craig
Murdoch Childrens Research Institute, Royal Children's Hospital, Parkville, Victoria 3052, Australia
Department of Paediatrics, The University of Melbourne, Parkville, Victoria 3052, Australia

Gilberto Paz-Filho, Claudio A. Mastronardi, Hardip R. Patel, Angad S. Johar, Aaron Chuah, Gavin A. Huttley and Mauricio Arcos-Burgos
Genome Biology Department, The John Curtin School of Medical Research, The Australian National University, Garran Rd, building 131, Acton, Canberra, ACT 0200, Australia

Margaret C.S. Boguszewski and Cesar L. Boguszewski
Endocrine Division (SEMPR), Department of Internal Medicine, Federal University of Parana, Avenida Agostinho Leão Junior, 285-Alto da Glória. CEP 80030-110, Curitiba-PR, Brazil

Ma-Li Wong and Julio Licinio
Mind and Brain Theme, South Australian Health and Medical Research Institute, and Department of Psychiatry, School of Medicine, Flinders University, Adelaide, Australia

Erin Sandford
Molecular & Behavioral Neuroscience Institute, University of Michigan, Ann Arbor, MI 48109, USA

Margit Burmeister
Molecular & Behavioral Neuroscience Institute, University of Michigan, Ann Arbor, MI 48109, USA
Departments of Human Genetics, University of Michigan, Ann Arbor, MI 48109, USA
Department of Computational Medicine & Bioinformatics, University of Michigan, Ann Arbor, MI 48109, USA
Department of Psychiatry, University of Michigan, Ann Arbor, MI 48109, USA

William G. Newman and Graeme C. Black
Manchester Centre for Genomic Medicine, University of Manchester, Manchester, M13 9WL, UK
Manchester Centre for Genomic Medicine, Central Manchester University Hospitals, NHS Foundation Trust, Manchester, M13 9WL, UK

Thiviyani Maruthappu, Claire A. Scott and David P. Kelsell
Centre for Cutaneous Research, Blizard Institute, Barts and The London School of Medicine and Dentistry, Queen Mary University of London, 4 Newark Street, London E1 2AT, UK

Scott D. Boyd and Jason D. Merker
Department of Pathology, Stanford University, Stanford, CA 94305, USA

Stephen J. Galli
Department of Pathology, Stanford University, Stanford, CA 94305, USA
Department of Microbiology and Immunology, Stanford University, Stanford, CA 94305, USA

Iris Schrijver
Department of Pathology, Stanford University, Stanford, CA 94305, USA
Department of Pediatrics, Stanford University, Stanford, CA 94305, USA

James L. Zehnder
Department of Pathology, Stanford University, Stanford, CA 94305, USA
Department of Medicine, Stanford University, Stanford, CA 94305, USA

Euan A. Ashley
Department of Medicine, Stanford University, Stanford, CA 94305, USA

Farshad Farshidfar
Department of Surgery, Tom Baker Cancer Center, University of Calgary, 1331 29th St NW, Calgary, AB T2N 4N2, Canada

Oliver F. Bathe
Department of Surgery, Tom Baker Cancer Center, University of Calgary, 1331 29th St NW, Calgary, AB T2N 4N2, Canada
Department of Oncology, Tom Baker Cancer Center, University of Calgary, 1331 29th St NW, Calgary, AB T2N 4N2, Canada

Alan F. Scott, David W. Mohr, Hua Ling and Peng Zhang
McKusick-Nathans Institute of Genetic Medicine, Johns Hopkins University School of Medicine, Baltimore, MD 21287, USA

Robert B. Scharpf
Department of Oncology, Johns Hopkins University School of Medicine, Baltimore, MD 21287, USA

Gregory S. Liptak
Department of Pediatrics, SUNY Upstate Medical Center, Golisano Children's Hospital, Syracuse, NY 13210, USA

Daniel F. Carr, Ana Alfirevic and Munir Pirmohamed
Wolfson Centre for Personalised Medicine, Department of Molecular and Clinical Pharmacology, University of Liverpool, Block A Waterhouse Buildings, 1-5 Brownlow Street, Liverpool, L69 3GL, UK

Lisa Smeester and Andrew E. Yosim
Department of Environmental Sciences and Engineering, Gillings School of Global Public Health, The University of North Carolina, 135 Dauer Drive, CB 7431, UNC, Chapel Hill, NC 27599, USA

Monica D. Nye
Lineberger Comprehensive Cancer Center, The University of North Carolina, 450 West Street, CB 7295, UNC, Chapel Hill, NC 27599, USA
Department of Obstetrics and Gynecology, Duke University Medical Center, B226 LSRC, Research Drive, Durham, NC 27708, USA

Susan K. Murphy
Department of Obstetrics and Gynecology, Duke University Medical Center, B226 LSRC, Research Drive, Durham, NC 27708, USA

Cathrine Hoyo
Department of Biological Sciences, Center for Human Health and Environment, NC State University, Raleigh, NC 27695, USA

Rebecca C. Fry
Department of Environmental Sciences and Engineering, Gillings School of Global Public Health, The University of North Carolina, 135 Dauer Drive, CB 7431, UNC, Chapel Hill, NC 27599, USA
Lineberger Comprehensive Cancer Center, The University of North Carolina, 450 West Street, CB 7295, UNC, Chapel Hill, NC 27599, USA
Curriculum in Toxicology, The University of North Carolina, 135 Dauer Drive, CB 7431, UNC, Chapel Hill, NC 27599, USA

Index

P

Q

R

S

T

U

V

W

X

Z

Printed in the USA
CPSIA information can be obtained
at www.ICGtesting.com
LVHW080044211023
761596LV00010BA/1908

9 781647 403935